DIGITAL SYSTEM DESIGN USING VHDL

DIGITAL SYSTEM DESIGN USING VHDL

Rishabh Anand

KHANNA BOOK PUBLISHING CO. (P) LTD.
Publisher of Engineering and Computer Books
4C/4344, Ansari Road, Darya Ganj, New Delhi-110002
Phone : 011-23244447-48 **Mobile:** +91-9910909320
E-mail : contact@khannabooks.com
Website : www.khannabooks.com

Price: ₹ 529.00

ISBN: 978-93-81068-78-6

Digital System Design Using VHDL
by Rishabh Anand

To view complete list of books, please scan the QR Code:

Published by:
Khanna Book Publishing Co. (P) Ltd.
CIN: U22110DL1998PTC095547

Visit us at: **www.khannabooks.com**
Write us at: **contact@khannabooks.com**

Printed in Bharat.

Dedicated to

MY MOTHER

whose love and affection made me
capable of writing this book

and

SHRI GURU MAHARAJ JI

of

SHRI ANANDPUR

MESSAGE TO MY STUDENTS

As you have signed up for this course and probably your teacher has adopted this book! You had no say about the choice, although you can form and express your opinion on the book after reading it. What I really meant is that although this is an easy book to study from (in my opinion), it is no-nonsense book. It does not have cartoons or nice looking photographs to amuse you. From here on it is all business and hard work. This book has many examples. Some of these are simple, for the purpose of illustrations of new ideas and subject matter. Some examples are made quite elaborate, in order to develop desired technical skill and confidence. Furthermore, the objective of this book is to present the complex subject in a clear and thorough manner. I hope that you will enjoy this book, and it will be remain as major text book in your college career. My advice to you is not sell it back at the end of semester. If you do, not only will you be violating the "buy low and sell high "axiom of finance but definitely be missing it during your higher semester or project assignments. When you finish the journey through this book, having variety of experiences, ideas and thoughts. You are welcome to express your opinion at rishabh.

Rishabh Anand

PREFACE

To keep up with advances in semiconductor technology, design complexity and design methodology have to be improved. The traditional gate-level schematic capture approach cannot meet the shortened design cycle and market window. A higher level of abstraction to capture designs is not an option, but a requirement. For this reason Hardware Description Languages (HDL) such as Verilog and VHDL were developed,, Synthesis tools take these HDLs and automatically translate them into schematics, thereby enabling designers to manage more complex and bigger designs in shorter design schedules. VHDL stands for VHSIC (Very High Speed Integrated Circuits) Hardware Description Language. Designers have interpreted VHDL as meaning Very Hard Description Language (VHDL). The major goal of this book is to define VHDL as Very Handy Description Language for designing and modelling digital circuits.

This book is intended for an introductory digital design course for students at the freshman Level. This book uses a spiral teaching approach by introducing a design problem and then, in the same or a later chapter, either (I) reemphasizing the same concepts when a different design problem is presented, or (2) working the same problem using a different technique. This is done to increase the likelihood of retention. There is no prerequisite for the book; however, computer familiarity and/ or a first programming course usually helps students to learn VHDL programming.

VHDL is a language for describing digital electronic systems. It arose out of the United States government's Very High Speed Integrated Circuits HSIC) program. In the course of this program, it became clear that there was a need for a standard Ian wage for .describing the structure and function of integrated circuits (ICs). Hence the VHSIC Hardware Description Language (VHDL) was developed. It was subsequently developed further under the auspices of the Institute of Electricals and Electronic Engineers (IEEE) and adopted in the form of the IEEE Standard, 1076, Standard VHDL Language Reference Manual, in 1987. This first standard version cf the language is often referred to as VHDL-87.

The book not only covers the entire scope of the subject but explains the philosophy of the subject. This makes the understanding of this subject but explains the philosophy of the subject. This makes the understanding of this subject more clear and makes it more interesting. The book will be very useful not only to the students but also to the subject teachers. This book allows the reader to apply what they've learned using realistic examples. Excellent for persons that want to introduce VHDL into their computer engineering sequence programs. The author hopes that the book will fulfill the pressing need of interested readers and welcome any suggestions towards the improvement of the book.

Rishabh Anand

ACKNOWLEDGEMENT

We all marvel at the beautiful rose. The rose, with all its beauty and grandeur. But seldom we pause and thank the kind of gardener who patiently manured and watered it. The book I have written is the fruit of such unseen hands. I wish to express my, profound thanks to the people around me who helped to make this book a reality. I am especially grateful to my teachers and my colleagues specially Mrs. Aparna Mahajan, Mr. Sanjeev Kumar, Mrs. Rashmi Mahajan, Wg.cdr. P. Patnayak, GpCapt. H.S. Dua, Wg.cdr. Onkar Singh, Mrs. Dimple Saproo, Mrs. Naresh Ktimari, Mrs. Chitra Kaul, Director M.M Jha, Mrs.Savita Khazanchi, Mrs. Meenakshi and other faculty members who influenced my academic grown during my study. There was, in and always will be the .love of my students whom I taught, for what is written in this book has been the fruit of experience that I taught them. Finally, thanks to my parents and to rest of my family for their patience and support during the long hours of writing this book and above all there is the one Almighty whose humble children we are. It is his blessings we cherish and pray for. It is the blessing I wish for you. I deeply express my heartful thanks to the publishers **Khanna Book Publishing Co. (P) Ltd.**, for publishing this book in such a beautiful get-up and well in time.

Rishabh Anand

You can also refer following book by the same author and publisher.

- Electromagnetic Field Theory.
- Digital Electronics.
- Mobile Computing
- Communication Systems (Analog and Digital)

CONTENT

UNIT 1

INTRODUCTION TO VHDL

1.1 VHDL TERMS

These are the basic VHDL building blocks that are used in almost every description, along with some terms that are redefined in VHDL to mean something different to the average designer.

Entity. All designs are expressed in terms of entities. An entity is the most basic building block in a design. The uppermost level of the design is the top-level entity. If the design is hierarchical, then the top-level description will have lower-level descriptions contained in it. These lower-level descriptions will be lower-level entities contained in the top-level entity description.

Architecture. All entities that can be simulated have an architecture description. The architecture describes the behavior of the entity. A single entity can have multiple architectures. One architecture might be behavioral while another might be a structural description of the design.

Configuration. A configuration statement is used to bind a component instance to an entity-architecture pair. A configuration can be considered like a parts list for a design. It describes which behavior to use for each entity, much like a parts list describes which part to use for each part in the design.

Package. A package is a collection of commonly used data types and subprograms used in a design. Think of a package as a toolbox that contains tools used to build designs.

Driver. This is a source on a signal. If a signal is driven by two sources, then when both sources are active, the signal will have two drivers.

Bus. The term "bus" usually brings to mind a group of signals or a particular method of communication used in the design of hardware. In VHDL, a bus is a special kind of signal that may have its drivers turned off.

Attribute. An attribute is data that are attached to VHDL objects or predefined data about VHDL objects. Examples are the current drive capability of a buffer or the maximum operating temperature of the device.

Generic. A generic is VHDL's term for a parameter that passes information to an entity. For instance, if an entity is a gate level model with a rise and a fall delay, values for the rise and fall delays could be passed into the entity with generics.

Process. A process is the basic unit of execution in VHDL. All operations that are performed in a simulation of a VHDL description are broken into single or multiple processes.

1.2 COMPUTER-AIDED DESIGN

Computer-aided design (CAD) tools have advanced significantly in the past decade, and nowadays, digital design is performed using a variety of software tools. Prototypes or even final designs can be created without discrete components and interconnection wires.

Figure 1.1 illustrates the steps in cccmodern digital system design. Like any engineering design, the first, step in the design flow is formulating the problem, stating the **design requirements** and arriving at the **design specification.** The next step is to **formulate the design** at a conceptual level, either at a block diagram level or at an algorithmic level.

Design entry is the next step in the design flow. In olden days, this would have been a hand-drawn schematic or blueprint. Now with CAD tools, the design conceptualized in the previous step needs to be entered into the CAD system in an appropriate manner. Designs can be entered in multiple forms. A few years ago, CAD tools used to provide a graphical method to enter designs. This was called **schematic capture.** The schematic editors typically were supplemented with a library of standard digital building blocks like gates, flip-flops, multiplexers, decoders, counters, registers, and so on. ORCAD (a company that produced design automation tools) provided a very popular schematic editor. Now a days, **hardware description languages** (HDLs) are used to enter designs. Two popular HDLs are VHDL and Verilog. The acronym VHDL stands for **VHSIC hardware description language,** and **VHSIC** in turn stands for **very high speed integrated circuit.**

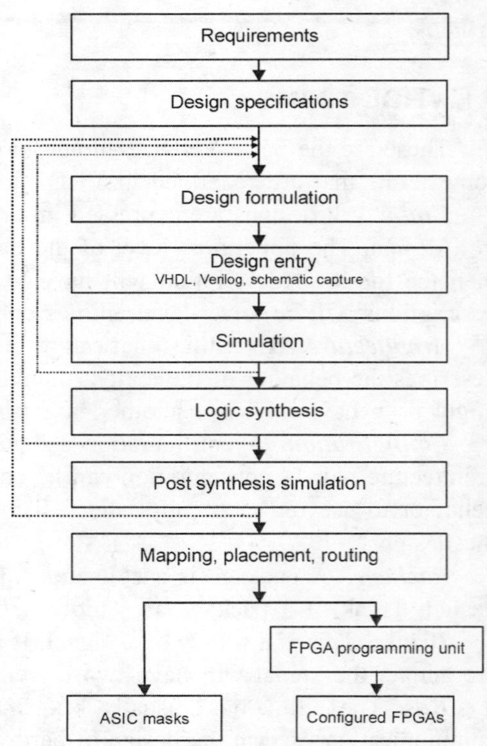

Figure 1.1: Design Flow in Modern Digital System Design

A hardware description language allows a digital system to be designed and debugged at a higher level of abstraction than schematic capture with gates, flip-flops, and standard MSI building blocks. The details of the gates and flip-flops do not need to be handled during early phases of design.

A design can be entered in what is called a **behavioral description** of the design. In a behavioral HDL description, one only specifies the general working of the design at a flow-chart or algorithmic level without associating to any specific physical parts, components, or implementations. Another method to enter a design in VHDL and Verilog is the **structural description** entry. In structural design, specific components or specific implementations of components are associated with the design. A structural VHDL or Verilog model of a design can be considered as a textual description of a schematic diagram that you would have drawn interconnecting specific gates and flip-flops.

Once the design has been entered, it is important to simulate it to confirm that the conceptualized design does function correctly. Initially, one should perform the **simulation** at the high-level behavioral model. This early simulation unveils problems in the initial design. If problems are discovered, the designer goes back and alters the design to meet the requirements.

Once the functionality of the design has been verified through simulation, the next step is synthesis. Synthesis mean "conversion of the higher-level abstract description of the design to actual components at the gate and flip-flop level." The output of the synthesis tool, consisting of a list of gates and a list of interconnections specifying how to interconnect them, is often referred to as a **netlist.** Synthesis is analogous to writing software programs in a high-level language such as C and then using a complier to convert the programs to machine language. Just like a C complier can generate optimized or unoptimized machine code, a synthesis tool can generate optimized or unoptimized hardware. The synthesis software generates different hardware implementations depending on algorithms embedded in the software to perform the translation and optimization techniques incorporated into the tools. A synthesis tool is nothing but a compiler to convert design description to hardware, and it is not unusual to name synthesis package with phrases similar to design compiler, silicon compiler, and so on.

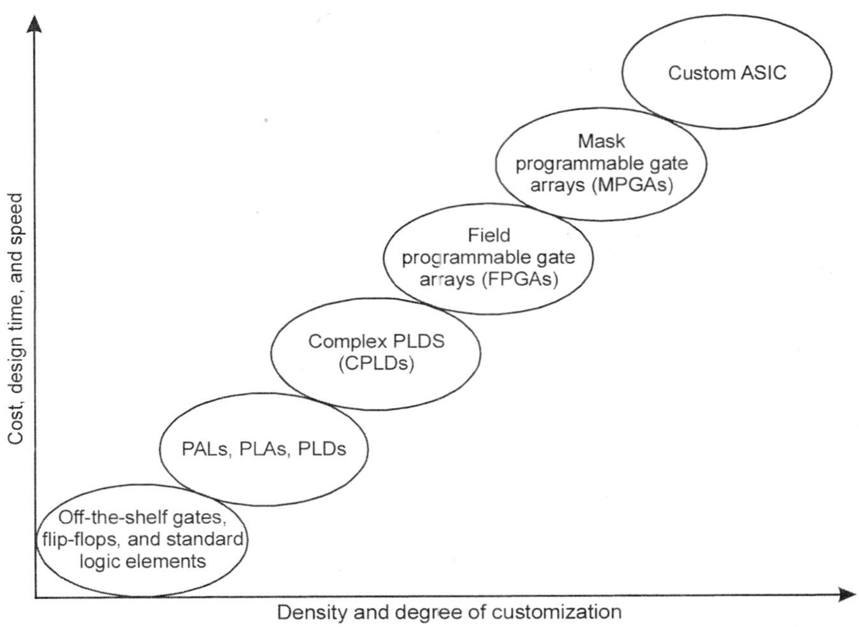

Figure 1.2: Spectrum of Design Technologies

The next step in the design flow is **post-synthesis simulation.** The earlier simulation at a higher level of abstraction does not take into account specific implementations of the hardware components that the design in using. If post-synthesis simulation unveils problems, one should go back and modify the design to meet timing requirements. Arriving at a proper design implementation is an iterative process.

Next, a designer moves into specific realization of the design. A design can be implemented in several different target technologies. It could be a completely custom IC or it could be implemented in a standard part that is easily available from a vendor. The target technologies that are commonly available now are illustrated in Figure 1.2.

At the lowest level of sophistication and density is an old-fashioned printed circuit board with off-the-shelf gates, flip-flops, and other standard logic building blocks. Slightly higher in density are programmable logic arrays (PLAs), programmable array logic (PAL), and simple programmable logic devices (SPLDs). PLDs with higher density and gate count are called complex programmable logic devices (CPLDs).Then there are the popular field programmable gate arrays (FPGAs) and mask programmable gate arrays (MPGAs), or simply gate arrays.The highest level of density and performance is a fully custom application-specific integrated circuit (ASIC).

Two most common target technologies nowadays are FPGAs and ASICs. The initial steps in the design flow are largely the same for either realization. Toward the final stages in the design flow different operations are performed depending on the target technology. This is indicated in Figure 1.1 The design is **mapped** into specific target technology and **placed** into specific parts in the target ASIC or FPGA. The paths taken by the connections between components are decided during the **routing.** If an ASIC is being designed, the routed design is used to generate a photomask that will be used in the IC manufacturing process. If a design is to be implemented in an FPGA, the design is translated to a format specifying what is to be done to various programmable points in the FPGA. In modern FPGAs, programming simply involves writing a sequence of 0's and l's into the programmable cells in the FPGA, and no specific programming unit other than a personal computer (PC) is required.

1.3 HARDWARE DESCRIPTION LANGUAGES

Hardware description languages (HDLs) are a popular mode of design entry. As mentioned previously, two popular HDLs are VHDL and Verilog.

VHDL is a hardware description language used to describe the behavior and structure of digital systems. VHDL is a general-purpose HDL that can be used to describe and simulate the operation of a wide variety of digital systems, ranging in complexity from a few gates to an interconnection of many complex integrated circuits. VHDL was originally developed under funding from the Department of Defense (DoD) to allow a uniform method for specifying digital systems. When VHDL was developed, the main purpose was to have a mechanism to describe and document hardware unambiguously. Synthesizing hardware from high-level descriptions was not one of the original purposes. The VHDL language has since become an IEEE (Institute of Electronic and Electrical Engineers) standard, and it is widely used in industry. IEEE created a VHDL standard in 1987 (VHDL-87) and later modified the standard in 1993 (VHDL-93). Further revisions were done to the standard in 2000 and 2002.

VHDL can describe a digital system at several different levels—**behavioral, data flow,** and **structural.** For example, a binary adder could be described at the behavioral level in terms of its function of adding two binary numbers without giving any implementation details. The same adder could be described at the data flow level by giving the logic equations for the adder. Finally, the adder could be described at the structural level by specifying the gates and the interconnections between the gates that comprise the adder.

VHDL leads naturally to a top-down design methodology, in which the system is first specified at a high level and tested using a simulator. After the system is debugged at this level, the design can gradually be refined, eventually leading to a structural description closely related to the actual hardware implementation. VHDL was designed to be technology independent. If a design is described in VHDL and implemented in today's technology, the same VHDL description could be used as a starting point for a design in some future technology. Although initially conceived as a hardware documentation language, most of VHDL can now be used for simulation and logic synthesis.

Verilog is another popular HDL. It was developed by the industry at about the same time the U.S. DoD was funding the creation of VHDL. Verilog was introduced by Gateway Design Automation in 1984 as a proprietary HDL. Synopsis created synthesis tools for Verilog around 1988. Verilog became an IEEE standard in 1995.

VHDL has its syntactic roots in ADA while Verilog has its syntactic roots in C. ADA was a general-purpose programming language, also sponsored by the Department of Defense. Due to the similarity with C, some find Verilog easier or less intimidating to learn. Many find VHDL to be excellent for supporting design and documentation of large systems. VHDL and Verilog enjoy approximately 50/50 market share. Both languages can accomplish most requirements for digital design rather easily. Often design companies continue to use what they are used to, and hence, Verilog users continue to use Verilog and VHDL users continue to use VHDL. If you know one of these languages, it is not difficult to transition to the other.

More recently, there also have been efforts in system design languages such as **System C, Handel-C,** and **System Verilog.** System C is created as an extension to C++ , and hence some who are very comfortable with general-purpose software development find it less intimidating. These languages are primarily targeted at describing large digital systems at a high level of abstraction. They are primarily used for verification and validation. When different parts of a large system are designed by different teams, one team can use a system level behavioral description of the block being designed by the other team during initial design. Problems that might otherwise become obvious only during system integration may become evident in early stages reducing the design cycle for large systems. System-level simulation languages are used during design of large systems.

1.3.1 Learning a Language

There are several challenges when you learn a new language, whether it be a language for common communication (English, Spanish, French, etc.), a computer language like C, or a special-purpose language such as VHDL. If it is not your first language, you typically have a tendency to compare it to a language you know. In the case of VHDL, if you already know another hardware description language, it is good to compare it with VHDL, but you should be careful when comparing it with languages like C. VHDL and Verilog have a very different purpose than languages like C, and a comparison with C is not a meaningful activity. We will be describing the language assuming it is your first HDL; however, we will assume basic knowledge of computer languages like C and the basic compilation and execution flow.

When one learns a new language, one needs to study the alphabet of the new language, its vocabulary, grammar, syntax rules, and semantics of language descriptions. The process of learning

VHDL is not much different. One needs to learn the alphabet, vocabulary or lexical elements of the language, syntax (grammar and rules), and semantics (meaning of descriptions). VHDL-87 uses the ASCII character set while VHDL-93 allows use of the full ISO character set. The ISO character set includes the ASCII characters and additionally includes accented characters. The ASCII character set only includes the first 128 characters of the ISO character set. The lexical elements of the language include various **identifiers, reserved words,** special symbols, and literals. We have listed these in Appendix A. The syntax or grammar determines what combinations of lexical elements can be combined to make valid VHDL descriptions. These are the rules that govern the use of different VHDL constructs. Then one needs to understand the semantics or meaning of VHDL descriptions. It is here that one understands what descriptions represent combinational hardware versus sequential hardware. And just like fluency in a natural language comes by speaking, reading, and writing the language, mastery of VHDL comes by repeated use of the language to create models for various digital systems.

Since VHDL is a hardware description language, it differs from an ordinary programming language in several ways. Most importantly, VHDL has statements that execute concurrently since they must model real hardware in which the components are all in operation at the same time. VHDL is popularly used for the purposes of describing, documenting, simulating, and automatically generating hardware. Hence, its constructs are tailored for these purposes.

Common Abbreviations

VHDL:	VHSIC hardware description language
VHSIC:	Very high speed integrated circuit
HDL:	Hardware description language
CAD:	Computer-aided design
EDA:	Electronic design automation
LSI:	Large scale integration
MSI:	Medium scale integration
SSI:	Small scale integration
VLSI:	Very large scale integration
ULSI:	Ultra large scale integration
ASCII:	American standard code for information exchange
ISO:	International Standards Organization
ASIC:	Application-specific integrated circuit
FPGA:	Field programmable gate array
PLA:	Programmable logic array
PAL:	Programmable array logic
PLD:	Programmable logic device
CPLD:	Complex programmable logic device

1.4 WHAT IS VHDL?

VHDL is an acronym for VHSIC Hardware Description Language (VHSIC is an acronym for Very High Speed Integrated Circuits). It is a hardware description language that can be used to model a digital system at many levels of abstraction, ranging from the algorithmic level to the gate level. The complexity of the digital system being modeled could vary from that of a simple gate to a complete digital electronic system, or anything in between. The digital system can also be described hierarchically. Timing can also be explicitly modeled in the same description.

The VHDL language can be regarded as an integrated amalgamation of the following languages:

sequential language +
concurrent language +
net-list language +
timing specifications +
waveform generation language =>VHDL

Therefore, the language has constructs that enable you to express the concurrent or sequential behavior of a digital system with or without timing. It also allows you to model the system as an interconnection of components. Test waveforms can also be generated using the same constructs. All the above constructs may be combined to provide a comprehensive description of the system in a single model.

The language not only defines the syntax but also defines very clear simulation semantics for each language construct. Therefore, models written in this language can be verified using a VHDL simulator. It is a strongly typed language and is often verbose to write. It inherits many of its features, especially the sequential language part, from the Ada programming language. Because VHDL provides an extensive range of modeling capabilities, it is often difficult to understand. Fortunately, it is possible to quickly assimilate a core subset of the language that is both easy and simple to understand without learning the more complex features. This subset is usually sufficient to model most applications. The complete language, however, has sufficient power to capture the descriptions of the most complex chips to a complete electronic system.

1.4.1 Capabilities

The following are the major capabilities that the language provides along with the features that differentiate it from other hardware description languages.

- The language can be *used as an exchange medium between chip vendors and CAD tool users.* Different chip vendors can provide VHDL descriptions of their components to system designers. CAD tool users can use it to capture the behavior of the design at a high level of abstraction for functional simulation.
- The language can also be *used as a communication medium between different CAD and CAE tools.* For example, a schematic capture program may be used to generate a VHDL description for the design, which can be used as an input to a simulation program.
- The language supports *hierarchy; that* is, a digital system can be modeled as a set of interconnected components; each component, in turn, can be modeled as a set of interconnected subcomponents.
- The language supports *flexible design methodologies:* top-down, bottom-up, or mixed.
- The language is *not technology-specific,* but is capable of supporting technology-specific features. It can also support various hardware technologies; for example, you may define new logic types and new components; you may also specify technology-specific attributes. By being technology-independent, the same model can be synthesized into different vendor libraries.
- It supports both *synchronous and asynchronous timing* models.

- Various digital modeling techniques, such as finite-state machine descriptions, algorithmic descriptions, and Boolean equations, can be modeled using the language.
- The language is publicly available, human-readable, machine-readable, and, above all, it is *not proprietary.*
- It is an *IEEE and ANSI standard;* therefore, models described using this language are portable. The government also has a strong interest in maintaining this as a standard so that reprocurement and second-sourcing may become easier.
- The language supports *three basic different description styles:* structural, dataflow, and behavioral. A design may also be expressed in any combination of these three descriptive styles.
- It supports a wide *range of abstraction levels* ranging from abstract behavioral descriptions to very precise gate-level descriptions. It does not, however, support modeling at or below the transistor level. It allows a design to be captured at a mixed level using a single coherent language.
- *Arbitrarily large designs can be modeled* using the language, and there are no limitations imposed by the language on the size of a design.
- The language has elements that make *large-scale design modeling easier;* for example, components, functions, procedures, and packages.
- *Test benches can be written* using the same language to test other VHDL models.
- Nominal propagation delays, min-max delays, setup and hold timing, timing constraints, and spike detection can all be described very naturally in this language.
- The use of generics and attributes in the models *facilitate back-annotation* of static information such as timing or placement information.
- Generics and attributes are also useful in describing *parameterized designs.*
- A model can not only describe the functionality of a design, but can also contain *information about the design itself* in terms of user-defined attributes, such as total area and speed.
- A common language can be used to describe library components from different vendors. Tools that understand VHDL models will have no difficulty in reading models from a variety of vendors since the language is a standard.
- Models written in this language can be verified by simulation since *precise simulation semantics are defined* for each language construct.
- Behavioral models that conform to a certain synthesis description style are *capable of being synthesized* to gate-level descriptions.
- The *capability of defining new data types* provides the power to describe and simulate a new design technique at a very high level of abstraction without any concern about the implementation details.

1.5 HARDWARE ABSTRACTION

VHDL is used to describe a model for a digital hardware device. This model specifies the external view of the device and one or more internal views. The internal view of the device specifies the functionality or structure, while the external view specifies the interface of the device through which it communicates with the other models in its environment. Figure 1.3 shows the hardware device and the corresponding software model.

The device-to-device model mapping is strictly one-to-many. That is, a hardware device may have many device models. For example, a device modeled at a high level of abstraction may not have a clock as one of its inputs, since the clock may not have been used in the description. Also, the data transfer at the interface may be treated in terms of, say, integer values, instead of logical values. In VHDL, each device model is treated as a distinct representation of a unique device, called an *entity*. Figure 1.4 shows the VHDL view of a hardware device that has multiple device models, with each device model representing one entity. Even though entity 1 through N represent N different entities from the VHDL point of view, in reality they represent the same hardware device.

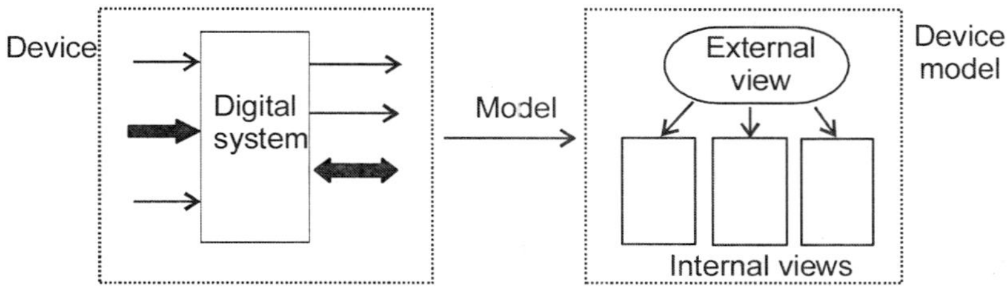

Figure 1.3: A VHDL view of a device

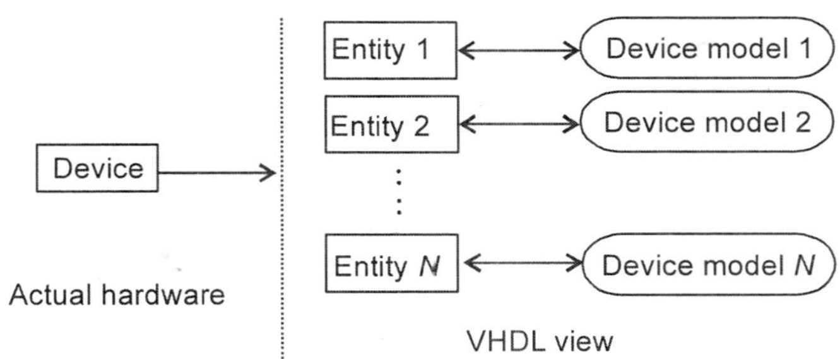

Figure 1.4: A VHDL view of a device.

The entity is thus a hardware abstraction of the actual hardware device. Each entity is described using one model, which contains one external view and one or more internal views. At the same time, a hardware device may be represented by one or more entities.

1.6 IDENTIFIERS

There are two kinds of identifiers in VHDL, basic identifiers and extended identifiers. A basic identifier in VHDL is composed of a sequence of one or more characters. A legal character is an uppercase letter (A ... Z), a lower-case letter (a ... z), a digit (0 ... 9), or the underscore (_) character. The first character in a basic identifier must be a letter, and the last character may not be an underscore. Lower-case and upper-case letters are considered to be identical when used in a basic identifier; as an example, Count, COUNT, and CouNT all refer to the same basic identifier. Also, two underscore characters cannot appear consecutively. Some more examples of basic identifiers are:

```
DRIVE_BUS        SelectSignal     RAM_Address
SET_CK_HIGH      CONST32_59       r2d2
```

An extended identifier is a sequence of characters written between two backslashes. Any of the allowable characters can be used, including characters like ., !, @,', and $. Within an extended identifier, lower-case and upper-case letters are considered to be distinct. Examples of extended identifiers are:

```
\TEST\                  -- Differs from the basic identifier TEST.
\–25\
\2FOR$\
\~Q\
\process\               -- Distinct from the keyword process.
\ .$*****\
\7400TTL\
\----\\----\            -- Two consecutive backslashes represents
                        -- one backslash.
\COunt\       \COUNT\   -- Each is a distinct identifier and both are
                        -- distinct from the basic identifier COUNT.
```

Comments in a description must be preceded by two consecutive hyphens (--); the comment extends to the end of the line. Comments can appear anywhere within a description. Examples are:

```
-- This is a comment; it ends at the end of this line.
-- To continue a comment onto a second line, a separate
-- comment line must be started.

entity UART is end; -- This comment starts after entity declaration.
```

These words, also called keywords, have a specific meaning in the language and therefore cannot be used as basic identifiers.

1.7 DATA OBJECTS

A data object holds a value of a specified type. It is created by means of an object declaration. An example is:

```
variable COUNT: INTEGER;
```

This results in the creation of a data object called COUNT, which can hold integer values. The object COUNT is also declared to be of variable class.

Every data object belongs to one of the following four classes:

1. *Constant*: An object of constant class (often called a constant) can hold a single value of a given type. This value is assigned to the constant before simulation starts, and the value cannot be changed during the course of the simulation. For a constant declared within a subprogram, the value is assigned to the constant every time the subprogram is called.
2. *Variable*: An object of variable class (often called a variable) can also hold a single value of a given type. However, in this case, different values can be assigned to the variable at different times using a variable assignment statement.
3. *Signal*: An object belonging to the signal class (often called a signal) holds a list of values, which include the current value of the signal, and a set of possible future values that are to appear on the signal. Future values can be assigned to a signal using a signal assignment statement.
4. *File*: An object belonging to the file class (often called a file) contains a sequence of values. Values can be read or written to the file using read procedures and write procedures, respectively.

Signals can be regarded as wires in a circuit, while variables and constants are analogous to their counterparts in a high-level programming language like C or Pascal. Signals are typically used to model wires and flip-flops, while variables and constants are typically used to model the behavior of the circuit. A file is used to model a file in the host environment.

An object declaration is used to declare an object, its type, and its class. A value can be optionally assigned to a signal, variable, or constant. For a file, the object declaration may specify information on how to open the file. Some examples of object declarations follow.

Constant Declarations

Examples of constant declarations are:

```
constant RISE_ITME: TIME := 10 ns;
constant BUS_WIDTH: INTEGER := 8;
```

The first declaration declares the object RISE_TIME, which can hold a value of type TIME (a predefined type in the language), and the value assigned to the object at the start of simulation is 10 ns. The second constant declaration declares a constant BUS_WIDTH of type INTEGER, with a value of 8.

An example of another form of constant declaration is:

```
constant NO_OF_INPUTS: INTEGER;
```

The value of the constant has not been specified in this case. Such a constant is called a *deferred constant*, and it can appear only inside a package declaration. The complete constant declaration with the associated value must appear in the corresponding package body.

Variable Declarations

Examples of variable declarations are:

```
variable CTRL_STATUS: BIT_VECTOR (10 downto 0);
variable SUM: INTEGER range 0 to 100 := 10;
variable FOUND, DONE: BOOLEAN;
```

The first declaration specifies a variable object CTRL_STATUS as an array of 11 elements, with each array element of type BIT. In the second declaration, an explicit initial value has been assigned to the variable SUM. When simulation starts, SUM will have an initial value of 10. If no initial value is specified for a variable object, a default value is used as an initial value. This default value is T'LEFT, where T is the object type and 'LEFT is a predefined attribute of a type that gives the leftmost value in the set of values belonging to type T. In the third declaration, the initial values assigned to FOUND and DONE at start of simulation is FALSE (FALSE is the leftmost value of the predefined type BOOLEAN). If the type of a variable is an array type or a record type, the initial value of each of the elements that form the variable is the initial value of each of its corresponding element types. Thus, the initial value for each of the array elements of CTRL_STATUS is '0'.

Signal Declarations

Here are some examples of signal declarations.

```
signal CLOCK: BIT;
signal DATA_BUS: BIT_VECTOR(0 to 7);
signal GATE_DELAY: TIME := 10 ns;
signal INIT_P: STD_LOGIC_VECTOR(7 downto 0) :=
        (0 =>' 1', others => 'U');
```

The interpretation of these signal declarations is very similar to that of the variable declarations. The first signal declaration declares the signal object CLOCK of type BIT and gives it an initial value of '0' ('0' being the leftmost value of type BIT). The third signal declaration declares a signal object GATE_DELAY of type TIME that has an initial value of 10 ns. The fourth signal declaration declares an 8-element signal array INIT_P of type STD_LOGIC_VECTOR; the 0th element is initialized to '1'and all other elements are initialized to the value 'U'.

File Declarations

A file is declared using a file declaration. The syntax of a file declaration is:

file *file-names : file-type-name* [[**open** *mode*] **is** *string-expression*];

The string expression is interpreted by the host environment as the physical name of a file. The mode specifies whether the file is to be used as a read-only or write-only, or in the append mode. Examples of file declarations are:

```
-- First two lines are file type declarations (explained later
--in this chapter):
type STD_LOGIC_FILE is file of STD_LOGIC_VECTOR;
type BIT_FILE is file of BIT_VECTOR;
-- File declarations are:
file STIMULUS: TEXT open READ_MODE
            is "/usr/home/jb/add.sti";
file VECTORS: BIT_FILE is "/usr/home/james/add.vec";
file FAT1, FAT2: STD_LOGIC_FILE;
```

In the first example of a file declaration, a file STIMULUS is declared to be of the predefined file type TEXT; that is, this file may contain an indefinite number of strings. The mode value, READ_MODE, specifies that the file will be opened in read-only mode, and the string expression after the keyword is specifies the path name to a physical file in the host environment. During elaboration, this declaration causes the specified file that STIMULUS is linked with "/usr/home/jb/add.sti", to be opened in read mode.

In the second example, VECTORS is declared as a file containing an indefinite number of bit vectors. This declaration also specifies the link to the file in the host environment. Since no mode is specified the default mode, READ _MODE, is used. The third file declaration declares two files, PAT1 and PAT2, each of type STD_LOGIC_FILE; that is, the files contain values of type STD_LOGIC_VECTOR. Since no link has been specified to a file in the host environment, the file is not opened during elaboration. The file will have to be explicitly opened using a file open procedure.

Other Ways to Declare Objects

Not all objects in a VHDL description are explicitly created using object declarations. These other objects are declared as one of the following:

1. Ports of an entity. All ports are signal objects.

2. Generics of an entity are constant objects.
3. Function parameters are constants or signals, while procedure parameters can belong to any object class.

There are two other types of objects that are implicitly declared. These are the indices of a for... loop statement and the generate statement (generate statements are discussed in Chapter 10). An example of such an implicit declaration for the loop index in a for... loop statement is shown.

```
for COUNT in 1 to 10 loop
  SUM := SUM + COUNT;
end loop;
```

In this for... loop statement, COUNT is a constant that has an implicit declaration of type INTEGER with range 1 to 10, and therefore cannot be explicitly declared. The constant COUNT is created when the loop is first entered and ceases to exist after the loop is exited.

1.8 DATA TYPES

Every data object in VHDL can hold a value that belongs to a set of values. This set of values is specified by using a **type declaration**. A type is a name that has associated with it a set of values and a set of operations. Certain types, and operations that can be performed on objects of these types, are predefined in the language. For example, INTEGER is a predefined type, with the set of values being integers in a specific range provided by the VHDL system. The minimum range that must be provided is $-(2^{31}-1)$ through $+(2^{31}-1)$. Some of the allowable and frequently used predefined operators are + for addition, – for subtraction, / for division, and * for multiplication. BOOLEAN is another predefined type that has the values FALSE and TRUE, and some of its predefined operators are **and, or, nor, nand, xor, xnor,** and **not**.

The language also provides the facility to define new types by using type declarations and also to define a set of operations on these types by writing functions that return values of this new type. All the possible types that can exist in the language can be categorized into the following four major categories:

1. *Scalar types*: Values belonging to these types appear in sequential order.
2. *Composite* types: These are composed of elements of a single type (an array type) or elements of different types (a record type).
3. *Access types*: These provide access to objects of a given type (via pointers).
4. *File types*: These provide access to objects that contain a sequence of values of a given type.

It is possible to derive subtypes from predefined or user-defined types.

1.8.1 Subtypes

A subtype is a type with a constraint. The constraint specifies the subset of values for the subtype. The type is called the base type of the subtype. An object is said to belong to a subtype if it is of the base type and if it satisfies the constraint. Subtype declarations are used to declare subtypes. An object can be declared to belong to either a type or a subtype.

The set of operations belonging to a subtype is the same as that associated with its base type. Subtypes are useful for range checking and for imposing additional constraints on types.

Examples of subtypes are:

```
subtype MY_INTEGER is INTEGER range 48 to 156;
type DIGIT is ('0', 1, '2, '3', '4', '5', '6', 7', '8', '9');
-- DIGIT is a type, not a subtype. It is the base type in the
-- following subtype declaration:
subtype MIDDLE is DIGIT range '3' to 7';
```

In the first example, MY_ INTEGER is a subtype of the INTEGER base type and has a range constraint with values ranging from 48 through 156. DIGIT is a user-defined enumeration type. The last subtype declaration declares a new subtype called MIDDLE, whose base type is DIGIT and has the values '3', '4', '5', '6', and 7'.

A subtype need not impose a constraint. In such a case, the subtype simply gives another name to an already existing type. For example,

```
subtype NUMBER is DIGIT;
```

NUMBER is a subtype with no constraint. Therefore, its set of values are the same as that for type DIGIT.

1.8.2 Scalar Types

The values belonging to these types are ordered, that is, relational operators can be used on these values. For example, BIT is a scalar type, and the expression '0' < '1' is valid and has the value TRUE. There are four different kinds of scalar types. These types are:

1. Enumeration
2. Integer
3. Physical
4. Floating point

Integer types, floating point types, and physical types are classified as numeric types because the values associated with these types are numeric. Enumeration and integer types are called discrete types because these types have discrete values associated with them. Every value belonging to an enumeration type, integer type, or a physical type has a position number associated with it. This number is the position of the value in the ordered list of values belonging to that type.

Enumeration Types

An enumeration type declaration defines a type that has a set of user-defined values consisting of identifiers and character literals. Examples are:

```
type MVL is ('U', '0', '1', 'Z');
type MICRO_OP is (LOAD, STORE, ADD, SUB, MUL, DIV);
subtype ARITH_OP is MICRO_OP range ADD to DIV;
```

Examples of objects defined for these types are:

```
signal CONTROL_A: MVL;
signal CLOCK: MVL range'0' to '1';     --Implicit subtype declaration.
variable IC: MICRO_OP := STORE;
 -- STORE is the initial value for IC.
variable ALU: ARITH_OP;
```

MVL is an enumeration type that has the set of ordered values, 'U', '0', '1', and 'Z'. ARITH__OP is a subtype of the base type MICRO_OP and has a range constraint specified to be from ADD to DIV; that is, the values ADD, SUB, MUL, and DIV belong to the subtype ARITH_OP. A range constraint can also be specified in an object declaration, as shown in the signal declaration for CLOCK; here, the value of signal CLOCK is restricted to '0' or '1'.

The order in which values appear in an enumeration type declaration defines their ordering. That is, when using relational operators, a value is always less than a value that appears to its right in the order. For instance, in the MICRO_OP type declaration, STORE < DIV is true, and SUB > MUL is false. Values of an enumeration type also have a position number associated with them. The position number of the leftmost element is 0. The position number of any other element is one more than the position number of the element to its left.

The values of an enumeration type are called enumeration literals. For example, consider the following enumeration type declaration.

```
type CAR_STATE is (STOP, SLOW, MEDIUM, FAST);
```

The enumeration literals specified in this type declaration are STOP, SLOW, MEDIUM, and FAST; therefore, objects of type CAR_STATE may be assigned only these values.

If the same literal is used in two different enumeration type declarations, the literal is said to be overloaded. In such a case, whenever such a literal is used, the type of the literal is determined from its surrounding context. The following example clarifies this.

```
type MVL is ('U','0','1', 'Z');                -- line 1
type TWO_STATE is ('0', '1');                  -- line 2
...
variable CLOCK: TWO_STATE;                     -- line 3
variable LATCH_CTRL: MVL;                      -- line 4
...
CLOCK := '0';                                  -- line 5
LATCH CTRL := LATCH CTRL xor '0';              -- line 6
...
```

Here, '0' and '1' are two literals that are overloaded since they appear in both the types, MVL and TWO_STATE. The value '0' being assigned to CLOCK in line 5 refers to the literal in the TWO_STATE type since variable CLOCK is of that type. The value '0' in the expression for LATCHLCTRL refers to the literal in type MVL since the first argument of the xor operator is of type MVL (this assumes that the xor operator is allowed for MVL type operands, which is achieved by overloading the operator; operator overloading is discussed in Chapter 8).

The predefined enumeration types of the language are CHARACTER, BIT, BOOLEAN, SEVERITY_LEVEL, FILE_OPEN_KIND, and FILE_OPEN_STATUS. Values belonging to the type CHARACTER constitute the 191 characters of the ISO eight-bit coded character set. These values are called character literals and are always written between two single quotes (' '). Examples are:

```
'A'.
' '
'''   -- the single quote character itself.
'3'        -- the character literal 3.
```

The predefined type BIT has the literals '0' and '1', while type BOOLEAN has the literals FALSE and TRUE. Type SEVERITY_LEVEL has the values NOTE, WARNING, ERROR, and FAILURE; this type is typically used in assertion statements. Type FILE_OPEN_KIND has the values READ_MODE, WRITE_MODE, and APPEND_MODE, while type FILE_OPEN_STATUS has the values OPEN_OK, STATUS_ERROR, NAME_ERROR, and MODE_ERROR; these two types are used exclusively with file operations.

Type STD_ULOGIC, defined in the package STD_LOGIC_1164, is an enumeration type. It is defined as:

```
type STD_ULOGIC is (
        'U',            -- Uninitialized
        'X',            -- Forcing unknown
        '0',            -- Forcing
        '1',            -- Forcing 1
        'Z',            -- High impedance
        'W',            -- Weak unknown
        'L',            -- Weak 0
        'H',            -- Weak 1
        '-'             -- Don't care
        );
```

Integer Types

An integer type defines a type whose set of values fall within a specified integer range. Examples of integer type declarations are:

```
type INDEX is range 0 to 15;
type WORD_LENGTH is range 31 downto 0;
subtype DATA_WORD is WORD_LENGTH range 15 downto 0;
type MY_WORD is range 4 to 6;
```

Some object declarations using these types are:

```
constant MUX_ADDRESS: INDEX := 5;
signal DATA_BUS: DATA_WORD;
```

INDEX is an integer type that includes the integer values from 0 through 15. DATA_WORD is a subtype of WORD_LENGTH that includes the integer values ranging from 15 through 0. The position of each Value Of an integer type is the value itself. For example, in the type declaration of WORD_LENGTH, value 31 is at position 31, value 14 is at position 14, and so on. In the declaration of MY_WORD, the position numbers of values 4, 5, and 6 are 4, 5, and 6, respectively. Contrast this with the position numbers of elements of an enumeration type; the position number in case of integer types does not refer to the index of the element in the range, but to its numeric value itself. The bounds of the range for an integer type must be constants or *locally static expressions*; a locally static expression is an expression that computes to a constant value at compile time (a *globally static expression* is an expression that computes to a constant value at elaboration time).

Values belonging to an integer type are called integer literals. Examples of integer literals are:

```
56349        6E2        0        98_71_28
```

Literal 6E2 refers to the decimal value 6 * (102) = 600. The underscore (_) character can be used freely in writing integer literals and has no impact on the value of the literal; 98_71_28 is same as 987128.

INTEGER is the only predefined integer type of the language. The range of the INTEGER type is implementation-dependent but must at least cover the range from $-(2^{31}-1)$ to $+(2^{31}-1)$.

Floating Point Types

A floating point type has a set of values in a given range of real numbers. Examples of floating point type declarations are:

```
typeTTL_VOLTAGE is range -5.5 to -1.4;
type REAL_DATA is range 0.0 to 31.9;
```

An example of an object declaration is:

```
variable LENGTH: REAL_DATA range 0.0 to 15.9;
...
variable L1, L2, L3: REAL_DATA range 0.0 to 15.9;
```

LENGTH is a variable object of type REAL_DATA that has been constrained to take real values in the range 0.0 through 15.9 only. Notice that in this case, the range constraint was specified in the variable declaration itself. Alternately, it is possible to declare a subtype and then use this subtype in the variable declarations as shown.

```
subtype RD16 is REAL_DATA range 0.0 to 15.9;
...
variable LENGTH: RD16;
...
variable L1, L2, L3: RD16;
```

The range bounds specified in a floating point type declaration must be constants or locally static expressions.

Floating point literals are values of a floating point type. Examples of floating point literals are:

```
16.26    0.0    0.002    3_1.4_2
```

Floating point literals differ from integer literals by the presence of the dot (.) character. Thus, 0 is an integer literal while 0.0 is a floating point literal.

Floating point literals can also be expressed in an exponential form. The exponent represents a power of 10 and the exponent value must be an integer. Examples are:

```
62.3E-2        5.0E+2
```

Integer and floating point literals can also be written in a base other than 10 (decimal). The base can be any value between 2 and 16. Such literals are called based *literals*. In this case, the exponent represents a power of the specified base. The syntax for a based literal is:

base # *based-value* # -- form 1
base # based-value # **E** *exponent* -- form 2

Examples are:

2#101_101_000#	represents $(101101000)_2$ = (360) in decimal,
16#FA#	represents $(FA)_{16}$ = (11111010)2 = (250) in decimal,
16#E#E1	represents $(E)_{16}$ * (16^1) = 14 * 16 = (224) in decimal,
2#110.01#	represents $(110.01)_2$ = (6.25) in decimal.

The base and the exponent values in a based literal must be in decimal notation.

The only predefined floating point type is REAL The range of REAL is again implementation-dependent, but it must at least cover the range –1.0E38 to +1.0E38, and it must provide at least six decimal digits of precision.

Physical Types

A physical type contains values that represent measurement of some physical quantity, like time, length, voltage, or current. Values of this type are expressed as integer multiples of a base unit. An example of a physical type declaration is:

```
type CURRENT is range 0 to 1E9
   units
      nA;                          -- (base unit) nano-ampere
      uA   = 1000 nA;              -- micro-ampere
      mA   = 1000 uA;              -- milli-ampere
      Amp  = 1000 mA;              -- ampere
   end units;

subtype FILTER_CURRENT is CURRENT range 10 uA to 5 mA;
```

CURRENT is defined to be a physical type that contains values from 0 nA to 109 nA. The base unit is a nano-ampere, while all others are derived units. The position number of a value is the number of base units represented by that value. For example, 2 uA has a position of 2000 while 100 nA has a position of 100. The range of values may include negative values, as shown in the next example.

```
type STEP_TYPE is range –10 to +10
   units
      STEP;                        -- base unit
      STEP2 = 2 STEP;              -- derived unit
      STEP5 = 5 STEP;              -- derived unit
end units;
```

Values in this type range from –10 STEP to +10 STEP. Other units of this type are STEP2 and STEP5.

Values of a physical type are called physical literals. Physical literals are written as an integer or a floating point literal followed by the unit name. For example, "10 nA" is a physical literal (note that a space between 10 and nA is essential), while "Amp" is also a literal that implies 1 Amp. Other examples are:

```
100 ns
10V
50 sec
Kohm                  -- implies 1 Kohm
5.2 mA                -- equivalent to 5200 uA
2.5 STEP2             -- equivalent to 5 STEP
5.6 nA                -- is 5 nA. Fractional part is truncated since nA is the
                      -- base unit for type CURRENT
7.2 STEP              -- is 7 STEP. Fractional part is truncated.
5.2643 uA             -- is 5264 nA
```

The only predefined physical type is TIME, and its range of base values, which again is implementation-dependent, must at least be $-(2^{31}-1)$ to $+(2^{31}-1)$. There is also a predefined physical subtype, DELAY_LENGTH, that represents non-negative time values.

1.8.3 A composite type

A composite type represents a collection of values. There are two composite types: array types and record types. An array type represents a collection of values all belonging to a single type; on the other hand, a record type represents a collection of values that may belong to different types. An object belonging to a composite type therefore represents a collection of subobjects, one for each element of the composite type. A composite type could have a value belonging to either a scalar type, a composite type, or an access type. For example, a composite type may be defined to represent an array of an array of records. This provides the capability of defining arbitrarily complex composite types.

Array Types

An object of an array type consists of elements that have the same type. Examples of array type declarations are:

```
type ADDRESS_WORD is array (0 to 63) of BIT;
type DATA_WORD is array (7 downto 0) of STD_ULOGIC;
type ROM is array (0 to 125) of DATA _ WORD;
type DECODE _MATRIX is array (POSITIVE range 15 downto 1,
NATURAL range 3 downto 0) of STD_ULOGIC;
-- POSITIVE and NATURAL are predefined subtypes; these are:
subtype NATURAL is INTEGER range 0 to INTEGER' HIGH;
subtype POSITIVE is INTEGER range 1 to INTEGER'HIGH;
-- T 'HIGH gives the highest value belonging to type T.
```

Examples of object declarations using these types are:

```
variable ROM_ADDR: ROM;
signal ADDRESS_BUS: ADDRESS_WORD;
constant DECODER: DECODE-MATRIX;      -- A deferred constant.
variable DECODE_VALUE: DECODE_MATRIX;
```

ADDRESS_BUS is a one-dimensional array object that consists of 64 elements of type BIT. ROM_ADDR is a one-dimensional array object that consists of 126 elements, each element being another array object consisting of eight elements of type STD_ULOGIC. We have thus created an array of arrays.

Elements of an array can be accessed by specifying the index values into the array. For example, ADDRESS_BUS(26) refers to the 27th element of ADDRESS_BUS array object; ROM_ADDR(10)(5) refers to the value (of type STD_ULOGIC) at index 5 of the ROM_ADDR(10) data object (of type DATA_WORD); DECODER(5, 2) refers to the value of the element at the 2nd column and 5th row of the two-dimensional object. Notice the difference in addressing for a two-dimensional array type and an array of a type that is an array of another type.

The language allows for an arbitrary number of dimensions to be associated with an array. It also allows an array object to be assigned to another array object of the same type using a single assignment statement. Assignment can be made to an entire array, or to an element of an array, or to a slice of an array. For example,

```
ROM_ADDR (5) := "01000100";          -- Assign to an element of an array.
DECODE_VALUE := DECODER;             -- An entire array is assigned.
ADDRESS_BUS (8 to 15) <=X"FF";       -- Assign to a slice of an array.
```

These examples of array types are constrained array declarations, since the number of elements in the type is explicitly specified. The language also allows array types to be unconstrained; in this case, the number of elements in the array is not specified in the type declaration. Instead, an object declaration for an object of that type declares the number of elements of the array. A subtype declaration constraining the array type may also specify the index constraint for an unconstrained array type. A subprogram parameter may be of an unconstrained type; in this case, the constraint is obtained from the actual parameter passed in during the subprogram call. Examples of unconstrained array declarations are:

```
-- The"<>"symbol is called"box".
type STACK_TYPE is array (INTEGER range <>)
  of ADDRESS_WORD;
subtype STACK is STACK_TYPE (0 to 63);
type OP_TYPE is (ADD, SUB, MUL, DIV);
type TIMING is array (OP_TYPE range <>, OP_TYPE range <>)
  of TIME;
```

Examples of object declarations using these types are:

```
variable FAST_STK: STACK_TYPE (–127 to 127);
constant ALU_TIMING: TIMING :=
--ADD, SUB, MUL
((10 ns, 20 ns, 45 ns),            -- ADD
(20 ns, 15 ns, 40 ns),             -- SUB
(45 ns, 40 ns, 30 ns));            -- MUL
```

STACK_TYPE is defined to be an unconstrained array type that specifies the index of the array as an integer type, and the element type as type ADDRESS_WORD. STACK is a subtype of the base type STACK_TYPE with the specified index constraint. The variable declaration for FAST_STK, defined to be of type STACK_TYPE, also specifies an index constraint. The constant ALU-TIMING specifies the timing for a two-operator ALU, where the operators could be ADD, SUB, or MUL; for example, an ALU that performs ADD and SUB operations has a delay of 20 ns. The declaration for ALU_TIMING is a special case of a constant declaration where no constraint need be specified for the unconstrained array type, since the size of the constant object is determined from the number of values in the constant.

There are two predefined one-dimensional unconstrained array types in the language, STRING and BIT_VECTOR. STRING is an array of characters, while BIT__VECTOR is an array of bits. Examples are:

```
variable MESSAGE: STRING(1 to 17) := "Hello, VHDL world";
signal RX_BUS: BIT_VECTOR (0 to 5) := 0"37";
-- 0"37" is a bit-string literal representing the octal value 37.
constant ADD_CODE: BIT_VECTOR := ('0', '1', '1', '1', '0');
```

There are two one-dimensional unconstrained array types defined in the package STD_LOGIC_1164. These are:

```
type STD_ULOGIC_VECTOR is array (NATURAL range <>)
                    of STD_ULOGIC;
type STD_LOGIC_VECTOR is array (NATURAL range <>)
                    of STD_LOGIC;

-- Type STD_LOGIC is a resolved subtype of STD-ULOGIC
-- (see Section 5.4 for what a resolved subtype means).
```

Examples of declarations using these types are:

```
signal RASTER_LINE: STD_LOGIC_VECTOR (0 to NBLITS);
variable REG_FILE: STD_ULOGIC.VECTOR
              (UPPER_BND downto LOWER_BND);
```

VHDL does not allow a type that is an unconstrained array of an unconstrained array.

```
type MEMORY is array (NATURAL range <>) of
          STD_LOGIC_VECTOR;        -- Not allowed in VHDL.
type REG_FILE is array (NATURAL range <>) of
BIT_VECTOR (0 to 7);               -- This is ok.
type DISPLAY_BANK is
     array (NATURAL range <>, NATURAL range <>) of BIT;
                            -- This is also ok.
```

The first type declaration, that of MEMORY, is not valid VHDL since STD_LOGIC_VECTOR is an unconstrained array type and therefore MEMORY is an unconstrained array of an unconstrained array, which is not allowed. Unconstrained arrays of constrained arrays are allowed as the example of type REG_FILE shows. The third type declaration, that of DISPLAY_BANK, shows that a two-dimensional (including multi-dimensional) unconstrained array type is supported by the language.

A value representing a one-dimensional array of characters is called a string literal. String literals are written by enclosing the sequence of characters within double quotes ("..."). Examples of string literals are:

```
"THIS IS A TEST"
"SPIKE DETECTED!"
"State " "READY"" entered!"      -- Two double quote characters in a
-- sequence represents one double quote character.
```

A string literal can be assigned to different types of objects; for example, to a STRING type object or a BIT_VECTOR type object. The type of a string literal is therefore determined from the context in which it appears. Here are some examples.

```
-- Example 1:
variable ERROR_MESSAGE: STRING(1 to 19);
ERROR_MESSAGE := "Fatal ERROR: abort!";
-- Example 2:
variable BUS_VALUE: BIT_VECTOR (0 to 3);
BUS_VALUE:="1101";
```

In the first example, the string literal is of type STRING, while in the second example, the string literal is of type BIT_VECTOR. The type of a string literal can also be explicitly stated by using a qualified expression (qualified expressions are described in Chapter 10).

A string literal that represents a sequence of bits (values of type BIT) can also be represented as a bit string literal. This sequence of bits, called bit strings, can be represented as either a binary value, an octal value, or a hexadecimal value. The underscore character can be freely used in bit string literals for clarity. Examples are:

```
X" FFO"                          -- X for hexadecimal.
B" 00_0011_1101"                 -- B for binary.
O"327"                           -- O for octal.
X"FF_F0_AB"
```

The type of a bit string literal is also determined from the context in which it appears. For example,

```
type MVL is ('X', '0', '1', 'Z');
type MVL_VECTOR is array (NATURAL range <>) of MVL;
variable FXA: MVL_VECTOR (0 to 3);
variable BRY: BIT_VECTOR (7 downto 0);
variable CLA: STD_LOGIC_VECTOR (1 to 9);
...
FXA := B"01 _00";-- Assign a binary value.
BRY:= X"AB";            -- Assign a hexadecimal value.
CLA:=O"304";       -- Assign an octal value.
```

The bit string literal in the first assignment statement is of type MVL_VECTOR, the bit string literal in the second assignment statement is of type BIT_VECTOR, while the bit string literal in the third assignment is of type STD_LOGIC_VECTOR.

There are many different ways to assign values to an array object. Here are some examples.

```
variable OP_CODES : BIT_VECTOR (1 to 5);
OP_CODES := "01001"; - A string literal is assigned.
OP_CODES := ('0', '1', '0', '0', '1');        -- Positional association is used;
-- first value is assigned to OP_CODES(l), second value to
-- OP_CODES (2), and so on. OP_CODES := (2 => '1', 5 => '1', others => '0'); -- Named association;
-- second and fifth element of OP_CODES get the value '1', and
-- the rest get a value of '0'.
OP_CODES := (others => '0');              -- All values set to '0'.
```

Notice the use of the keyword others to mean all previously un-assigned values; when used, it must be the last association. The expressions used in the last three assignments to OP_CODES are examples of array aggregates. An aggregate is a set of comma-separated elements enclosed within parenthesis.

An aggregate can also be used to provide an initial value for an array object in its declaration. Here are two examples.

```
variable OP_CODES : BIT_VECTOR(l to 5) :=
(2 | 5 => '1', others => '0');
signal CG_PORT: STD_LOGIC_VECTOR(0 to 15) := (others => '0');
```

In the variable declaration, the 2nd and the 5th bits of the variable OP_CODES are set to '1', while all other bits are set to '0'. In the signal declaration, all elements of the array CG_PORT are initialized to the value'0'.

Here is an example of an aggregate specified in a constant declaration.

```
constant CLN: STD_LOGIC_VECTOR (2downto 0) := "001";
```

CLN(2) is '0', CLN(1) is '0' and CLN (0) is '1'. When an aggregate has only one element, then named association must be used to specify the value.

```
constant TIK: STD_LOGIC_VECTOR (2 downto 2) := (2 => '1');
-- It is illegal to write:
-- constant TIK: STD_LOGIC_VECTOR (2 downto 2) := ('1');
```

Record Types

An object of a record type is composed of elements of same or different types. A record type is analogous to the record data type in Pascal and the struct declaration in C. An example of a record type declaration is:

```
type PIN_TYPE is range 0 to 10;
type MODULE is
  record
SIZE                    : INTEGER range 20 to 200;
CRITICAL_DLY            : TIME;
NO_INPUTS               : PIN_TYPE;
NO_OUTPUTS              : PIN_TYPE;
end record;
```

Values can be assigned to a record object using aggregates. For example,

```
variable NAND_COMP: MODULE;
  -- NAND_COMP is an object of record type MODULE.
NAND_COMP := (50, 20 ns, 3, 2);
  -- Implies 50 is assigned to SIZE, 20 ns is assigned
  -- to CRITICAL_DLY, etc.
```

Values can be assigned to a record object from another record object of the same type using a single assignment statement. In the following example, each element of NAND_GENERIC is assigned the value of the corresponding element in NAND_COMP.

```
signal NAND_GENERIC: MODULE;

NAND_GENERIC <= NAND_COMP;
```

Each element of a record object can also be assigned individually by using selected names. For example,

```
NAND_COMPNO_INPUTS := 2;
```

Aggregate values can be assigned to record objects using both positional and named associations. Since the type of an aggregate is determined from the context in which it is used, the aggregate can be either an array or a record aggregate, depending on its usage. For example,

```
CAB := (others => '0');
```

If CAB is an array type object, the aggregate is treated as an array aggregate; on the other hand, if CAB is a record type object, the aggregate is a record aggregate where **others** refers to all the elements in the record.

1.8.4 Access Types

Values belonging to an access type are pointers to a dynamically allocated object of some other type. They are similar to pointers in Pascal and C languages. Examples of access type declarations are:

```
-- MODULE is a record type declared in the previous sub-section.
type PTR is access MODULE;
type FIFO is array (0 to 63, 0 to 7) of BIT;
type FIFO_PTR is access FIFO;
```

PTR is an access type whose values are addresses that point to objects of type MODULE. Every access type may also have the value **null,** which means that it does not point to any object. Objects of an access type can only belong to the variable class. When an object of an access type is declared, the default value of this object is **null.** For example,

```
variable MOD1PTR, MOD2PTR: PTR;          -- Default value is null.
```

Objects to which access types point can be created using allocators. Allocators provide a mechanism to dynamically create objects of a specific type.

```
MOD1PTR := new MODULE;
```

The **new** in this assignment causes an object of type MODULE to be created, and the pointer to this object is returned. The values of the elements of the MODULE record are the default values of each element; these are 20 (the leftmost value of the implied subtype) for the SIZE element, TIME'LEFT for the CRITICAL_DLY element, and the value 0 (this is PIN_TYPE'LEFT) for the NO_INPUTS and NO-OUTPUTS elements. Initial values can also be assigned to a newly created object by explicitly specifying the values, as shown in the following example.

```
MOD2PTR := new MODULE'(25,10 ns, 4, 9);
```

Objects of an access type can be referenced as:

1. *obj-ptr*.**all**: Accesses the entire object pointed to by obj_ptr, where *obj-ptr* is a pointer to an object of any type
2. *array-obj-ptr* (*element-index*): Accesses the specified array element, where array-*obj-ptr* is a pointer to an array object
3. *record-obj-ptr.element-name*: Accesses the specified record element, where record-obj-ptr is a pointer to a record object

The elements of the object to which MOD2PTR points can be accessed as MOD2PTR.SIZE, MOD2PTR.CRITICAL_DLY, MOD2PTR.NO_OUTPUTS, and MOD2PTR.NO_OUTPUTS, provided the pointer is not null.

For every access type, a procedure DEALLOCATE is implicitly declared. This procedure, when called, returns the storage occupied by the object to the host environment. For the PTR and FIFO_PTR access types declared earlier, the following DEALLOCATE procedures are implicitly declared:

```
procedure DEALLOCATE (P: intout PTR);
procedure DEALLOCATE (P: inout FIFO_PTR);.
```

The execution of the following statement:

```
DEALLOCATE (MOD2FTR);
```

causes the storage occupied by the object to which MOD2PTR points, to be deallocated, and MOD2PTR is set to **null.**

Pointers can be assigned to other pointer variables of the same access type. Therefore,

```
MOD1PTR := MOD2PTR;
```

is a legal assignment. Both MOD1PTR and MOD2PTR now point to the same object. Consider the following example.

```
type BITVEC_PTR is access BIT_VECTOR;
variable BLTVEC1: BITVEC_PTR := new BIT_VECTOR'("1001");
```

BITVEC1 points to an object that is constrained to be of four bits. The bits can be accessed as BITVEC1(0), which is 1, BITVEC1(1), which is 0, and so on. Here is another example in which values are assigned to a dynamically allocated object using named association.

```
MOD2PTR := new MODULE' (CRITICAL_DLY => 10 ns,
NO_INPUTS => 2, NO_OUTPUTS => 3, SIZE => 100);
```

Access types are useful in modeling high-level behavior, especially that of regular structures like RAMs and FIFOs, where pointers can be used to access objects one at a time in a sequence.

1.8.5 Incomplete Types

It is possible to have an access type that points to an object which has elements that are also access types. This can lead to mutually dependent or recursive access types. Since a type must be declared before it is used, an incomplete type declaration can be used to solve this problem. An incomplete type declaration has the form:

```
type type-name;
```

Once an incomplete type has been declared, the type-name can now be used in any mutually dependent or recursive access type. However, a corresponding full type declaration must follow later. An example of a mutually dependent access type is:

```
type COMP;  -- Record contains component name and list of
            -- nets its' connected to
type NET;   -- Record contains net name and list of
            -- components it is connected to.
type COMP_PTR is access COMP;
type NET_TR is access NET;
constant MODMAX: INTEGER := 100;
constant NETMAX: INTEGER := 2500;
type COMP_LIST is array (1 to MODMAX) of COMP_PTR;
type NET_LIST is array (1 to NETMAX) of NET_PTR;
type COMPLIST_PTR is access COMP_LIST;
type NETLIST_PTR is access NET_LIST;
— Full type declarations for COMP and NET follow:
type COMP is
  record
    COMP_NAME          : STRING(1 to 10);
```

```
        NETS                  : NETLIST_PTR;
    end record;

    type NET is
      record
        NET_NAME              : STRING(1 to 10);
        COMPONENTS            : COMPLIST_PTR;
    end record;
```

Here, COMP and NET have elements that access objects of type NET and COMP, respectively. An example of a recursive access type is:

```
type DFG;              --A data flow graph node.
type OP_TYPE is (ADD, SUB, MUL, DIV, SHIFT, ROTATE);
type PTR is access DFG;
type DFG is
record
OP_CODE     : OP-TYPE;
SUCC        : PTR;          -- Successor node list.
PRED        : PTR;          -- Predecessor node list.
end record;
```

PTR is an access type of DFG. It is also the type of an element in DFG. Here is another example.

```
type MEM_RANGE is range 0 to 16383;
type HOLE;                  -- Represents available memory.
type HOLE_PTR is access HOLE;
type HOLE is
   record
     START_LOC, END_LOC    : MEM_RANGE;
     PREV_HOLE, NEXT.HOLE  : HOLE_PTR;
end record;
```

1.8.6 File Types

Objects of file types represent files in the host environment. They provide a mechanism by which a VHDL design communicates with the host environment. The syntax of a file type declaration is:

type *file-type-name* **is file of** *type-name;*

The type-name is the type of values contained in the file. Here are two examples.

```
type VECTORS is file of BIT_VECTOR;
type NAMES is file of STRING;
```

A file of type VECTORS has a sequence of values of type BIT_VECTOR; a file of type NAMES has a sequence of strings as values in it.

A file can be opened, closed, read, written to, or tested for an end-of-file condition by using special procedures and functions that are implicitly declared for every file type. These are:

```
procedure FILE_OPEN (file F: file-type-name;
      EXTERNAL_NAME: in STRING;
      OPEN_KIND: in FILE_OPEN_KIND := READ_MODE);
  -- Opensthe file F that points to the physical file specified in the
  -- string EXTERNAL_NAME, with the specified mode.
  -- Type FILE _OPEN_KIND has values:
```

```
  -- READ_MODE (default), WRITE _MODE, and APPEND_MODE.
procedure FILE_OPEN (STATUS: out FILE_OPEN_STATUS;
    file F: file-type-name;
      EXTERNAL_NAME: in STRING;
      OPEN_KIND: in FILE_OPEN_KIND := READ_MODE);
  -- Same as first procedure, except that this procedure returns the file
  -- open status. Type FILE_OPEN_STATUS has the following values:
  -- OPEN_OK      : file open was successful.
  -- STATUS_ERROR : file was already open.
  -- NAME_ERROR   : file not found or not accessible.
  -- MODE_ERROR   : could not open file with specified
                  : access mode.
procedure FILE_CLOSE (file F: file-type-name);
  -- Closes the specified file.
procedure READ (file F: file-type-name;VALUE: out type-name);
  -- Gets the next value in VALUE from file F.
procedure WRITE (file F: file-type-name; VALUE: in type-name);
  -- Appends a given value in VALUE to file F.
function ENDFILE (file F: file-type-name) return BOOLEAN;
  -- Returns false if a read on a file F will be successful in getting
  -- another value, otherwise it returns true.
```

If type-name is an unconstrained array type, a different READ procedure is implicitly declared, which is of the form:

```
procedure READ (file V: file-type-name; VALUE: out type-name;
    LENGTH: out NATURAL);
  -- LENGTH returns the number of elements of the array that
  -- was read.
```

Values within a file can only be accessed sequentially. A file can be opened implicitly by specifying the file open information in the file declaration. Some examples are shon next.

```
file DUMP: NAMES open APPEND_MODE is "top.dump";
  -- An implicit call to FILE_OPEN is performed at elaboration time:
  -- FILE_OPEN (DUMP, "top.dump", APPEND_MODE);
file PATTERNS: VECTORS is "uart.pat";
  -- An implicit call to FILE_OPEN is performed at elaboration time:
  -- FILE_OPEN (PATTERNS, "uart.pat", READ_MODE);
file TMP: VECTORS;
  -- Since no file open information is provided, an explicit call to
  -- FILE_OPEN needs to be performed before the file TMP can be
  -- accessed.
```

Here is a complete example of a test bench that reads vectors from a file, "fadd.vec", applies these vectors to the test component, a one-bit full adder, and then writes the results into another file, "fadd.out".

```
entity FA_TEST is end;
architecture IO_EXAMPLE of FA_TEST is
  component FULL_ADD
    port (CIN, A, B: in BIT; COUT, SUM: out BIT);
  end component;
  subtype STRING3 is BIT _VECTOR (0 to 2);
  subtype STRING2 is BIT_VECTOR (0 to 1);
  type IN_TYPE is file of STRING3;
```

```
   type OUT_TYPE is file of STRING2;

  fileVEC_FILE: IN_TYPE
    open READ_MODE is "/usr/home/jb/vhdl_ex/fadd.vec";
  file RESULT_FILE: OUT_TYPE
    openWRITE_MODE is "/usr/home/jb/vhdl_ex/fadd.out"; signal S: STRING3; signal Q: STRING2;
  begin

  FA: FULL_ADD port map (S(0), S(l), S(2), Q(0), Q(l));

   process
    constant PROPAGATION_DELAY: TIME := 25 ns;
    variable IN_STR: STRING3;
    variable OUT_STR: STRING2;
    begin
      while not ENDFILE (VEC_FILE) loop
          READ (VEC_FILE, IN_STR);
          S <= IN_STR;
          wait for PROPAGATION_DELAY;
          OUT_STR := Q;
          WRITE (RESULT_FILE, OUT_STR);
      end loop;
      report "Completed processing all vectors.";
      wait;                         -- Stop simulation.
    end process;
  end IO_ EXAMPLE;
```

Two files, VEC_FILE and RESULT_FILE, are declared. VEC_FILE is an input file and contains three-bit strings, and RESULT_FILE is an output file into which two-bit strings are written. Input vectors are read one at a time until end of file is reached. Each vector is applied, and then the process waits for the full-adder circuit to stabilize before sampling the adder output value which is written to the output file. The report statement, when executed, simply prints the report message. The wait statement causes the process to suspend indefinitely.

One file type, TEXT, is predefined in the language; this file type represents a file consisting of variable-length text strings.

```
type TEXT is file of STRING;
  -- A file of variable-length strings.
```

An access type, LINE, is provided to point to such strings.

```
type LINE is access STRING;     -- A line is a pointer to a STRING.
```

Operations to read and write data from a single line are provided. Operations to read and write entire lines are also provided. The definitions for all these types and operations appear in the predefined package, TEXTIO.

Here is an example that reads from each line, a last name (a string), an ID number (an integer) and a dollar amount (a real number), and rewrites out to another file in the reverse order, that is, dollar amount first, ID second and then the last name.

```
use STD.TEXTIO.all;          -- Include the predefined package TEXTIO.
entity REVERSE_RECORDS is end;

architecture SIMPLE of REVERSE_RECORDS is
```

```
  file FROM_FILE: TEXT open READ_MODE
                          is " /home/jb/vhdLex/master_records.in";
file TO_FILE: TEXT open WRITE_MODE
                    is "/home/jb/vhdl_ex/master_records.out";
begin
  process
    variable BUF_IN, BUF_OUT: LINE;
    variable LNAME: STRING(1 to 20);
    variable ID_NUM: INTEGER;
    variable DOLLARS: REAL;
begin
  while not ENDFILE(FROM_FILE) loop
    READLINE (FROM_FILE/ BUF_IN);       -- Read line from file.
    READ (BUF_IN, LNAME);               -- Read string.
    READ (BUF_IN, ID_NUM);              -- Read integer.
    READ (BUF_IN, DOLLARS);             -- Read real number.
    WRITE (BUF_OUT, DOLLARS);           -- Write real number.
    WRITE (BUF_OUT,' ');                – Write space character.
    WRITE (BUF_OUT, ID _NUM);           – Write integer.
    WRITE (BUF_OUT, '');                – Write space character.
    WRITE (BUF_OUT, LNAME);             – Write string.
    WRLTELINE (TO_FILE, BUF_OUT);       – Write line to file.
   end loop;
   wait;
  end process;
end SIMPLE;
```

Function ENDFILE and procedures READLINE, READ, WRITELINE and WRITE are predefined in the package TEXTIO. The while loop goes through all the lines in the file, one at a time. The READLINE procedure reads one line into the buffer BUF_IN. The READ procedure reads each field in the buffer. The WRITE procedure writes out the values into the output buffer. The WRITELINE procedure writes the output buffer to the output file. Here is a test input to this program.

```
Soccer       4569889 566Q.23
Basketball   1890942 3561.99
Hockey       8766567 8009.12
```

Here is the output produced in the output file.

```
5.660230e+0   4569889 Soccer
3.561990e+03  1890942 Basketball
8.009120e+03  8766567 Hockey
```

1.9 OPERATORS

The predefined operators in the language are classified into the following six categories:

1. Logical operators
2. Relational operators
3. Shift operators
4. Adding operators
5. Multiplying operators
6. Miscellaneous operators

The operators have increasing precedence going from category 1 to 6. Operators in the same category have the same precedence, and evaluation is done left to right. Parentheses may be used to override the left to right evaluation.

1.9.1 Logical Operators

The seven logical operators are:

and **or** **nand** **nor** **xor** **xnor** **not**

These operators are defined for the predefined types BIT and BOOLEAN. They are also defined for one-dimensional arrays of BIT and BOOLEAN. During evaluation of logical operators, bit values '0' and '1' are treated as FALSE and TRUE values of the BOOLEAN type, respectively. The result of a logical operation has the same type as its operands. The not operator is a unary logical operator and has the same precedence as that of miscellaneous operators.

The operators nand and nor are not associative; therefore, the syntax of an expression with a sequence of nand or nor operators is illegal. For example, the following expression is illegal.

```
A nand B nand C        -- is illegal.
```

Parentheses can be used to avoid this problem.

1.9.2 Relational Operators

These are:

```
=    /=    <    <=    >    >=
```

The result type for all relational operations is always the predefined type BOOLEAN. The = (equality) and the /= (inequality) operators are predefined on any type except file types. The remaining four relational operators are predefined on any scalar type (e.g., integer or enumerated types) or discrete array type (i.e., arrays in which element values belong to a discrete type). When operands are discrete array types, comparison is performed one element at a time from left to right. For example,

```
BIT_VECTOR'('0', '1', 1') < BIT_VECTOR'('1', '0', '1')
```

is true, since the first element in the first array aggregate is less than the first element in the second aggregate. Similarly, if

```
type MVL is ('U', '0', 1', 'Z');
```

then

```
MVL'('U') < MVL'('Z')
```

is true, since 'U' occurs to the left of 'Z'. Note that it is necessary to qualify the aggregates and literals used in these examples since the literals are overloaded and their type cannot be determined from its use.

Here is an example in which operands are arrays of different lengths.

```
"VHDL" < "VHDL92"
```

is true. Comparison is again performed one element at a time from left to right. However, if a corresponding element does not exist in an operand, it is considered to be null, and null is always considered to be less than any character.

1.9.3 Shift Operators

These are:

sll **srl** **sla** **sra** **rol** **ror**

Each of the operators takes an array of BIT or BOOLEAN as the left operand and an integer value as the right operand and performs the specified operation. If the integer value is a negative number, the opposite action is performed, that is, a left shift or rotate becomes a right shift or rotate, respectively, and vice versa.

The **sll** operator (shift left logical) and the **srl** operator (shift right logical) fill the vacated bits with left-operand-type'LEFT. The **sla** operator (shift left arithmetic) fills the vacated bits with the rightmost bit of the left operand, while the **sra** operator (shift right arithmetic) fills the vacated bits with the leftmost bit of the left operand. The rotate operators cause the vacated bits to be filled with the displaced bits in a circular fashion. Here are some examples.

```
-- Assume that all left operands are BIT_VECTORS:
"1001010" sll 2    is "0101000"  -- filled with BIT'LEFT, which is '0'.
"1001010" srl 3    is "0001001"  -- filled with '0'.
"1001010" sla 2    is "0101000"  -- filled with rightmost bit which is '0'.
"1001010" sra 3    is "1111001"  -- filled with '1' which is the leftmost bit.
"1001010" rol 2    is "0101010"  -- rotate left.
"1001010" ror 3    is "0101001"  -- rotate right
"1001010" ror –4   is "0101001"  -- "rol 4"operation performed.
"1001010" srl –5   is "1000000"  -- "sll 5"operation performed.
"1001010" sla –2   is "1110010"  -- "sra 2"operation performed.
"1001010" sra -4   is "0100000"  -- "sla 4"Operation performed.
"1001010" rol –1   is "0100101"  -- "ror 1"operation performed.
"1001010" sll –5   is "0000010"  -- "srl 5"operarion performed.
```

1.9.4 Adding Operators

These are:

\+ - &

The operands for the + (addition) and - (subtraction) operators must be of the same numeric type, with the result being of the same numeric type. The addition and subtraction operators may also be used as unary operators, in which case the operand and the result type are the same. The operands for the & (concatenation) operator can be either a one-dimensional array type or an element type. The result is always an array type. For example,

'0'& '1'

results in an array of characters "01".

'C & 'A' & T

results in the value "CAT".

"BA" & "LL"

creates an array of characters "BALL".

A good example of an use of a concatenation operator is when many separate signals are to be grouped into one bus. Here is an example.

```
signal DBUS: STD_LOGIC_VECTOR (0 to 7);
signal CTRL: STD_LOGIC_VECTOR (1 downto 0);
signal ENABLE, RW: STD_LOGIC;
signal COUNT: STD_LOGIC_VECTOR (0 to 3);
-- This is the intention:
-- DBUS (0 to 1) <= CTRL;
-- DBUS (2) <= ENABLE;
-- DBUS (3) <= RW;
-- DBUS (4 to 7) <= COUNT;
-- Using the concatenation operator, we get:
DBUS <= CTRL & ENABLE & RW & COUNT;
```

1.9.5 Multiplying Operators

These are:

```
*    mod    rem
```

The * (multiplication) and / (division) operators are predefined for both operands being of the same integer or floating point type. The result is also of the same type. The multiplication operator is also defined for the case when one of the operands is of physical type and the second operand is of integer or real type. The result is of physical type.

For the division operator, division of a value of physical type by either an integer or a real value is allowed, and the result type is the physical type. Division of a value of physical type by another object of the same physical type is also defined, and it yields an integer value as a result.

The **rem** (remainder) and **mod** (modulus) operators operate on operands of integer types, and the result is also of the same type. The result of a **rem** operation has the sign of its first operand and is defined as:

```
A rem B = A- (A/B) * B
```

The result of a **mod** operator has the sign of the second operand and is defined as:

```
A mod B = A-B * N          -- For some integer N.
```

Here are some examples using the **mod** and **rem** operators.

```
7 mod 4 -- has value 3
(–7) rem 4        -- has value –3
7 mod (–4)        -- has value –1
(–7) rem (–4)     -- has value –3
```

1.9.6 Miscellaneous Operators

The miscellaneous operators are:

```
abs **
```

The **abs** (absolute) operator is defined for any numeric type.

The ** (exponentiation) operator is defined for the left operand to be of integer or floating point type, and for the right operand (i.e., the exponent) to be of integer type only.

The **not** logical operator has the same precedence as the above two operators.

1.10 SIGNAL DRIVERS

What if there is more than one assignment to the same signal within a process? To understand how the waveform elements contribute to the effective value of a signal, it is important to understand the concept of a driver. A driver is created for every signal that is assigned a value in a process. The driver of a signal holds its current value and all its future values as a sequence of one or more transactions, where each transaction identifies the value to appear on the signal along with the time at which the value is to appear.

Consider the following signal assignment statement with three waveform elements and the driver which is created for that signal.

```
process
begin
   ...
   RESET <= 3 after 5 ns, 21 after 10 ns, 14 after 17 ns;
end process;
```

RESET ←	curr@now	3@T+5 ns	21@T+10 ns	14@T+17ns

All transactions on the driver are ordered in increasing order of time. A driver always contains at least one transaction, which could be the initial value of the signal. In the previous example, when the signal assignment statement is executed, say, at time 7, three new transactions are added to the driver for the RESET signal. The first transaction is the current value of the signal.

When simulation time advances to 7+5 ns, the first transaction is deleted from the driver, and RESET gets the value of 3. When time advances to 7+10 ns, the second transaction is deleted, and RESET gets the value of 21. When time advances to 7+17 ns, the third transaction is deleted, and RESET gets the value of 14.

What if there is another signal assignment to RESET in the same process? Since a signal inside a process has only one driver, the transactions of the second signal assignment modify the transactions already present on the driver, depending on whether inertial or transport delay model is used. We consider the transport delay case first.

Effect of Transport Delay on Signal Drivers

Here is an example of a process having three signal assignments to the same signal RX_DATA.

```
signal RX_DATA: NATURAL;
...
process
begin
   ...
   RX_DATA <= transport 11 after 10 ns;
   RX_DATA <= transport 20 after 22 ns;
   RX_DATA <= transport 35 after 18 ns;
end process;
```

Assume that the statements are executed at time 7. The transactions on the driver for RX_DATA are created as follows. When the first signal assignment is executed, the transaction 11@7+10 ns is added to the driver. After the second signal assignment is executed, the transaction 20@7+ 22 ns is appended to the driver since the delay of this transaction (= 22 ns) is larger than the delay of the pending transactions on the driver. The driver for RX_DATA looks like this:

RX_DATA ⟵	curr@now	11@T+10 ns	20@T+22 ns

When the third signal assignment statement is executed, the new transaction 35@T+18 ns causes the 20@ T+22 ns transaction to be deleted, and the new transaction is appended to the driver. The reason for this is that the delay for the new transaction (=18 ns) is less than the delay of the last transaction sitting on the driver (=22 ns). This effect is caused because transport delay is used. In general, a new transaction will delete all transactions sitting on a driver that are to occur at or later than the delay of the new transaction. Therefore, the driver for RX_DATA is changed to:

RX_DATA ⟵	curr@now	11@T+10 ns	35@T+18 ns

Here is another example.

```
process
    -- DATA_BUS is a signal of type BIT_VECTOR.
begin
    ...
    DATA_BUS <= transport X"01" after 5 ns, X"FA" after 10 ns,
                        X"E8" after 15 ns;
    DATA_BUS <= transport X"B5" after 12 ns;
end process;
```

When these statements are executed, the status of the driver for DATA_BUS after the first signal assignment is:

DATA_BUS ⟵	curr@now	X"01"@T+5 ns	X"FA"@T+10 ns	X"E8"@T+15 ns

When the second signal assignment is executed, the new transaction X"B5" @T+ 12 ns causes the X"E8" @T+15 ns transaction to be deleted from the driver (because its delay value is larger than the delay of the new transaction). The remaining transactions on the driver are not deleted, since their delays are less than the delay of the new transaction. Therefore, the final state of the driver after both the statements are executed is:

DATA_BUS ⟵	curr@now	X"01"@T+5 ns	X"FA"@T+10ns	X"B5"@T+12ns

The following summarizes the rules for adding transactions from a signal assignment, using the transport delay mode, to a driver:

1. All transactions on the driver that occur at or after the delay time of the first new transaction are deleted.
2. All the new transactions are added at the end of the driver.

Effect of Inertial Delay on Signal Drivers

When inertial delays are used, both the signal value being assigned and the delay value affect the deletion and addition of transactions. If the delay of the first new transaction is earlier than an existing transaction, the latter is deleted regardless of the signal values of the two transactions. Note that this is the same as rule 1 for the transport delay case. On the other hand, existing transactions that fall within the window of the time of the first new transaction, say, F and (F - pulse rejection limit), are checked for their values. If their values are different, these already existing transactions are deleted; otherwise, they are retained. The following examples will help clarify this.

Consider the following process statement.

```
process
begin
  TX_DATA <= 11 after 10 ns;                    -- Pulse rejection limit is 10 ns.
  TX_DATA <= reject 15 ns inertial 22 after 20 ns;
  TX_DATA <= 33 after 15 ns;                    -- Pulse rejection limit is 15 ns.
  wait;                                         -- Suspends indefinitely.
end process;
```

The transaction 11@10 ns first gets added to the driver. The second transaction, 22@20 ns, causes the 11@10 ns transaction on the driver to be deleted. This is because the only transaction that falls in the window of 20 ns (the time of the first new transaction) and (20 ns -15 ns) (the time of the first new transaction minus the pulse rejection limit) is the 11@10 ns transaction, and its value (which is 11) is different from the value of the first new transaction being added (which is 22). The state of the driver at this point is:

```
TX_DATA  <----| curr@now | 22@20 ns |
```

The execution of the third signal assignment causes the transaction 22@20 ns to be deleted from the driver, since the delay of the new transaction (=15 ns) is less than the delay of the transaction on the driver (similar to the transport delay case). The final status of the driver is:

1.11 TRANSPORT VERSUS INERTIAL DELAY

In VHDL, there are two types of delay that can be used for modeling behaviours. Inertial delay is the most commonly used, while transport delay where a wire delay model is required.

1.11.1 Inertial delay models

Inertial delay models the delays often found in switching circuits. An input value must be stable for a specified pulse rejection limit duration before the value is allowed to propagate to the output. In addition, the value appears at the output after the specified inertial delay. If the input is not stable for the specified limit, no output change occurs. When used with signal assignments, the input value is represented by the value of the expression on the right-hand-side and the output is represented by the target signal. The syntax of such a specification in a signal assignment is:

signal-object <= [[**reject** *pulse-rejection-limit*] **inertial**] *expression*
after ***inertial-delay-value;***

If no pulse rejection limit is specified, the default pulse rejection limit is the inertial delay value itself. The pulse rejection limit cannot be negative or greater than the value of the inertial delay.

Figure 1.5 shows a simple example of a non-inverting buffer with an inertial delay of 10 ns and a pulse rejection limit of 4 ns.

```
Z <= reject 4 ns inertial A after 10 ns;
```

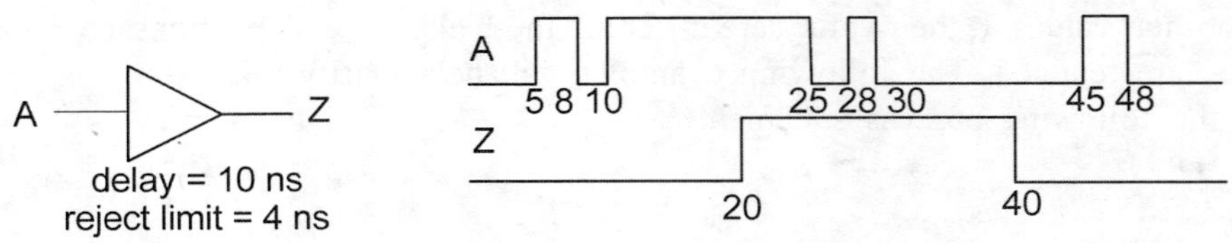

Figure 1.5: Inertial delay example.

Events on signal A that occur at 5 ns and 8 ns are not stable for the duration of the pulse rejection limit, and hence do not propagate to the output. The event on A at 10 ns remains stable for more than the pulse rejection limit; therefore, the value is propagated to the target signal Z after the inertial delay - Z gets the value '1' at 20 ns. Events On signal A at 25 ns and 28 ns do not affect the output, since they are not stable for the duration of the pulse rejection limit. The '1' to '0' transition at time 30 ns on signal A remains stable for at least the duration of the pulse rejection limit, and therefore a '0' is propagated to signal Z with a delay of 10 ns; Z gets the new value at 40 ns. Other

TX_DATA ←

curr@now	33@15ns

Let us consider another example.

```
process
begin
  ADDR_BUS <= 1 after 5 ns, 21 after 9 ns, 6 after 10 ns,
        12 after 19 ns;            -- Pulse rejection limit is 5 ns, the
                                   -- delay of the first waveform element.
  ADDR_BUS <= reject 4 ns inertial 6 after 12 ns, 20 after 19 ns;
  wait;
end process;
```

The status of the driver for the ADDR_BUS signal after the first statement is executed is:

ADDR_BUS ←

curr@now	1@5ns	21@9ns	6@10ns	12@19ns

The execution of the second statement causes all transactions on the driver later than 12 ns (the inertial delay of the first waveform element) to be deleted, which would be the 12@19 ns transaction. The two new transactions, 6@12 ns and 20@19 ns, are added to the driver. The old transaction, 21@9 ns, also gets deleted since its delay falls in the window of time 12 ns and (12 ns - 4 ns), and its value (which is 21) is not the same as the value of the first new transaction (which is 6).

The resulting driver is:

ADDR_BUS ←

curr@now	1@5 ns	6@10ns	6@12ns	20@19ns

The summary of rules for adding transactions from a signal assignment using inertial delays are the following:

1. All transactions on a driver that are scheduled to occur at or after the delay of the first new transaction are deleted (as in the transport case).
2. Add all the new transactions to the driver.
3. For all the old transactions on the driver that occur at times between the time of the first new transaction (say F) and F minus the pulse rejection limit, delete the old transactions whose value is different from the value of the first new transaction.

events on A do not affect the target signal Z. In this example, if the assignment had no reject clause specified, such as

```
Z <= A after 10 ns;
```

then the pulse rejection limit is also 10 ns, which is the inertial delay value.

Since inertial delay is most commonly found in digital circuits, it is the default delay model; that is, no keyword need be explicitly specified. It is possible, however, to explicitly identify the delay in a signal assignment as an **inertial** delay, even though it is not necessary, by using the keyword inertial explicitly, such as:

```
Z <= inertial A after 10 ns;
-- Same as: Z <= A after 10 ns;
```

This delay model is often used to filter out unwanted spikes and transients on signals.

1.11.2 Transport Delay Model

Transport delay models the delays in hardware that do not exhibit any inertial delay. This delay represents pure propagation delay; that is, any changes on an input are transported to the output, no matter how small, after the specified delay. To use a transport delay model, the keyword **transport** must be used in a signal assignment statement. Figure1.6 shows an example of a non-inverting buffer using a transport delay of 10 ns.

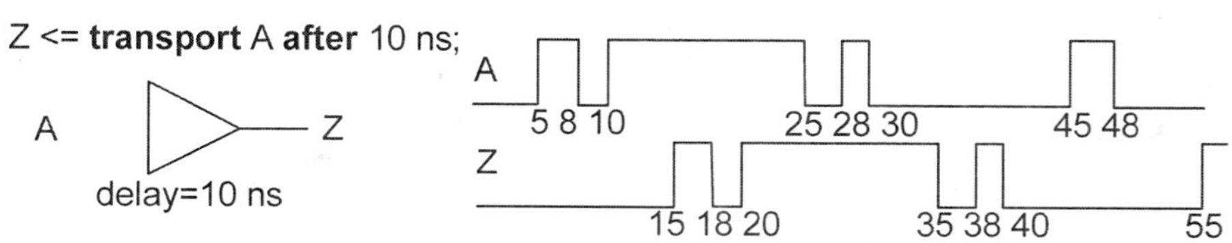

Figure 1.6: Transport delay example.

Ideal delay modeling can be obtained by using this delay model. In this case, spikes would be propagated through instead of being ignored, as in the inertial delay case. Routing delays can be modeled using transport delay. An example of a routing delay model is:

```
entity WIRE14 is
  port (A: in BIT; Z: out BIT);
end WIRE14;

architecture WIRE14_TRANSPORT of WIRE14 is
begin
```

```
    process (A)
    begin
      Z <= transport A after 0.1 ms;
    end process;
end WIRE14_TRANSPORT;
```

1.11.3 Creating Signal Waveforms

In all the examples of signal assignment statements that we have seen so far, we have always assigned a single value to a signal; this need not be so. It is possible to assign multiple values to a signal, each with a different delay value. For example,

```
PHASEl <= '0', '1' after 8 ns, '0' after 13 ns, '1' after 50 ns;
```

When this signal assignment statement is executed, say, at time *T*, it causes four values to be scheduled for signal PHASE1; the value '0' is scheduled to be assigned at time $T+\Delta$, '1' at T+8 ns, '0' at *T*+13 ns, and '1' at *T*+50 ns. Thus, a waveform appears on the signal PHASE1, as shown in Figure 1.7.

A more general syntax of a signal assignment statement is:.

```
signal-object<=¦[transport | [reject pulse-rejection-limit] inertial]
                          waveform-element, waveform-element,
                          waveform-element, waveform-element ;
-- Can have any number of waveform elements.
```

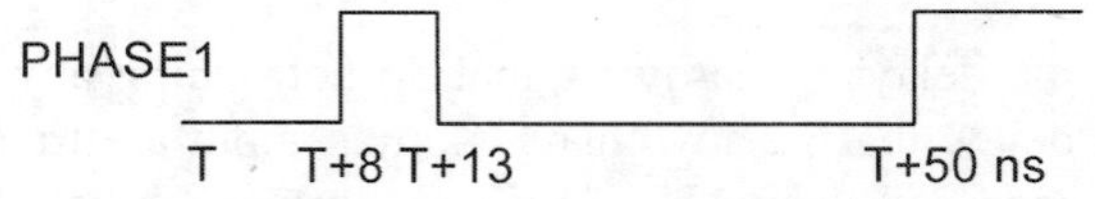

Figure 1.7: A signal waveform.

Each waveform element has a value part, specified by an expression (called the waveform expression), and a delay part, specified by an after clause that specifies the delay. The delays in the waveform elements must appear in increasing order. A waveform element is of the form:

expression [**after** *time-expression*]

If no pulse rejection limit is specified in an assignment using interial dealys, the delay, the delay of the first waveform element becomes the default pulse rejection limit. Any arbitry waveform can therefore easily be created using a signal assignment statement.

1.12 BASIC TERMINOLOGY

VHDL is a hardware description language that can be used to model a digital system. The digital system can be as simple as a logic gate or as complex as a complete electronic system. A hardware abstraction of this digital system is called an entity in this text. An entity X, when used in another entity Y, becomes a component for the entity Y. Therefore, a component is also an entity depending on the level at which you are trying to model.

To describe an entity, VHDL provides five different types of primary constructs, called design units. They are:

1. Entity declaration
2. Architecture body
3. Configuration declaration
4. Package declaration
5. Package body

An entity is modeled using an entity declaration and at least one architecture body. The entity declaration describes the external view of the entity; for example, the input and output signal names. The are-chitecture body contains the internal description of the entity; for example, as a set of interconnected components that represents the structure of the entity, or as a set of concurrent or sequential statements that represents the behavior of the entity. Each style of representation can be specified in a different architecture body or mixed within a single architecture body. Figure 1.8 shows an entity and one possible model.

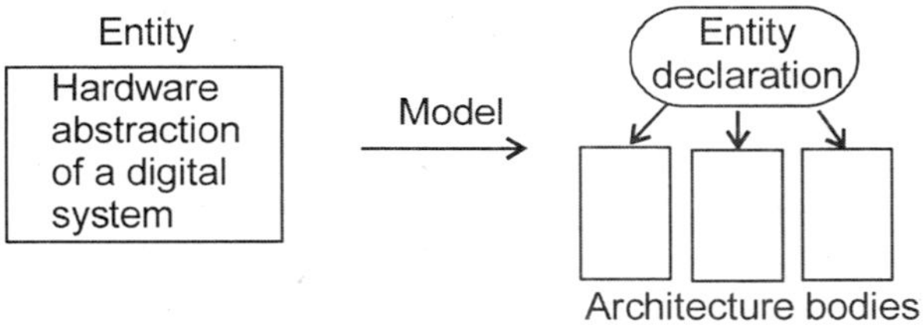

Figure 1.8: An entity and its model.

A configuration declaration is used to create a configuration for an entity. It specifies the binding of one architecture body from the many architecture bodies that may be associated with the entity. It may also specify the bindings of components used in the selected architecture body to other entities. An entity may have any number of different configurations.

A package declaration encapsulates a set of related declarations, such as type declarations, subtype declarations, and subprogram declarations, which can be shared across two or more design units. A package body contains the definitions of subprograms declared in a package declaration.

Figure 1.9 shows three entities called E1,E2, and E3. Entity E1 has three architecture bodies, E1_A1, E1_A2, and E1_A3. Architecture body E1_A1 is a purely behavioral model without any hierarchy. Architecture body E1_A2 uses a component called BX, while architecture body E1_A3 uses a component called CX. Entity E2 has two architecture bodies, E2_A1 and E2_A2, and architecture body E2_A1 uses a component called M1. Entity E3 has three architecture bodies, E3_A1, E3_A2, and E3_A3. Notice that each entity has a single entity declaration but more than one architecture body.

The dashed lines represent the binding that may be specified in a configuration for entity E1. There are two types of binding shown: binding of an architecture body to its entity and the binding of components used in the architecture body with other entities. For example, architecture body E1_A3 is bound to entity E1, while architecture body E2_A1 is bound to entity E2. Component M1 in architecture body E2_A1 is bound to entity E3. Component CX in the architecture body E1_A3 is bound to entity E2. However, one may choose a different configuration for entity E1 with the following bindings:

- Architecture E1_A2 is bound to its entity E1
- Component BX is bound to entity E3
- Architecture E3_A1 is bound to its entity E3

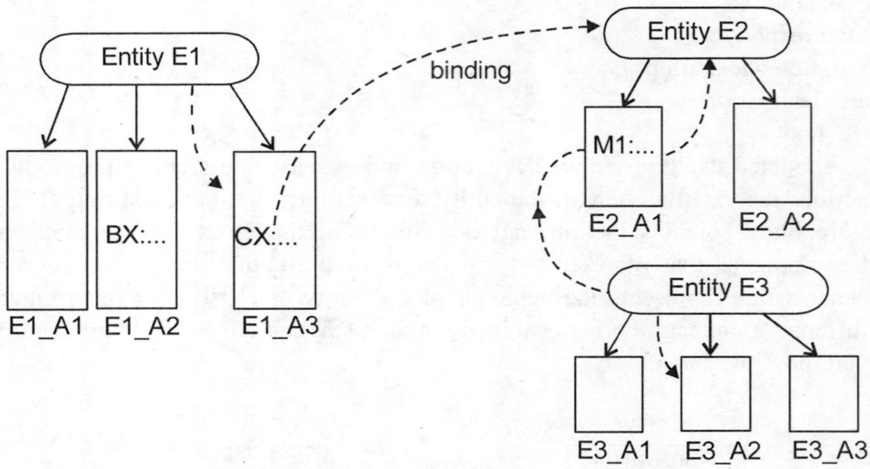

Figure 1.9: A configuration for entity E1.

Once an entity has been modeled, it needs to be validated by a VHDL system. A typical VHDL system consists of an analyzer and a simulator. The analyzer reads in one or more design units contained in a single file and compiles them into a design library after validating the syntax and performing some static semantic checks. The design library is a place in the host environment (that is, the environment that supports the VHDL system) where compiled design units are stored.

The simulator simulates an entity, represented by an entity-architecture pair or by a configuration, by reading in its compiled description from the design library and then performing the following steps:

1. Elaboration
2. Initialization
3. Simulation

A note on the language syntax. The language is case-insensitive; that is, lower-case and upper-case characters are treated alike (except in extended identifiers, string literals, and character literals). For example, CARRY, CarrY, or carrY all refer to the same name. The language is also free-format, very much like in Ada and Pascal. Comments are specified in the language by preceding the text with two consecutive dashes (--). All text between the two dashes and the end of that line is treated as a comment.

1.13 ENTITY DECLARATION

The entity declaration specifies the name of the entity being modeled and lists the set of interface ports. Ports are signals through which the entity communicates with the other models in its external environment.

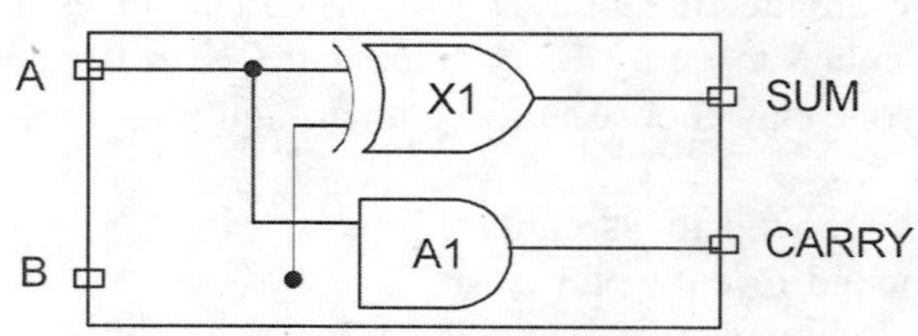

Figure 1.10: A half-adder circuit

Here is an example of an entity declaration for the half-adder circuit shown in Figure 1.10.

```
entity HALF_ADDER is
  port (A, B: in BIT; SUM, CARRY: out BIT);
end HALF_ADDER;
--This is a comment line.
```

The entity, called HALF_ADDER, has two input ports, A and B (the mode **in** specifies input port), and two output ports, SUM and CARRY (the mode **out** specifies output port). BIT is a predefined type of the language; it is an enumeration type containing the character literals '0' and '1'.The port types for this entity have been specified to be of type BIT, which means that the ports can take the values '0' or '1'.

The following is another example of an entity declaration for the 2-to-4 decoder circuit shown in Figure 1.11

```
entity DECODER2x4 is
  port (A, B, ENABLE: in BIT; Z: out BIT_VECTOR(0 to 3));
end DECODER2x4;
```

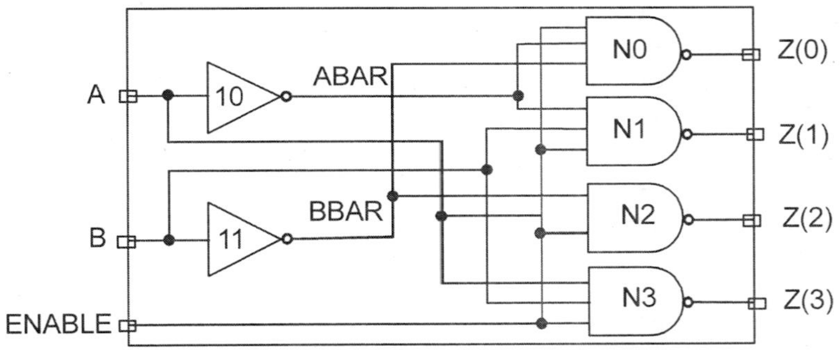

Figure1.11: A 2-to-4 decoder circuit.

This entity, called DECODER2x4, has three input ports and four output ports. BIT VECTOR is a predefined unconstrained array type of BIT. An unconstrained array type is a type in which the size of the array is not specified. The range "0 **to** 3" for port Z specifies the array size.

From the last two examples of entity declarations, we see that the entity declaration does not specify anything about the internals of the entity. It only specifies the name of the entity and the interface ports.

1.14 ARCHITECTURE BODY

The internal details of an entity are specified by an architecture body using any of the following modeling styles:

1. As a set of interconnected components (to represent structure)
2. As a set of concurrent assignment statements (to represent dataflow)
3. As a set of sequential assignment statements (to represent behavior)
4. As any combination of the above three

1.14.1 Structural Style of Modeling

In the structural style of modeling, an entity is described as a set of interconnected components. Such a model for the HALF_ADDER entity, shown in Figure 1.10 is described in an architecture body as shown below.

```
architecture HA_STRUCTURE of HALF_ADDER is
  component XOR2
    port (X,Y: in BIT; Z: out BIT);
end component;
component AND2
  port (L, M: in BIT; N: out BIT);
  end component;
begin
  X1: XOR2 port map (A, B, SUM);
  Al: AND2 port map (A, B, CARRY);
end HA_STRUCTURE;
```

The name of the architecture body is HA_STRUCTURE. The entity declaration for HALF_ADDER specifies the interface ports for this architecture body. The architecture body is composed of two parts: the declarative part (before the keyword begin) and the statement part (after the keyword **begin**). Two component declarations are present in the declarative part of the architecture body. These declarations specify the interface of components that are used in the architecture body. The components XOR2 and AND2 may either be predefined components in a library or, if they do not exist, they may later be bound to other components in a library.

The declared components are instantiated in the statement part of the architecture body using component instantiation statements. X1 and A1 are the component labels for these component instantiations. The first component instantiation statement, labeled X1, shows that signals A and B (the input ports of the HALF_ADDER), are connected to the X and Y input ports of a XOR2 component, while output port Z of this component is connected to output port SUM of the HALF_ADDER entity.

Similarly, in the second component instantiation statement, signals A and B are connected to ports L and M of the AND2 component, while port N is connected to the CARRY port of the HALF_ADDER. Note that in this case, the signals in the port map of a component instantiation and the port signals in the component declaration are associated by position (called positional association). The structural representation for the HALF_ADDER does not say anything about its functionality. Separate entity models would be described for the components XOR2 and AND2, each having its own entity declaration and architecture body.

A structural representation for the DECODER2x4 entity, shown in Figure 1.11 is shown next.

```
architecture DEC_STR of DECODER2x4 is
  component INV
    port (PIN: in BIT; POUT: out BIT);
end component;
component NAND3
  port (DO, Dl, D2: in BIT; DZ: out BIT);
end component;
signal ABAR, BBAR: BIT;  .
begin
```

```
    V0: INV port map (A, ABAR);
    V1: INV port map (B, BBAR);
    NO: NAND3 port map (ENABLE, ABAR, BBAR, Z(0)); -
    N1: NAND3 port map (ABAR, B, ENABLE, Z(l));
    N2: NAND3 port map (A, BBAR, ENABLE, Z(2));
    N3: NAND3 port map (A, B, ENABLE, Z(3));
end DEC.STR;
```

In this example, the name of the architecture body is DEC_STR, and it is associated with the entity declaration with the name

DECODER2x4; therefore, it inherits the list of interface ports from that entity declaration. In addition to the two component declarations (for INV and NAND3), the architecture body contains a signal declaration that declares two signals, ABAR and BBAR, of type BIT. These signals, which represent wires, are used to connect the various components that form the decoder. The scope of these signals is restricted to the architecture body; therefore, these signals are not visible outside the architecture body. Contrast these signals with the ports of an entity declaration, available for use within any architecture body associated with the entity declaration.

A component instantiation statement is a concurrent statement. Therefore, the order of these statements is not important. The structural style of modeling describes only an interconnection of components (viewed as black boxes), without implying any behavior of the components themselves nor of the entity that they collectively represent. In the architecture body DEC_STR, the signals A, B, and ENABLE, used in the component instantiation statements, are the input ports declared in the DECODER2x4 entity declaration. For example, in the component instantiation labeled N3, port A is connected to input DO of component NAND3, port B is connected to input port D1 of component NAND3, port ENABLE is connected to input port D2, and port Z(3) of the DECODER2x4 entity is connected to the output port DZ of component NAND3. Again, positional association is used to map signals in a port map of a component instantiation with the ports of a component specified in its declaration. The behavior of the components NAND3 and INV are not apparent, nor is the behavior of the decoder entity that the structural model represents.

1.14.2 Dataflow Style of Modeling

In this modeling style, the flow of data through the entity is expressed primarily using concurrent signal assignment statements. The structure of the entity is not explicitly specified in this modeling style, but it can be implicitly deduced. Consider the following alternate architecture body for the HALF_ADDER entity that uses this style.

```
architecture HA_CONCURRENT of HALF_ADDER is
begin
  SUM <= A xor B after 8 ns;
  CARRY <= A and B after 4 ns;
end HA_CONCURRENT;
```

The dataflow model for the HALF_ADDER is described using two concurrent signal assignment statements. In a signal assignment statement, the symbol <= implies an assignment of a value to a signal. The value of the expression on the right-hand-side of the statement is computed and is assigned to the signal on the left-hand-side, called the target signal. A concurrent signal assignment statement is executed only when any signal used in the expression on the right-hand-side has an event on it, that is, the value for the signal changes.

Delay information is included in the signal assignment statements using after clauses. If either signal A or B, which are input port signals of HALF_ADDER entity, has an event, say at time T, the right-hand-side expressions of both signal assignment statements are evaluated. Signal SUM is scheduled to get the new value after 8 ns, while signal CARRY is scheduled to get the new value after 4 ns. When simulation time advances to (T+4) ns, CARRY will get its new value, and when simulation advances to (T+8) ns, SUM will get its new value. Thus, both signal assignment statements execute concurrently.

Concurrent signal assignment statements are concurrent statements, and therefore, the ordering of these statements in an architecture body is not important. Note again that this architecture body, with name HA_CONCURRENT, is also associated with the same HALF_ADDER entity declaration.

Here is a dataflow model for the DECODER2x4 entity.

```
architecture DEC_DATAFLOW of DECODER2x4 is
  signal ABAR, BBAR: BIT;
begin
    Z(3) <= not (A and B and ENABLE);               -- statement 1
    Z (0) <= not (ABAR and BBAR and ENABLE);        -- statement 2
    BBAR <= not B;                                  -- statement 3
    Z(2) <= not (A and BBAR and ENABLE);            -- statement 4
    ABAR <= not A;                                  -- statement 5
    Z(l) <= not (ABAR and B and ENABLE);            -- statement 6
end DEC_DATAFLOW;
```

The architecture body consists of one signal declaration and six concurrent signal assignment statements. The signal declaration declares signals ABAR and BBAR to be used locally within the architecture body. In each of the signal assignment statements, no after clause was used to specify delay. In all such cases, a default delay of 0 ns is assumed. This delay of 0 ns is also known as delta delay, and it represents an infinitesimally small delay.

To understand the behavior of this architecture body, consider an event happening on one of the input signals, say, input port B at time T. This would cause the concurrent signal assignment statements 1, 3, and 6 to be triggered. Their right-hand-side expressions would be evaluated, and the corresponding values would be scheduled to be assigned to the target signals at time (T+Δ). When simulation time advances to (T+Δ), new values are assigned to signals Z(3), BBAR, and Z(1). Since the value of BBAR changes, this will in turn trigger signal assignment statements 2 and 4. Eventually, at time (T+2Δ), signals Z(0) and Z(2) will be assigned their new values.

The semantics of this concurrent behavior indicate that the simulation, as defined by the language, is event-triggered, and that simulation time advances to the next time unit when an event is scheduled to occur. Simulation time could also advance a multiple of delta time units. For example, events may have been scheduled to occur at times 1, 3, 4, 4+Δ, 5, 6, 6+Δ, 6+2Δ, 6+3Δ, 10, 10+Δ, 15, 15+Δ time units.

The after clause may be used to generate a clock signal, as shown in the following concurrent signal assignment statement.

```
CLK <= not CLK after 10 ns;          -- Signal CLK is of type BIT.
```

This statement creates a periodic waveform on the signal CLK with a time period of 20 ns, as shown in Figure 1.12

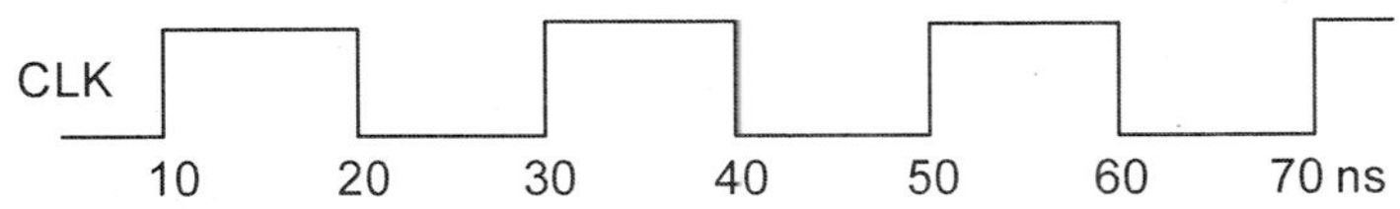

Figure 1.12: A clock waveform with constant on-off period.

1.14.3 Behavioral Style of Modeling

In contrast to the styles of modeling described earlier, the behavioral style of modeling specifies the behavior of an entity as a set of statements that are executed sequentially in the specified order. This set of sequential statements, which are specified inside a process statement, do not explicitly specify the structure of the entity but merely its functionality. A process statement is a concurrent statement that can appear within an architecture body. For example, consider the following behavioral model for the DECODER2x4 entity.

```
architecture DEC_SEQUENTIAL of DECODER2x4 is
begin
  process (A, B, ENABLE)
    variable ABAR, BBAR: BIT;
begin
  ABAR := not A; -- statement 1
  3AR := not B;   -- statement 2
  if ENABLE = '1' then      -- statement 3
    Z(3) <= not (A and B); -- statement 4
    Z(0) <= not (ABAR and BBAR); -- statement 5
    Z(2) <= not (A and BBAR); -- statement 6
    Z(1) <= not (ABAR and B); -- statement 7
  else
    Z<="1111";     -- statement 8
  end if;
 end process;
end DEC_SEQUENTIAL;
```

A process statement also has a declarative part (before the keyword **begin**) and a statement part (between the keywords **begin** and **end process**). The statements appearing within the statement part are sequential statements and are executed sequentially. The list of signals specified within the parentheses after the keyword process constitutes a sensitivity list, and the **process** statement is invoked whenever there is an event on any signal in this list. In the previous example, when an event occurs on signals A, B, or ENABLE, the statements appearing within the process statement are executed sequentially.

The variable declaration (starts with the keyword **variable**) declares two variables called ABAR and BBAR. A variable is different from a signal in that it is always assigned a value instantaneously, and the assignment operator used is the := compound symbol; contrast this with a signal that is assigned a value always after a certain delay (user-specified or the default delta delay), and the assignment operator used to assign a value to a signal is the <= compound symbol. Variables declared within a process have their scope limited to that process. Variables declared outside of a process or a subprogram are called shared variables. These variables can be updated and read by more than one process. Note that signals cannot be declared within a process.

Signal assignment statements appearing within a process are called sequential signal assignment statements. Sequential signal assignment statements, including variable assignment statements, are executed sequentially independent of whether an event occurs on any signals in its right-hand-side expression; contrast this with the execution of concurrent signal assignment statements in the dataflow modeling style. In the previous architecture body, if an event occurs on any signal, A, B, or ENABLE, statement 1, which is a variable assignment statement, is executed, then statement 2 is executed, and so on. Execution of the third statement, an if statement, causes control to jump to the appropriate branch based on the value of the signal ENABLE. If the value of ENABLE is '1', the next four signal assignment statements, 4 through 7, are executed independent of whether A, B, ABAR, or BBAR changed values, and the target signals are scheduled to get their respective values after delta delay. If ENABLE has a value '0', a value of '1' is assigned to each of the elements of the output array Z. When execution reaches the end of the process, the process suspends itself and waits for another event to occur on a signal in its sensitivity list.

It is possible to use case or loop statements within a process. The semantics and structure of these statements are very similar to those in other high-level programming languages like C or Pascal. An explicit wait statement can also be used to suspend a process. It can be used to wait for a certain amount of time, until a certain condition becomes true, or until an event occurs on one or more signals. Here is an example of a process statement that generates a clock with a different on-off period. Figure 1.13 shows the generated waveform.

```
process
begin
  CLK <= '0';
  wait for 20 ns;
  CLK<='1';
  wait for 12 ns;
end process;
```

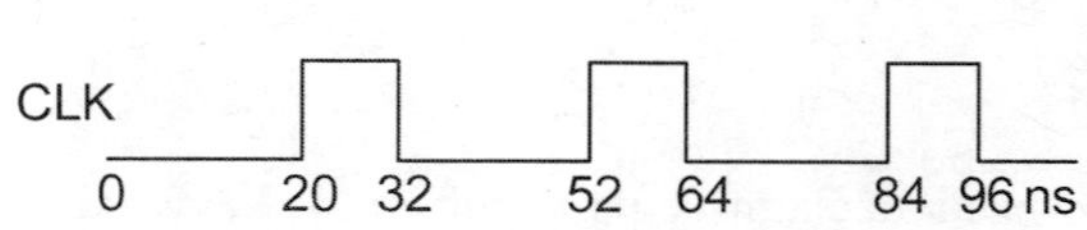

Figure 1.13: A clock waveform with varying on-off period.

This process does not have a sensitivity list because explicit wait statements are present inside the process. It is important to remember that a process never terminates. It is always either being executed or in a suspended state. All processes are executed once during the initialization phase of simulation until they get suspended. Therefore, a process with no sensitivity list and no explicit wait statements will never suspend itself.

A signal can represent not only a wire but also a place holder for a value, that is, it can be used to model a flip-flop. Here is such an example. Port signal Q models a level-sensitive flip-flop.

```
entity LS_DFF is
  port (Q: out BIT; D, CLK: in BIT);
end LS_DFF;

architecture LS_DFF_BEH of LS_DFF is
begin
```

```
    process (D, CLK)
     begin
       if CLK = '1' then
        Q <= D;
     end if;
    end process;
end LS_DFF_BEH;
```

The process executes whenever there is an event on signal D or C.L.K. If the value of CLK is '1', the value of D is assigned to Q. If C LK is '0', then no assignment to Q takes place. Thus, as long as CLK is '1', any change on D will appear cn Q. Once C LK becomes '0', the value in Q is retained.

1.14.4 Mixed Style of Modeling

It is possible to mix the three modeling styles that we have seen so far in a single architecture body. That is, within an architecture body, we could use component instantiation statements (that represent structure), concurrent signal assignment statements (that represent dataflow), and process statements (that represent behavior). Here is an example of a mixed style model for the one-bit full-adder shown in Figure 1.14

```
entity FULL_ADDER is
  port (A, B, CIN: in BIT; SUM, COUT: out BIT);
end FULL_ADDER;

architecture FA_MIXED of FULL_ADDER is
  component XOR2
     port (P1, P2: in BIT; PZ: out BIT);
  end component;
  signal S1: BIT;
begin
  X1: XOR2 port map (A, B, S1);                -- structure
  process (A, B, CIN)                          -- behaviour
    variable T1,T2,T3: BIT;
begin
  Tl := A and B;
  T2 := B and CIN;
  T3 := A and CIN;
  COUT <=Tl or T2 or T3;
 end process;
 SUM <= S1 xor CIN;                            -- dataflow
end FA_MIXED;
```

The full-adder is represented using one component instantiation statement, one process statement, and one concurrent signal as-signment statement. All of these statements are concurrent state-ments; therefore, their order of appearance within the architecture body is not important. Note that a process statement itself is a concurrent statement; however, statements within a process statement are always executed sequentially. S1 is a signal locally declared within the architecture body and is used to pass the value from the output of the component X1 to the expression for signal SUM.

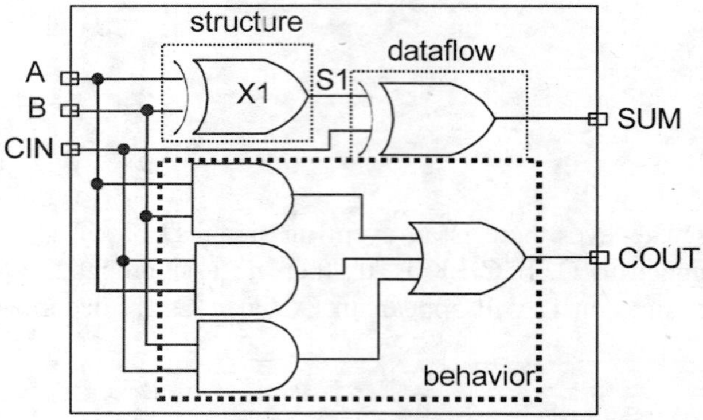

Figure1.14 A 1-bit full-adder.

1.15 CONFIGURATION DECLARATION

A configuration declaration is used to select one of the possibly many architecture bodies that an entity may have, and to bind components, used to represent structure in that architecture body, to entities represented by an entity-architecture pair or by a configuration, which reside in a design library. Consider the following configuration declaration for the HALF_ADDER entity.

```
library CMOS_LIB, MY_LIB;
configuration HA_BINDING of HALF_ADDER is
  for HA_STRUCTURE
   for X1:XOR2
     use entity CMOS_LIB.XOR_GATE(DATAFLOW);
   end for;
   for Al:AND2
     use configuration MY_LIB.AND_CONFIG;
   end for;
 end for;
end HA_BINDING;
```

The first statement is a library clause that makes the library names CMOS_LIB and MY_LIB visible within the configuration declaration. The name of the configuration is HA_BINDING, and it specifies a configuration for the HALF_ADDER entity. The next statement specifies that the architecture body HA_STRUCTURE selected for this configuration. Since this architecture body contains two component instantiations, two component bindings are required. The first statement **(for** X1: ... **end for)** binds the component instantiation with label X1 to an entity represented by the entity-architecture pair, the XOR_GATE entity declaration, and the DATAFLOW architecture body, which resides in the CMOS_LIB design library. Similarly component instantiation A1 is bound to a configuration of an entity defined by the configuration declaration, with name AND_CONFIG, residing in the MY_LIB design library.

There are no behavioral or simulation semantics associated with a configuration declaration. It merely specifies a binding that is used to build a configuration for an entity. These bindings are performed during the elaboration phase of simulation when the entire design to be simulated is being assembled. Having defined a configuration for the entity, the configuration can then be simulated.

An architecture body that does not contain any component instantiations, for example, when dataflow style is used, can also be selected to create a configuration. For example, the DEC_DATAFLOW architecture body can be selected for the DECODER2x4 entity using the following configuration declaration.

```
configuration DEC_CONFIG of DECODER2x4 is
  for DEC_DATAFLOW
  end for;
end DEC_CONFIG;
```

DEC_CONFIG defines a configuration that selects the DEC_DATAFLOW architecture body for the DECODER2x4 entity. The configuration DEC_CONFIG, which represents one possible configuration for the DECODER2x4 entity, can now be simulated.

1.16 PACKAGE DECLARATION

A package declaration is used to store a set of common declarations, such as components, types, procedures, and functions. These declarations can then be imported into other design units using a use clause. Here is an example of a package declaration.

```
package EXAMPLE_PACK is
  type SUMMER is (MAY, JUN, JUL, AUG, SEP);
  component D_FLIP_FLOP
     port (D, CK: in BIT; Q, QBAR: out BIT)
  end component-constant PIN2PIN_DELAY: TIME := 125 ns;
  function INT2BIT_VEC (INT_VALUE: INTEGER)
    return BIT_VCTOR;
end EXAMPLE_PACK;
```

The name of the package declared is EXAMPLE_PACK. It contains type, component, constant, and function declarations. Notice that the behavior of the function INT2BIT_VEC does not appear in the package declaration; only the function interface appears.

Assume that this package has been compiled into a design library called DESIGN_LIB. Consider the following clauses associated with an entity declaration.

```
library DESIGN_LIB;                     -- This is a library clause.
use DESIGN_LIB.EXAMPLE_PACK.all;        -- This is a use clause.
entity RX is ...
```

The library clause makes the name of the design library DESIGN_LIB visible within this description; that is, the name DESIGN_LIB can be used within the description. This is followed by a use clause that imports all declarations in package EXAMPLE_PACK into the entity declaration of RX.

It is also possible to selectively import declarations from a package declaration into other design units. For example,

```
library DESIGN_LIB;
useDESIG LIB.EXAMPLE_PACK.D_FLIP_FLOP;
use DESIGN_LIB.EXAMPLE_PACK.PIN2PIN_DELAY;
architecture RX_STRUCTURE of RX is ...
```

The two use clauses make the component declaration for D_FLIP_FLOP and the constant declaration for PIN2PIN_DELAY visible within the architecture body.

Another approach to selectively import items declared in a package is by using selected names. For example,

```
library DESIGN_LIB;           -- Package EXAMPLE_PACK has been
  -- compiled into a design library called DESIGN_LIB.
package ANOTHER_PACKAGE is
  function POCKET_MONEY
    (MONTH: DESIGN_LIB.EXAMPLE_PACK.SUMMER)
    return INTEGER;
  constant TOTAL_ALU: INTEGER;          -- A deferred constant.
end ANOTHER_PACKAGE;
```

The type SUMMER declared in package EXAMPLE_PACK is used in this new package by specifying a selected name. In this case, a use clause was not necessary. Package ANOTHER_PACKAGE also contains a constant declaration with the value of the constant not specified; such a constant is referred to as a *deferred constant.* The value of this constant is supplied in a corresponding package body.

This is the package STD_LOGIC_1164 which is an IEEE standard. Here is an example that uses this package.

```
library IEEE; -- Makes name IEEE visible.
use IEEE.STD_LOGIC_1164.all;   -- Makes all declarations,
  -- including types STD_LOGIC and STD_LOGIC_VECTOR,
  -- declared in package STD_LOGIC_1164 visible within
  -- entity declaration.
entity DEGODER2x4 is
  port (A, B, ENABLE: in STD_LOGIC;
          Z: out STD_LOGIC_VECTOR (0 to 3));
end DECODER2x4;
architecture DEC_DATAFLOW of DECODER2x4 is
  signal ABAR, BBAR: STD_LOGIC;
begin
  Z(3) <= not (A and B and ENABLE);
  Z(0) <= not (ABAR and BBAR and ENABLE);
  BBAR <= not B;
  Z(2) <= not (A and BBAR and ENABLE);
  ABAR <= not A;
  Z(l) <= not (ABAR and B and ENABLE);
end DEC_DATAFLOW;
```

The library clause declares the design library IEEE. The use clause imports all the declarations within the package STD_LOGIC_1164 into the entity declaration including the type declarations for STD_LOGIC and STD_LOGIC_VECTOR.

1.17 PACKAGE BODY

A package body is used to store the definitions of functions and procedures that were declared in the corresponding package declaration, and also the complete constant declarations for any deferred constants that appear in the package declaration. Therefore, a package body is always associated with a package declaration. Furthermore, a package declaration can have at most one package body associated with it. Contrast this with an architecture body and an entity declaration, where multiple architecture bodies may be associated with a single entity declaration.

Here is the package body for the package EXAMPLE_PACK, declared in the previous section.

```
package body EXAMPLE_PACK is
  function INT2BIT_VEC (INT_VALUE: INTEGER)
     return BIT_VECTOR is
begin
  -- Behavior of function described here.
 end INT2BIT_VEC;
end EXAMPLE_PACK;
```

The name of the package body must be the same as that of the package declaration with which it is associated. It is important to note that a package body is not necessary if the corresponding package declaration has no function or procedure declarations and no deferred constant declarations.

```
package body ANOTHER_PACKAGE is
  -- A complete constant declaration:
  constant TOTAL_ALU: INTEGER := 10;
  function POCKET_MONEY    - - Function body.
    (MONTH: DESIGN_LIB.EXAMPLE_PACK.SUMMER)
    return INTEGER is
  begin
    case MONTH is
      when MAY => return 5;
      when JUL | SEP => return 6;              -- When JUL or SEP.
      when others => return 2;                 -- When JUN or AUG.
    end case;
  end POCKET_MONEY;
end ANOTHER_PACKAGE;
```

1.18 MODEL ANALYSIS

Once an entity is described in VHDL, it can be validated using an analyzer and a simulator that are part of a VHDL system. The first step in the validation process is analysis. The analyzer takes a file that contains one or more design units (remember that a design unit is an entity declaration, an architecture body, a configuration declaration, a package declaration, or a package body) and compiles them into an intermediate form. The format of this compiled intermediate representation is not defined by the language. During compilation, the analyzer validates the syntax and performs static semantic checks. The generated intermediate form is stored in a specific design library that has been designated as the working library.

A design library is a location in the host environment (the computer that supports the VHDL system) where compiled descriptions are stored. Each design library has a logical name that is used when referring to a library in a VHDL description. The mapping of the logical library name to a physical storage location is provided externally by the host environment and is not defined by the language. As an example, a design library can be implemented as a directory in the host environment with the compiled design units being stored as files within this directory. One possible way of providing the mapping of physical names to logical names is by specifying the mappings in a special file that the VHDL system can interpret.

An arbitrary number of design libraries may exist simultaneously. Of all the design libraries that may coexist, one particular library is designated as the working library with the logical name WORK. The language analyzer always compiles descriptions into this library; therefore, at any given time, only one library is updated. Figure 1.15 shows the compilation process.

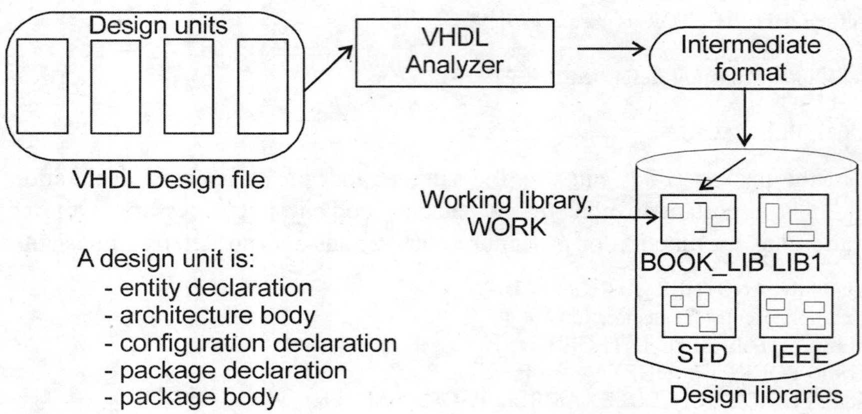

Figure1.15: The compilation process.

If compiled descriptions need to be stored in a different design library, the reference to WORK is changed to point to this design library prior to analysis. The mapping of WORK to a design library is again provided by the host environment. Items compiled in one design library can be imported into design units compiled in a different design library by using library and use clauses, or by accessing them with a selected name.

There is a design library with the logical name *STD* predefined by the VHDL language environment. This library contains two packages: *STANDARD* and *TEXTIO* The STANDARD package contains declarations for all the predefined types of the language (BIT, TIME, INTEGER, etc.). The TEXTIO package contains procedures and functions that are necessary for supporting formatted text read and write operations.

There also exists an IEEE standard package called *STD_LOCIC_1164.* This package defines a nine-value logic type, called *STD_ULOCIC,* and contains its associated subtypes, overloaded operator functions, and other useful utilities. This standard is called the IEEE Std 1164-1993. If available, a host environment must provide this package in a design library called *IEEE.*

1.19 SIMULATION

Once the model description is successfully compiled into one or more design libraries, the next step in the validation process is simulation. For a hierarchical entity to be simulated, all of its lowest-level components must be described at the behavioral level. A simulation can be performed on either one of the following:

- An entity declaration and an architecture body pair
- A configuration

Preceding the actual simulation are two major steps:

1. *Elaboration phase:* In this phase, the hierarchy of the entity is expanded and linked, components are bound to entities in a library, and the top-level entity is built as a network of behavioral models that is ready to be simulated. Also, storage is allocated for signals, variables, and constants declared in the design units. Ini- tial values are also assigned to variables and constants. Files are opened if so indicated in their declarations.
2. *Initialization phase:* Driving and effective values for all explicitly declared signals are computed, implicit signals (discussed in later chapters) are assigned val ues, processes are executed once until they suspend, and simulation time is set to 0 ns.

Simulation commences by advancing time to that of the next event. Values that are scheduled to be assigned to signals at this time are assigned. If the value of a signal changes, and if that signal is present in the sensitivity list of a process, the process is executed until it suspends. Simulation stops when an assertion violation occurs, depending on the implementation of the VHDL system or when the maximum time as defined by the language is reached.

PROBLEMS

1.1. What is the role of HDL in digital system design? How do HDL differ from other software language ?

1.2. List out the capabilities of VHDL.

1.3. What is the Data Object? Explain its all categories.

1.4. Define Data type and discuss its types in brief.

1.5. What is the operator ? Define its all types. Explain shift operators in detail.

1.6. What is overloading ? Define all types of overloading.

1.7. Why do we require signal driver ? Explain with example. 1.8. How does transport delay differ from inertial delay ?

1.9. Explain entity with syntax and example.

1.10. What is the architecture declaration? Define its all types.

1.11. What are various design stages in designing a digital system? Discuss specification and implementation of digital system. Prove how VHDL is capable of releasing the purpose of each design stage ?

1.12. What are the steps involved in implementation and analysis of digital system ?

1.13. What are the different specifications for the design of digital system?

1.14. What is VHDL ? Who formulated the rules for VHDL? What are its specific capabilities?

1.15. What are identifiers ? How many types of identifiers are there ? How do we use identifiers? Explain with example.

1.16. Enumerate the difference between Data Object and Data Types?

1.17. What are Data Types ? Explain all its types with program examples.

1.18. Differentiate between sub and scalar types of data types.

1.19. Write short note on subprogram overloading.

1.20. What is subprogram overloading. Explain with example.

1.21. Differentiate between operator and subprogram overloading.

1.21. In behavioral modeling, explain inertial and transport delay.

1.23. Differentiate between behavioral and structural style of modeling.

1.24. While designing a digital system, what possibly could be various stages. Also prove that how VHDL is capable of realizing the purpose of each design stage.

1.25. Explain the following terms:

(i)	Bus	(ii)	Generic
(iii)	Configuration	(iv)	Package
(v)	Drivers	(vi)	Attribute
(vii)	Process	(viii)	Entity
(xi)	Simultation.		

UNIT 2

VHDL STATEMENTS

2.1 ENTITY DECLARATION

An entity declaration describes the external interface of the entity,that is, it gives the black-box view. It specifies the name of the entity, the names of interface ports, their mode (i.e., direction), and the type of ports. The syntax for an entity declaration is:

entity *entity-name* **is**
  [**generic** (*list-of-generics-and-their-types*);]
  [**port** (*list-of-interface-port-names-and-their-types*);]
  [*entity-item-dedarations*]
[**begin**
  entity-statements]
end [**entity**] [*entity-name*];

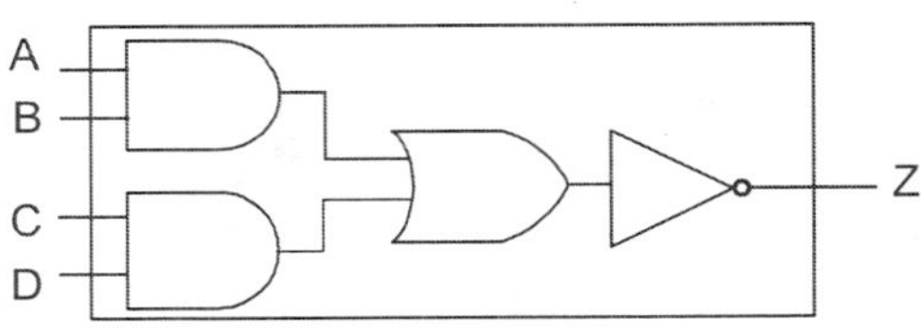

Figure 2.1: An And-Or-Invert circuit.

The entity-name is the name of the entity, and the interface ports are the signals through which the entity passes information to and from its external environment. Each interface port can have one of the following modes:

1. **in:** The value of an input port can only be read within the entity model.
2. **out:** The value of an output port can only be updated within the entity model; it cannot be read.
3. **inout:** The value of a bidirectional port can be read and updated within the entity model.
4. **buffer:** The value of a buffer port can be read and updated within the entity model. However, it differs from the **inout** mode in that it cannot have more than one source, and the only kind of signal that can be connected to it can be another buffer port or a signal with at most one source.
5. **linkage:** The value of a linkage port can be read and updated. This can be done only by another port of mode **linkage.** The usage of linkage ports is not very clear and is thus not recommended. In the past, it has been used to interface with foreign-language models and foreign simulators.

An And-Or-lnvert circuit is shown in Figure 2.1, and its corresponding entity declaration is shown next.

```
entity AOI is
port (A, B, C, D: in BIT; Z: out BIT);
end AOI;
```

The entity declaration specifies that the name of the entity is AOI, and that has four input signals of type BIT and one output signal of type BIT. Note that it does not specify the composition or functionality of the entity.

2.2 ARCHITECTURE BODY

An architecture body describes the internal view of an entity. It describes the functionality or the structure of the entity. The syntax of an architecture body is:

```
architecture architecture-name of entity-name is
[ architecture-item-declarations ]
begin
concurrent-statements; these are -->
process-statement
block-statement
concurrent-procedure-call-statement
concurrent-assertion-statement
concurrent-signal-assignment-statement
component-instantiation-statement
generate-statement
end [ architecture ] [ architecture-name ];
```

The concurrent statements describe the internal composition of the entity. All concurrent statements execute in parallel; therefore, their textual order of appearance within the architecture body has no impact on the implied behavior. The internal composition of an entity can be expressed in terms of structure, dataflow, and sequential behavior. These are described using concurrent statements. For example, component instantiations are used to express structure, concurrent signal assignment statements are used to express dataflow, and process statements are used to express sequential behavior. Each concurrent statement is a different element operating in parallel, similar to individual gates of a design operating in parallel.

The item declarations declare items that are available for use within the architecture body. The names of items declared in the entity declaration, including ports and generics, are available for use within the architecture body due to the association of the entity name with the architecture body by the statement:

```
architecture architecture-name of entity-name is ...
```

An entity can have many internal views, each of which is described using a separate architecture body. In general, an entity is represented using one entity declaration (which provides the external view) and one or more architecture bodies (which provide the internal view). Here are two examples of architecture bodies for the same AOI entity.

```
architecture AOI_CON CURRENT of AOI is
begin
  Z <= not ((A and B) or (C and D));
end AOL_CONCURRENT;
architecture AOI_SEQUENTIAL of AOI is
begin
  process (A, B, C, D)
    variable TEMPI, TEMP2: BIT;
begin
  TEMP1 := A and B;                      -- statement 1
  TEMP2 := C and D;                      -- statement 2
  TEMP1 := TEMP1 or TEMP2;               -- statement 3
    --Z <= not TEMPI;                    -- statement 4
-- It is possible to replace the above four statements with just
-- one statement inside the process:
  -- Z <= not ((A and B) or (C and D));
-- However, it has been used as such here to explain the
-- sequential nature of the statements within a process.
end process;
end AOL_SEQUENTIAL;
```

The first architecture body, AOI_CONCURRENT, describes the AOI entity using the dataflow style of modeling; the second architecture body, AOL_SEQUENTIAL, uses the behavioral style of modeling. In this chapter, we are concerned with describing an entity using the behavioral modeling style. A process statement, which is a concurrent statement, is the primary mechanism used to describe the functionality of an entity in this modeling style.

2.3 PROCESS STATEMENT

A process statement contains sequential statements that describe the functionality of a portion of an entity in sequential terms. The syntax of a process statement is:

```
[ process-label: ] process [ (sensitivity-list) ] [ is ]
  [ process-item-declarations ]
begin
sequential-statements; these are —>
  variable-assignment-statement
  signal-assignment-staiement
  wait-statement
  if-statement
  case-statement
```

```
        loop-statement
        null-statement
        exit-statement
        next-statement
        assertion-statement
        report-statement
        procedure-call-statement
        return-statement
    end process [ process-label ];
```

A set of signals to which the process is sensitive is defined by the sensitivity list. In other words, each time an event occurs on any of the signals in the sensitivity list, the sequential statements within the process are executed in a sequential order, that is, in the order in which they appear (similar to statements in a high-level programming language like C or Pascal). The process then suspends after executing the last sequential statement and waits for another event to occur on a signal in the sensitivity list. Items declared in the item declarations part are available for use only within the process. Every sequential statement can optionally have a label.

The architecture body, AOI_SEQUENTIAL, presented earlier, contains one process statement. This process statement has four signals in its sensitivity list and one variable declaration. If an event occurs on any of the signals, A, B, C, or D, the process is executed. This is accomplished by executing statement 1 first, then statement 2, followed by statement 3, and then statement 4. After this, the process suspends (simulation does not stop, however) and waits for another event to occur on a signal in the sensitivity list.

2.4 VARIABLE ASSIGNMENT STATEMENT

Variables can be declared and used inside a process statement. A variable is assigned a value using the variable assignment statement that typically has the form:

```
variable-object := expression;
```

The expression is evaluated when the statement is executed, and the computed value is assigned to the variable object instantaneously, that is, at the current simulation time.

Variables are created at the time of elaboration and retain their values throughout the entire simulation run (like static variables in C). This is because a process is never exited; it is either in an active state, that is, being executed, or in a suspended state, that is, waiting for a certain event to occur. A process is first entered at the start of simulation (actually, during the initialization phase of simulation), at which time it is executed until it suspends because of a wait statement or a sensitivity list.

Consider the following process statement.

```
process (A)
   variable EVENTS_ON_A: INTEGER := -1;
begin
   EVENTS_ON_A := EVENTS_ON_A+ 1;
end process;
```

At the start of simulation, the process is executed once. The variable EVENTS_ON_A gets initialized to -1 and then incremented by 1. After that, any time an event occurs on signal A, the process is activated and the single variable assignment statement is executed. This causes the variable EVENTS_ON_A to be incremented. At the end of simulation, variable EVENTS_ON_A contains the total number of events that occurred on signal A.

Here is another example of a process statement.

```
signal A, Z: INTEGER;
...
PZ: process (A)                       -- PZ is a label for the process.
    variable V1, V2: INTEGER;
begin
  V1:=A–V2;              -- statement 1
  Z <= –V1;              -- statement 2
  V2:= Z +V1 * 2;        -- statement 3
end process PZ;
```

If an event occurred on signal A at time T1,and variable V2 was assigned a value, say 10, in statement 3, then the next time an event occurs on signal A, say, at time T2, the value of V2 used in statement 1 would still be 10.

A variable can also be declared outside of a process or a subprogram. Such a variable can be read and updated by more than one process. These variables are called shared variables.

2.5 SIGNAL ASSIGNMENT STATEMENT

Signals are assigned values using a signal assignment statement. The simplest form of a signal assignment statement is:

signal-object <= *expression* [**after** *delay-value*];

A signal assignment statement can appear within a process or outside of a process. If it occurs outside of a process, it is considered to be a concurrent signal assignment statement. When a signal assignment statement appears within a process, it is considered to be a sequential signal assignment statement and is executed in sequence with respect to the other sequential statements which appear within that process.

When a signal assignment statement is executed, the value of the expression is computed, and this value is scheduled to be assigned to the signal after the specified delay. ft is important to note that the expression is evaluated at the time the statement is executed (which is the current simulation time) and not after the specified delay. If no after clause ("**after** delay-value") is specified, the delay is assumed to be a default delta delay.

Some examples of signal assignment statements are:

```
COUNTER <= COUNTER + "0010";                 -- Assign after a delta delay.
PAR <= PAR xor DIN after 12 ns;
Z <= (A0 and A1) or (BO and B1) or (CO and C1) after 6 ns;
```

Delta Delay

A delta delay is a very small delay (infinitesimally small). It does not correspond to any real delay, and actual simulation time does not advance. This delay models hardware where a minimal amount of time is needed for a change to occur, for example, in performing zero delay simulation. Delta delay allows for ordering of events that occur at the same simulation time during a simulation. Each unit of simulation time can be considered to be composed of an infinite number of delta delays. Therefore, an event always occurs at a real simulation time plus an integral multiple of delta delays. For example, events can occur at 15 ns, 15 ns +1Δ, 15 ns + 2Δ, 15 ns + 3Δ, 22 ns, 22 ns + Δ, 27 ns, 27 ns + Δ, and so on.

Let us assume that an event occurs on input signal D (that is, there is a change of value on signal D) at simulation time T. Statement 1 is executed first, and TEMP1 is assigned a value immediately because it is a variable. Statement 2 is executed next, and TEMP2 is assigned a value immediately. Statement 3 is executed next; it uses the values of TEMP1 and TEMP2 computed in statements 1 and 2, respectively, to determine the new value for TEMP1. Finally, statement 4 is executed, which causes signal Z to get the value of its right-hand-side expression after a delta delay, that is, signal Z gets its value only at time T+Δ. This is shown in Figure 2.2

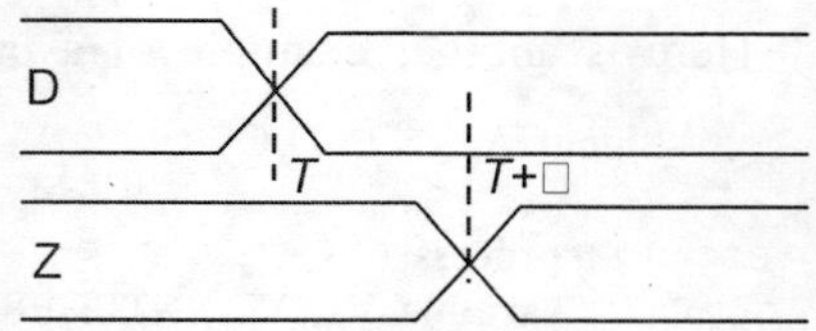

Figure 2.2 Delta delay.

Consider the process PZ described in the previous section. If an event occurs on signal A at time 7, execution of statement 1 causes V1 to get a value, signal Z is then scheduled to get a value at time 7+Δ, and finally, statement 3 is executed in which the old value of signal Z is used, that is, its value at time 7, not the value that was scheduled to be assigned in statement 2. The reason for this is that simulation time is still at time 7 and has not advanced to time 7+Δ. Later when simulation time advances to 7+Δ, signal Z will get its new value. This example shows the important distinction between a variable assignment and a signal assignment statement. Variable assignments cause variables to get their values instantaneously, while signal assignments always cause signals to get their values at a later time (at least a delta delay later).

2.6 WAIT STATEMENT

The wait statement provides an alternate way to suspend the execution of a process. There are three basic forms of the wait statement.

```
Wait on sensitivity- list;
Wait until boolean- expression;
Wait for time-expression;
```

They may also be combined in a single wait statement. For example,

```
wait on sensitivity-list until boolean-expression for time-expression;
```

Some examples of wait statements are:

```
wait on A, B, C;                    -- statement 1
wait until A = B;                   -- statement 2
wait for 10 ns;                     -- statement 3
wait on CLOCK for 20 ns;            -- statement 4
wait until SUM > 100 for 50 ms;     -- statement 5
wait on CLOCK until SUM > 100;      -- statement 6
```

The execution of the wait statement in statement 1 causes the enclosing process to suspend and then wait for an event to occur on signal A, B, or C. Once that happens, the process resumes execution from the next statement onwards. If the wait statement was the last statement in the process, the process resumes execution from the first statement.

The execution of the wait statement in statement 2 causes the enclosing process to suspend until the specified condition becomes true. When an event occurs on signal A or B, the condition is evaluated. If it is true, the process resumes execution from the next statement onwards, otherwise, it suspends again. When the wait statement in statement 3 is executed, say, at time 7, the enclosing process suspends for 10 ns, and when simulation time advances to T+10 ns, the enclosing process resumes execution from the statement following the wait statement.

The execution of statement 4 causes the enclosing process to suspend and then wait for an event to occur on signal CLOCK for a time-out interval of 20 ns. If no event occurs within 20 ns, the process resumes execution with the statement following the wait. In statement 5, the process suspends for a maximum of 50 ms until the value of signal SUM is greater than 100. The Boolean condition is evaluated every time there is an event on signal SUM. If the Boolean condition is not satisfied for 50 ms, the process resumes from the statement following the wait statement.

In the last example of the wait statement (statement 6), it means to wait until an event occurs on CLOCK, then to check if SUM > 100. If SUM <= 100, continue to wait. If SUM > 100, resume from the statement following the wait statement.

It is possible for a process not to have an explicit sensitivity list. In such a case, the process may have one or more wait statements. It must have at least one wait statement; otherwise, the process will never get suspended and would remain in an infinite loop during the initialization phase of simulation. It is an error if both a sensitivity list and a wait statement are present within a process.

The presence of a sensitivity list in a process implies the presence of an implicit "**wait on** *sensitivity-list*" statement as the last statement in the process. An equivalent process statement for the process statement in the AOI_SEQUENTIAL architecture body is:

```
process                                  -- No sensitivity list.
  variable TEMP1, TEMP2: BIT;
begin
  TEMP1 := A and B;
  TEMP2 := C and D;
  TEMP1 := TEMP1 or TEMP2;
  Z<= not TEMP1;
  wait on A, B, C,D;                     -- Replaces the sensitivity list.
end process;
```

Therefore, a process with a sensitivity list always suspends at the end of the process and, when reactivated due to an event, resumes execution from the first statement in the process.

wait for 0

What does the following wait statement imply?

```
wait for 0 ns;
```

It means to wait for one delta cycle. This statement is useful when you want the process to be delayed so that delta-delayed signal assignments within a process can take effect. For example,

```
WAIT0: process
begin
  wait on DATA;
  SIG_A <= DATA;
  wait for 0 ns;
  SIG_B <= SIG_A;
end process;
```

If DATA changes at 10 ns, SIG_A is scheduled to get the new value of DATA at 10+1Δ. The wait statement ("**wait for** 0 ns") causes the process to suspend for one delta. SIG_A gets updated with it's new value. Process resumes at 10 + 1Δ. SIG_B gets scheduled to get the new value of SIG_A at10+2Δ.

If the "**wait for** 0 ns" statement was not present, both the signalassignments are executed sequentially at time 10 ns, and SIG_B is scheduled to get the current value of SIG_A (which is not yet the new value of DATA) at time 10+1 Δ. SIG_B always gets the value of SIG_A at the time DATA changes but after a delta delay. This effect is further elaborated in Figure 2.3.

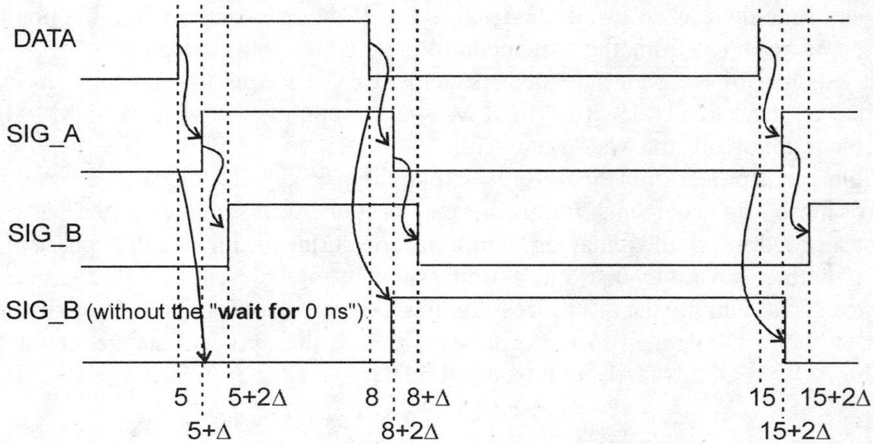

Figure 2.3: Effect of "**wait for** 0 ns"

wait until

The behavior of the "**wait until**" form is worth a second explanation. So let us look at an example.

wait until CLOCK = '1',

When the wait statement is executed, the process (enclosing) must first suspend, then wait for an event to occur on CLOCK, and then check the value of the boolean expression. It does not matter if the boolean expression is true at the time the wait statement is executed. So if CLOCK is already '1' when the wait statement is executed, the boolean expression is not checked until another event occurs on signal CLOCK.

For the same reason,

wait until TRUE;

when executed will suspend the process indefinitely (since no signal is present in the boolean expression and hence no event can occur).

2.7 IF STATEMENT

An if statement selects a sequence of statements for execution based on the value of a condition. The condition can be any expression that evaluates to a Boolean value. The general form of an if statement is:

if *boolean-expression* **then**
  sequential-statements
{**elsif** *boolean-expression* **then**   -- elsif clause; if statement can

```
  sequential-statements}              -- have 0 or more elsif clauses.
[else                                 -- else clause (optional).
  sequential-statements ]
end if;
```

The if statement is executed by checking each condition sequentially until the first true condition is found; then, the set of sequential statements associated with this condition is executed. The if statement can have zero or more elsif clauses and an optional else clause. An if statement is also a sequential statement, and therefore, the previous syntax allows for arbitrary nesting of if statements. Here are some examples.

```
if SUM <= 100 then                    -- "<="is a less-than-or-equal-to operator.
  SUM := SUM + 10; end if;
if NICKEL_IN then
  DEPOSITED <= TOTAL_10;              -- This"<=" is a signal assignment
                                      -- operator.
elsif  DIME_IN then
  DEPOSITED <=TOTAL_15;
elsif QUARTER_IN then
  DEPOSITED <= TOTAL_30;
else
  DEPOSITED <=TOTAL_ERROR;
end if;
if CTRL1 = '1' then
  if CTRL2 = '0' then
    MUX_OUT <= "0010";
  else
    MUX_OUT<="0001";
  end if;
else
  if CTRL2 ='0' then
    MUX_OUT<= "1000";
  else
    MUX_OUT <= "0100";
  end if;
end if;
```

A complete example of a two-input nor gate entity using an if statement is shown next.

```
entity NOR2 is
  port (A, B: in BIT; Z: out BIT);
end NOR2;
architecture NOR2 of NOR2 is          -- Architecture body can have
                                      -- same name as entity.
begin
  P1: process (A, B)
  constant RISE_TIME: TIME := 10 ns;
  constant FALL_TIME: TIME := 8 ns;
begin
  if (A nor B) = '1' then
    Z<='1' after RISE_TIME;
  else
```

```
      Z <= '0' after FALL_TIME;
    end if;
  end process P1;
end NOR2;
```

2.8 CASE STATEMENT

The format of a case statement is:

```
case expression is
  when choices => sequential-statements              -- branch #1
  when choices => sequential-statements              -- branch #2
  -- Can have any number of branches.
  [ when others => sequential-statements ]           -- last branch
end case;
```

The case statement selects one of the branches for execution based on the value of the expression. The expression value must be of a discrete type or of a one-dimensional array type. Choices may be expressed as single values, as a range of values, by using | (vertical bar: represents an "or" choice), or by using the others clause ("**when others**"). All possible values of the expression must be covered in the case statement exactly once. The others clause can be used as a choice to cover the "catch-all" values and, if present, must be the last branch in the case statement. An example of a case statement is:

```
type WEEK_DAY is (MON, TUE, WED, THU, FRI, SAT, SUN);
type DOLLARS is range 0 to 10;
variable DAY: WEEK_DAY;
variable POCKET_MONEY: DOLLARS;
case DAY is
  when TUE         => POCKET_MONEY := 6;      -- branch 1
  when MON | WED   => POCKET_MONEY := 2;      -- branch 2
  when FRI to SUN  => POCKET_MONEY := 7;      -- branch 3
  when others      => POCKET_MONEY := 0;      -- branch 4
end case;
```

Branch 2 is chosen if DAY has the value of either MON or WED. Branch 3 covers the values FRI, SAT, and SUN, while branch 4 covers the remaining value, THU. The case statement is also a sequential statement, and it is therefore possible to have nested case statements. A model for a 4-by-1 multiplexer using a case statement is shown next.

```
library IEEE;
use IEEE.STD_LOGIC_1164.all;
entity MUX is
  port ( A, B, C, D   : in STD_LOGIC;
         CTRL         : in STD_LOGIC_VECTOR(0 to 1);
         Z            : out STD_LOGIC);
end MUX;

architecture MUX_BEHAVIOR of MUX is
  constant MUX_DELAY: TIME := 10 ns;
begin
```

```
PMUX: process (A, B, C, D, CTRL)
  variable TEMP: STD_LOGIC;
begin
 case CTRL is
   when "00" => TEMP := A;
   when "01" => TEMP := B;
   when "10" => TEMP := C;
   when "11" => TEMP := D;
   when others => TEMP := 'X';     -- Covers values such as
                  -- "0Z","ZU", "Ul",etc.
end case;
Z <= TEMP after MUX_DELAY;
end process PMUX;
end MUX_BEHAVIOR;
```

2.9 NULL STATEMENT

The statement

null;

is a sequential statement that does not cause any action to take place; execution continues with the next statement. One example of this statement's use is in an if statement or in a case statement where, for certain conditions, it may be useful or necessary to explicitly specify that no action needs to be performed.

2.10 LOOP STATEMENT

A loop statement is used to iterate through a set of sequential statements. The syntax of a loop statement is:

[*loop-label*:] *iteration-scheme* **loop**
  sequential-statements
end loop [*loop-label*];

There are three types of iteration schemes. The first is the for iteration scheme, which has the form:

for *identifier* **in** *range*

An example of this iteration scheme is:

FACTORIAL := 1;

```
for NUMBER in 2 to N loop
  FACTORIAL := FACTORIAL * NUMBER;
end loop;
```

In this example, the body of the for loop is executed (N-1) times, with the loop identifier, NUMBER, being incremented by 1 at the end of each iteration. The object NUMBER is implicitly declared within the for loop to belong to the integer type whose values are in the range 2 to N. Therefore, no explicit declaration for the loop identifier is necessary. The loop identifier also cannot be assigned any value inside the for loop. If another variable with the same name exists outside the for loop, these two variables are treated separately, and the variable used inside the for loop refers to the loop identifier.

The range in a for loop can also be a range of an enumeration type, such as the one shown in the following example.

```
type HEXA is ('0', '1,' '2', '3', '4', '5', '6', 7', '8', '9', 'A'. 'B', 'C','D', 'E', 'F');
...
for NUM in HEXA ('9') downto HEXA'('0') loop
  -- NUM will take values in type HEXA from '9' through '0'.
  ...
end loop;

for CHAR in HEXA loop
  -- CHAR will take all values in type HEXA from '0' through 'F'.
end loop;
```

Notice that it is necessary to qualify the values being used for NUM (e.g., HEXA'('0')) since the literals '0' through '9' are overloaded, being defined once in type HEXA and the second time in the predefined type CHARACTER.

The second form of the iteration scheme is the while scheme, which has the form:

```
while boolean-expression
```

An example of the while iteration scheme is:

```
J := 0; SUM := 10;
WH_LOOP: while J < 20 loop          -- This loop has a label, WH_LOOP.
  SUM := SUM * 2;
  J:=J + 3;
end loop;
```

The statements within the body of the loop are executed sequentially and repeatedly as long as the loop condition, J < 20, is true. If the loop condition becomes false, execution continues with the statement following the loop statement.

The third and final form of the iteration scheme is one where no iteration scheme is specified. In this form of loop statement, all statements in the loop body are repeatedly executed until some other action causes the loop to exit. This action can be caused by an exit statement, a next statement, or a return statement. Here is an example.

```
SUM := 1; J := 0;
L2: loop                  -- This loop also has a label.
J:=J + 21;
SUM := SUM * 10;
exit when SUM > 100;
end loop L2;              -- This loop label, if present, must be the same
                          -- as the initial loop label.
```

In this example, the exit statement causes the execution to jump out of loop L2 when SUM becomes greater than 100. If the exit statement were not present, the loop would execute indefinitely.

2.11 EXIT STATEMENT

The exit statement is a sequential statement that can be used only inside a loop. It causes execution to jump out of the innermost loop or the loop whose label is specified. The syntax for an exit statement is:

```
exit [ loop-label] [ when condition ];
```

If no loop label is specified, the innermost loop is exited. If the when clause ("**when** *condition*") is used, the specified loop is exited only if the given condition is true; otherwise, execution continues with the next statement. An alternate form for loop L2 described in the previous section is:

```
SUM := 1; J := 0;
L3: loop
  J : = J + 21;
  SUM := SUM * 10;

  if SUM > 100 then
     exit L3;                    --"exit;"is also sufficient.
  end if;
end loop L3;
```

Here is another example of an exit statement.

```
loop
  wait on A, B;
  exit when A = B;               -- Since no label, innermost loop is exited.
end loop;
-- This loop behaves exactly like the wait statement:
-- wait until A = B;
```

2.12 NEXT STATEMENT

The next statement is also a sequential statement that can be used only inside a loop. The syntax is the same as that for the exit statement except that the keyword **next** replaces the keyword **exit**. Its syntax is:

```
next [ loop-label ] [ when condition ];
```

The next statement results in skipping the remaining statements in the current iteration of the specified loop; execution resumes with the first statement in the next iteration of this loop, if one exists. If no loop label is specified, the innermost loop is assumed. In contrast to the exit statement, which causes the loop to be terminated (i.e., exits the specified loop), the next statement causes the current loop iteration of the specified loop to be prematurely terminated; execution resumes with the next iteration. Here is an example.

```
for J in 10 downto 5 loop
  if SUM < TOTAL_SUM then
     SUM := SUM + 2;
  elsif  SUM = TOTAL_SUM then
     next;
  else
     null;
  end if;
  K:=K + 1;
end loop;
```

When the next statement is executed, execution jumps to the end of the loop (the last statement, K := K + 1, is not executed), decrements the value of the loop identifier, J, and resumes loop execution with this new value of J.

The next statement can also cause an inner loop to be exited. Here is such an example.

```
L4: for K in 10 downto 1 loop
  -- statements section 1
  L5: loop
    -- statements section 2
    next L4 when WR_DONE = '1',
    -- statements section 3
  end loop L5;
  -- statements section 4
end loop L4;
```

When WR_DONE = '1' becomes true, statements sections 3 and 4 are skipped, and execution jumps to the beginning of the next iteration of loop L4. Notice that the loop L5 was terminated because of the result of the next statement.

2.13 ASSERTION STATEMENT

Assertion statements are useful in modeling constraints of an entity. For example, you may want to check if a signal value lies within a specified range, or check the setup and hold times for signals arriving at the inputs of an entity. If the check fails, a message is reported. The syntax of an assertion statement is:

```
assert boolean-expression
  [ report string-expression ]
  [ severity expression ];
```

If the value of the Boolean expression is false, the report message is printed along with the severity level. The expression in the severity clause must be a value of type SEVERITY_LEVEL (a predefined enumeration type with values NOTE, WARNING, ERROR, and FAILURE). The severity level is typically used by a simulator to initiate appropriate actions depending on its value. For example, if the severity level is ERROR, the simulator may abort the simulation process and provide relevant diagnostic information. At the very least, the severity level is displayed.

Here is a model of a D-type rising-edge-triggered flip-flop that uses assertion statements to check for setup and hold times.

```
entity DFF is
  port (D, CK: in BIT; Q, NOTQ: out BIT);
end DFF;

architecture CHECK_TMES of DFF is
  constant HOLD_TIME: TIME := 5 ns;
  constant SETUP_ TIME: TIME := 3 ns;
begin
  process (D, CK)
    variable LastEventOnD, LastEventOnCk: TIME;
  begin
   -- Check for hold time:
```

```
        if D'EVENT then
            assert NOW = 0 ns or
                (NOW - LastEventOnCk) >= HOLD_TIME
              report "Hoid time too short!"
              severity FAILURE;
             LastEventOnD := NOW;
        end if;

        -- Check for setup time:
        if CK = '1' and CK' EVENT then
           assert NOW = 0 ns or
                (NOW - LastEventOnD) >= SETUP_ TIME
              report "Setup time too short!"
              severity FAILURE;
           LastEventOnCk := NOW;

        end if;
        -- Behavior of FF:
        if CK = '1'and CK'EVENT then
           Q <= D;
           NOTQ <= not D;
      end if;
    end process;
  end CHECK_TIMES;
```

'EVENT is a predefined attribute of a signal and is true if an event (a change of value) occurred on that signal at the time the value of the attribute is determined.NOW is a predefined function that returns the current simulation time. In the previous example, the process is sensitive to signals D and CK. When an event occurs on either of these signals, the first if statement is executed. This checks to see if an event occurred on D. If so, the assertion is checked to make sure that the difference between the current simulation time and the last time an event occurred on signal CK is greater than a constant HOLD_TIME delay. If not, a report message is printed and the severity level is returned to the simulator. Similarly, the next if statement checks for the setup time. The last if statement describes the latch behavior of the D-type flip-flop. The setup and hold times have been modeled as constants in this example. These could also be modeled as generic parameters of the flip-flop. The expression "(NOW = 0 ns)" is used in the assertion statements to prevent misleading messages appearing during the initialization phase of simulation (remember that all processes are executed until they suspend during the initialization phase).

Here is another example that uses an assertion statement to check for spikes at the input of an inverter.

```
package PACK1 is
  constant MIN_PULSE: TIME := 5 ns;
  constant PROPAGATE_DLY: TIME := 10 ns;
end PACK1;

useWORK.PACK1.all;
entity INV is
   port (A: in BIT; NOT_A: out BIT);
end INV;
```

```
architecture CHECK_INV of INV is
begin
  process (A)
   variable LastEventOnA: TIME := 0 ns;
  begin
    assert NOW = 0 ns or
          (NOW - LastEventOnA) >= MIN_PULSE
     report "Spike detected on input of inverter"
     severity WARNING;
   LastEventOnA := NOW;
   NOT_A <= not A after PROPAGATE_DLY;
  end process;
end CHECK_INV;
```

Some other examples of assertion statements are:

```
assert DATA <= 255
  report "Data out of range.";
assert (CLK = '0') or (CLK = '1');                 -- CLK is of type STD_ULOGIC.
```

In the last assertion statement example, the default report message "Assertion Violation" is printed. The default severity level is ERROR if the serverty clause is not specified as in the previous two exmples.

2.14 REPORT STATEMENT

A report statement can be used to display a message. It is similar to an assertion statement, but without the assertion check. The syntex is of the from :

```
report string- experssion
  [Serverity expression];
```

The experssion in the severity clause if present, must be of the pre- defiend type SEVERITY_ LEVEL. If not present, the default serverity level of NOTE is used.

When a report statement is executed, it causes the specified string to be printed and the serveity level to be reported to the simulator for appropriate action. Here are some examples.

```
              if CLR = 'Z' then
report "Signal CLR has a high-impedance Value.";
       -- Default serverity level is NOTE.
       end if;
   if CLK/ = '0' and CLK/ = '1' then
report " CLK is neither a '0' nor a '1'!!! "
      serverity ERROR;
      end if;
```

2.15 CONCURRENT SIGNAL ASSIGNMENT STATEMENT

One of the primary mechanisms for modeling the dataflow behavior of an entity is using the concurrent signal assignment statement. An example of a dataflow model for a two-input or gate, shown in Figure 2.4 follows.

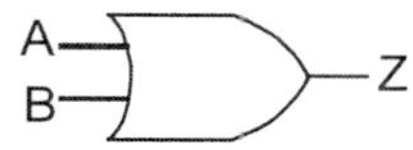

Figure 2.4: An or gate.

```
entity OR2 is
 port (signal A, B: in BIT; signal Z: out BIT);
end OR2;
architecture OR2 of OR2 is
begin
  Z <= A or B after 9 ns;
end OR2;
```

The architecture body contains a single concurrent signal assignment statement that represents the dataflow of the or gate. The interpretation of this statement is that whenever there is an event (a change of value) on either signal A or B (A and B are signals in the expression for Z), the expression on the right is evaluated, and its value is schedulec to appear on signal Z after a delay of 9 ns. The signals in the expression, A and B, form the "sensitivity list" for the signal assignment statement.

There are two other points to mention about this example. First, the 0input and output ports have their object class "signal" explicitly specified in the entity declaration. If it were not so, the ports would still have been signals, since this is the default and the only object class that is allowed for ports. The second point to note is that the architecture name and the entity name are the same. This is not a problem, since architecture bodies are secondary units while entity declarations are primary units, and the language allows secondary units to have the same names as primary units.

An architecture body can contain any number of concurrent signal assignment statements. Since they are concurrent statements, the ordering of the statements is not important. Concurrent signal assignment statements are executed whenever events occur on signals that are used in their expressions. An example of a dataflow model for a 1-bit full-adder, whose external view is shown in Figure 2.5, is presented next.

```
entity FULL_ADDER is
  port (A, B, CIN: in BIT; SUM, COUT: out BIT);
end FULL_ADDER;
architecture FULL.ADDER of FULL_ADDER is
begin
  SUM <= A xor B xor CIN after 15 ns;
  COUT <= (A and B) or (B and CIN) or (CIN and A) after 10 ns;
end FULL_ADDER;
```

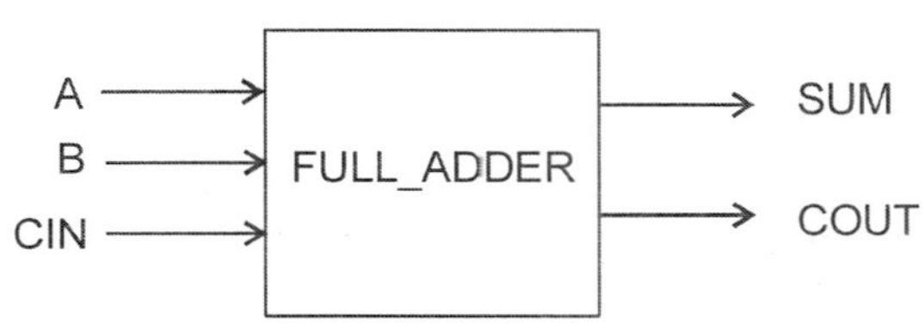

Figure 2.5: External view of a 1-bit full-adder.

Two signal assignment statements are used to represent the dataflow of the FULL_ADDER entity. Whenever an event occurs on signals A, B, or CIN, expressions of both statements are evaluated, and the value to SUM is scheduled to appear after 15 ns, while the value to COUT is scheduled to appear after 10 ns. The after clause models the delay of the logic represented by the expression.

Contrast this with the statements that appear inside a process statement. Statements within a process are executed sequentially, while statements in an architecture body are ail concurrent statements and order-independent. A process statement is itself a concurrent statement. What this means is that if there were any concurrent signal assignment statements and process statements within an architecture body, the order of these statements also would not matter.

2.16 CONCURRENT VERSUS SEQUENTIAL SIGNAL ASSIGNMENT

We saw that signal assignment statements can also appear within the body of a process statement. Such statements are called sequential signal assignment statements, while signal assignment statements that appear outside of a process are called concurrent signal assignment statements. Concurrent signal assignment statements are event-triggered, that is, they are executed whenever there is an event on a signal that appears in its expression, while sequential signal assignment statements are not event-triggered and are executed in sequence in relation to the other sequential statements that appear within the process. To further understand the difference between these two kinds of signal assignment statements, consider the following two architecture bodies.

```
architecture SEQ_SIG_ASG of FRAGMENT1 is
  -- A, B and Z are signals.
begin
  process (B)
  begin        -- Following are sequential signal assignment statements:
     A<=B;
     Z<=A;
  end process;
end;

architecture CON_SIG_ASG of  FRAGMENT2 is
begin    -- Following are concurrent signal assignment statements:
  A <= B;
  Z<=A;
end;
```

In architecture SEQ_SIG_ASG, the two signal assignments are sequential signal assignments. Therefore, whenever signal B has an event (say, at time T), the first signal assignment statement is executed and then the second, both in zero time. However, signal A is scheduled to get its new value of B only at time 7+Δ (the delta delay is implicit), and Z is scheduled to be assigned the old value of A (not the value of B) at time 7+Δ also.

In architecture CON_SIG_ASG, the two statements are concurrent signal assignment statements. When an event occurs on signal B (say, at time 7), signal A gets the value of B after delta delay, that is, at time 7+Δ. When simulation time advances to 7+Δ, signal A will get its new value, and this event on A (assuming there is a change of value on signal A) will trigger the second signal assignment statement, which will cause the new value of A to be assigned to Z after another delta delay, that is, at time 7+2Δ.

For every concurrent signal assignment statement, there is an equivalent process statement with the same semantic meaning. The concurrent signal assignment statement:

```
CLEAR <= RESET or PRESET after 15 ns;
  -- RESET and PRESET are signals.
```

is equivalent to the following process statement.

```
process
begin
  CLEAR <= RESET or PRESET after 15 ns;
  wait on RESET; PRESET;
end process;
```

An identical signal assignment statement (now a sequential signal assignment) appears in the body of the process statement along with a wait statement whose sensitivity list comprises signals used in the expression of the concurrent signal assignment statement.

A concurrent signal assignment statement can be marked as postponed, such as:

```
CLEAR <= postponed RESET or PRESET after 15 ns;
```

The semantics of such a statement is equivalent to the semantics of its equivalent process statement, which is a postponed process.

2.17 DELTA DELAY REVISITED

In a signal assignment statement, if no delay is specified or a delay of 0 ns is specified, a delta delay is assumed. Delta delay is an infini-tesimally small amount of time. It is not a real-time delta and doesnot cause real simulation time to change. The delta delay mechanism provides for ordering of events on signals that occur at the same simulation time. Consider the circuit shown in Figure 2.6 and the corresponding model.

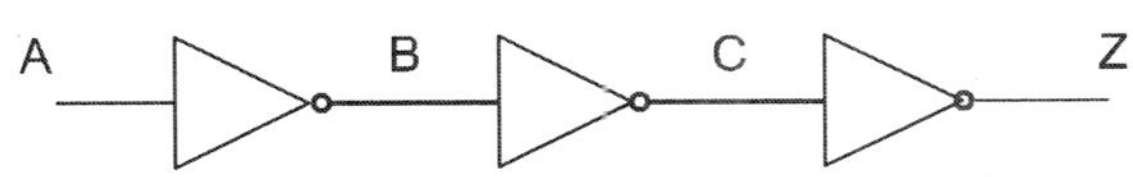

Figure 2.6: Three inverting buffers in series.

```
entity FAST_INVERTER is
   port (A: in BIT; Z: out BIT);
end;

architecture DELTA_DELAY of FAST_INVERTER is
  signal B, C: BIT;
begin      -- Following statements are order-independent:
  Z <= not C;                          -- signal assignment #1
  C <= not B;                          -- signal assignment #2
  B <= not A;                          -- signal assignment #3
end;
```

The three signal assignments in the FAST_INVERTER entity use delta delays. When an event occurs on signal A, say at 20 ns, the third signal assignment is triggered, which causes signal B to get the inverted value of A at 20 ns+1Δ. When time advances to 20 ns+1Δ, signal B changes. This triggers the second signal assignment, causing signal C to get the inverted value of B after another delta delay, that is, at 20 ns+2Δ. When simulation time advances to 20 ns+2Δ, the first signal assignment is triggered, causing Z to get a new value at time 20 ns+3Δ. Even though the real simulation time stayed at 20 ns, Z was updated with the correct value through a sequence of delta-delayed events. This sequence of waveforms is shown in Figure 2.7.

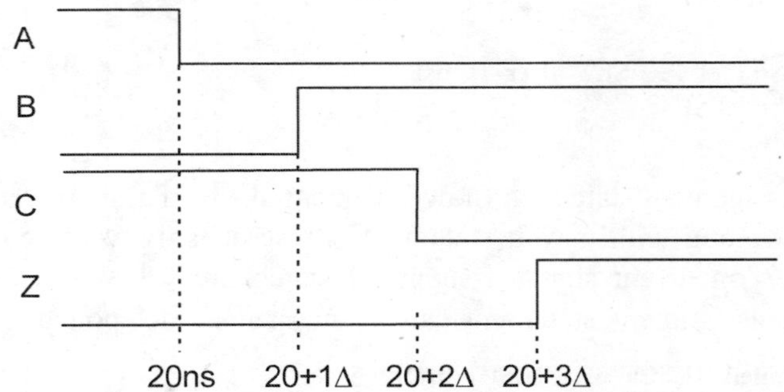

Figure 2.7: Delta delays in concurrent signal assignment statements.

A typical VHDL simulator maintains a list of events that is to occur in an event queue. The events in this queue are ordered not only on the real simulation time, but also on each delta delay. Figure 2.8 shows a snapshot of an event queue in a VHDL simulator during simulation. Each event has associated with it a list of signal-value pairs that are to be scheduled. For example, the value '0' is to be assigned to signal Z when simulation time advances to 10 ns+2Δ. Any postponed process is evaluated only at the end of all the deltas of a simulation time, that is, in the marked shaded regions. For example, if a postponed process activates at time 10+1Δ, the process will actually resume execution only after events at time 10+2Δ is processed.

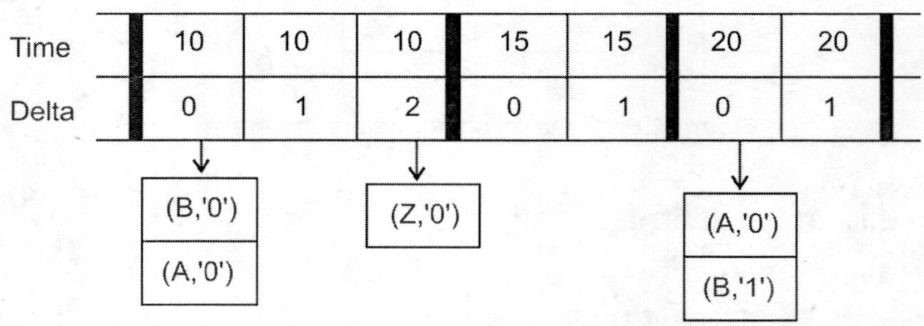

Figure 2.8: An event queue in a VHDL simulator.

Let us consider another example of delta delay. The dataflow model for the RS latch shown in Figure 2.9 appears next.

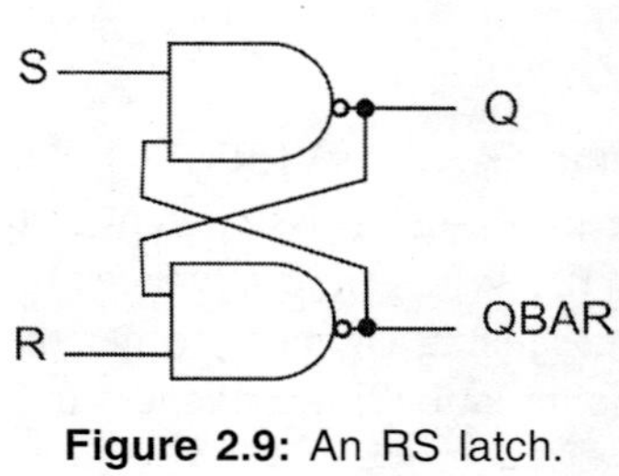

Figure 2.9: An RS latch.

```
entity RS_LATCH is
   port (R, S: in BIT; Q: buffer BIT;
      QBAR: buffer BIT);
end RS_LATCH;

architecture DELTA of RS_LATCH is
begin
  QBAR<=R nand Q;
  Q <= S nand QBAR;
end DELTA;
```

Assume both R and S have value '1' and Q and QBAR are at '1' and '0', respectively. Say signal R changes from '1' to '0' at 5 ns. Figure 2.10 shows the sequence of events that occur as a result. After two delta delays, the circuit stabilizes with the final values of Q and QBAR being '0' and '1', respectively.

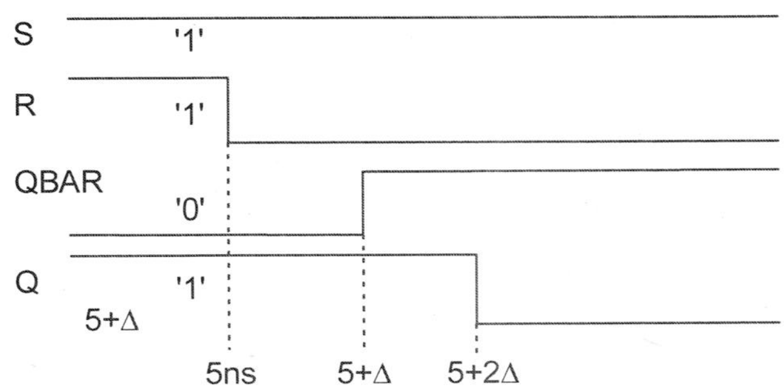

Figure 2.10: Sequence of events in the RS latch.

2.18 RESOLUTION FUNCTIONS

Each concurrent signal assignment statement creates a driver for the signal being assigned. What happens when there is more than one assignment to the same signal? In this case, the signal has more than one driver, and a mechanism is needed to compute the effective value of the signal. Consider the circuit shown in Figure 2.11 and its corresponding dataflow model.

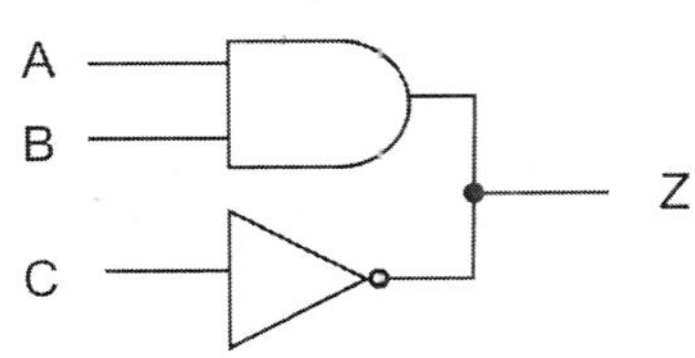

Figure 2.11: Two drivers driving signal Z.

```
entity TWO_DR_EXAMPLE is
  port (A, B, C: in BIT; Z: out BIT);
end TWO_DR_EXAMPLE;

architecture NOT_LEGAL of TWO_DR_EXAMPLE is
begin
```

```
    Z <= A and B after 10 ns;
    Z <= not C after 5 ns;
end;                                  -- Effective value for signal Z has to be
                                      -- determined: this is not a legal VHDL model.
```

In this example, there are two gates driving the output signal Z. How is the value of Z determined? It is determined by using a user-defined resolution function that considers the values of both the drivers for Z and determines the effective value. Consider the following architecture body.

```
architecture NO_ENTITY of DUMMY is
begin
    Z <= '1' after 2 ns, '0' after 5 ns, '1' after 10 ns;
    Z <= '0' after 4 ns, '1' after 5 ns, '0' after 20 ns;
    Z <= '1' after 10 ns, '0' after 20 ns;
end NO_ENTITY;
```

In this case, there are three drivers for signal Z. Each driver has a sequence of transactions where each transaction defines the value to appear on the signal and the time at which it is to appear. The resolution function resolves the value for the signal Z from the current value of each of its drivers. This is shown pictorially in Figure 2.12

The value of each driver is an input to the resolution function and, based on the computation performed within the resolution function, the value returned by this function becomes the resolved value for the signal. The resolution function is user-written and may perform any function. The function is not restricted to perform a wired-and or wired-or operation; for example, it could be used to count the number of events on a signal.

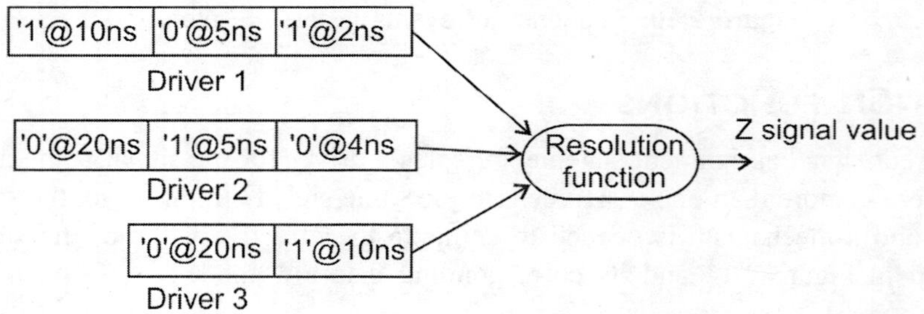

Figure 2.12: Resolving signal drivers.

A signal with more than one driver must have a resolution function associated with it; otherwise, it is an error. Such a signal is calleda resolved signal. A resolution function is associated with a signal by specifying its name in the signal declaration. For example,

```
signal BUSY: WIRED_OR BIT;
```

is one way of associating the resolution function WIRED_OR with the signal BUSY. No arguments can be specified, since the arguments for the function are the current values of all the drivers for that signal. The TWO_DR_EXAMPLE entity previously described is therefore incorrect; a resolution function must be specified for port Z, such as:

```
port (A, B, C: in BIT; Z: out WIRED_OR BIT);
```

Another way is by declaring a resolved subtype, that is, by including the name of the resolution function in the subtype declaration and then declaring the signal to be of that subtype. For example,

```
subtype RESOLVED_BIT is WIRED_OR BIT;
signal BUSY: RESOLVED_BIT;
```

The resolved signal Z in the TWO_DR_EXAMPLE entity can now be specified as:

```
port (A, B, C: in BIT; Z: out RESOLVED_BIT);
```

The semantics of when a resolution function is invoked are as follows. Whenever an event occurs on any source of a resolved signal, the resolution function associated with that signal is called with the values of all its drivers. The return value from the resolution function becomes the value for the resolved signal. In the example of architecture NO_ENTITY, the resolution function is invoked at time 2 ns with driver values '1', '0', and '0' (drivers 2 and 3 have '0' because that is assumed to be the initial value of Z). The function WIRED_OR is performed, and the resulting resolved value of '1' is assigned to Z at 2 ns. Signal Z is scheduled to have another event at 4 ns, at which time the driver values '1', '0', and '0' are passed to the resolution function, which returns the value of '1' for signal Z. At time 5 ns, the driver values, '0', '1', and '0' are passed to the resolution function which returns the value T. At 10 ns, the driver values '1', '1', and '1' are passed to the resolution function. Finally, at time 20 ns, the driver values '1', '0', and '0' are passed to the resolution function to determine the effective value for signal Z, which is '1'.

The resolution function has only one input parameter, a constant one-dimensional unconstrained array. The array element type and the return type of the function are the same type as the signal. The function typically computes a value from the various driver values, each element of the input array corresponding to one of the driver values. It should be noted that the identity of the driver is lost in the input array, that is, there is no way of knowing which driver is associated with which element of the input array. Here is an example of a WIRED_OR function that can be used as a resolution function.

```
function WIRED_OR (INPUTS: BIT_VECTOR) return BIT is
begin
  for J in INPUTS' RANGE loop
    if INPUTS(J) = '1', then
       return '1';
    end if;
  end loop;
  return '0';
  end WIRED_OR;
```

There is nothing special about the syntax for a resolution function. A function is recognized as a resolution function if it is associated with a signal in the signal declaration or a subtype declaration. In the previous example, we introduced a predefined attribute 'RANGE of an array object. This attribute returns the range of the elements of the specified array object. For example, if there are four drivers when the WIRED_OR function is called, INPUTS' RANGE returns the range "0 to 3".

If a signal KCQ is of type STD_ULOGIC (type STD_ULOGIC is declared in package STD_LOGIC_1164) and has multiple drivers, then a resolution function must be used. A resolution function called RESOLVED is predefined in the STD_LOGIC_1164 package. Here is a possible declaration for signal KCQ.

```
signal KCQ: RESOLVED STD_ULOGIC;
```

Alternately, the resolved subtype STD_LOGIC (subtype STD_LOGIC is declared in package STD_LOGIC_164) can be used leading to a more compact signal declaration for KCQ.

```
signal KCQ: STD_LOGIC;
```

Any multiple drivers for KCQ will imply a call to the function RESOLVED to determine the effective value for signal KCQ.

Drivers are also created for signals that are assigned within a process. The one difference is that irrespective of how many times a signal is assigned a value inside a process, there is only one driver for that signal in that process. Therefore, each process will create at most one driver for a signal. If a signal is assigned a value using multiple concurrent signal assignment statements (which can only appear outside a process), an equal number of drivers are created for that signal.

2.19 CONDITIONAL SIGNAL ASSIGNMENT STATEMENT

The conditional signal assignment statement selects different values for the target signal based on the specified, possibly different, conditions. (It is like an if statement.) A typical syntax for this statement is:

```
target-signal<=     [ waveform-elements when condition else ]
                    [ waveform-elements when condition else ]
                    ...
                    waveform-elements [ when condition ];
```

The semantics of this concurrent statement are as follows. Whenever an event occurs on a signal used in either any of the waveform expressions (recall that a waveform expression is the value expression in a waveform element) or any of the conditions, the conditional signal assignment statement is executed by evaluating the conditions one at a time. For the first true condition found, the corresponding value (or values) of the waveform is scheduled to be assigned to the target signal. For example,

```
Z <=     IN0 after 10 ns when S0 = '0' and S1 = '0' else
         IN1 after 10 ns when S0 = '1' and S1 = '0' else
         IN2 after 10 ns when S0 = '0' and S1 = '1' else
         IN3 after 10 ns;
```

In this example, the statement is executed any time an event occurs on signals IN0, IN1, IN2, IN3, S0, or S1. The first condition (S0 = '0' **and** S1 = '0') is checked; if false, the second condition (S0 = '1' **and** S1 = '0') is checked; if false, the third condition is checked; and so on. Assuming S0 = '0' and S1 = '1', then the value of IN2 is scheduled to be assigned to signal Z after 10 ns.

For a given conditional signal assignment statement, there is an equivalent process statement that has the same meaning. Such a process statement has exactly one If statement and one wait statement within it. The signals in the sensitivity list for the wait statement are the signals in all the waveform expressions and the signals referenced in all the conditions. The equivalent process statement for these conditional signal assignment statement example is:

```
process
begin
  if S0 = '0' and S1 = '0' then
```

```
    Z <= IN0 after 10 ns;
  elsif  S0 = '1' and S1='0' then
    Z <= IN1 after 10 ns;
  elsif  S0 = 0' and S1 = '1', then
    Z <= IN2 after 10 ns;
  else
    Z <= IN3 after 10 ns;
  end if;
  wait on IN0, IN1, IN2, IN3, S0, S1;
end process;
```

2.20 SELECTED SIGNAL ASSIGNMENT STATEMENT

The selected signal assignment statement selects different values for a target signal based on the value of a select expression (it is like a case statement). A typical syntax for this statement is:

```
with expression select                          -- This is the select expression.
  target-signal <=        waveform-elements when choices,
                          waveform-elements when choices,
                          ...
                          waveform-elements when choices;
```

The semantics of a selected signal assignment statement is very similar to that of the conditional signal assignment statement. Whenever an event occurs on a signal in the select expression or on any signal used in any of the waveform expressions, the statement is executed. Based on the value of the select expression that matches the choice value specified, the value (or values) of the corresponding waveform is scheduled to be assigned to the target signal. Note that the choices are not evaluated in sequence. All possible values of the select expression must be covered by the choices exactly once. Values not covered explicitly may be covered by an "others" choice, which covers all such values. The choices may also be an "or" of several values or may be specified as a range of values.

Here is an example of a selected signal assignment statement.

```
type OP is (ADD, SUB, MUL, DIV);
signal OP_CODE: OP;
...
with OP_CODE select
  Z <=  A + B after ADD_PROP_DLY when  ADD,
        A – B after SUB_PROP_DLY when SUB,
        A* B  after MUL_PROP_DLY when MUL,
        A/ B after DIV_PROP_DLY when DIV;
```

In this example, whenever an event occurs on signals OP_CODE, A, or B, the statement is executed. Assuming the value of the select expression OP_CODE is SUB, the expression "A - B" is computed, and its value is scheduled to be assigned to signal Z after SUB_PROP_DLY time.

For every selected signal assignment statement, there is an equivalent process statement with the same semantics. In the equivalent process statement, there is one case statement that uses the select expression to branch. The list of signals in the sensitivity list of the wait statement comprises all signals in the select expression and in the waveform expressions. The equivalent process statement for the previous example is:

```
process
begin
  case OP_CODE is
    when ADD    => Z <= A + B after ADD_PROP_DLY;
    when SUB    => Z <= A – B after SUB_PROP_DLY;
    when MUL    => Z <= A * B after MUL_PROP_DLY;
    when DIV.   => Z <= A / B after DIV_PROP_DLY;
  end case;
  wait on OP„CODE, A, B;
end process;
```

2.21 THE UNAFFECTED VALUE

It is possible to assign a value of **unaffected** to a signal in a concurrent signal assignment statement. Such an assignment causes no change to the driver for the target signal. Here are some examples.

```
type STATE_TYPE is (RESET, APPLY, WAITS, HOLD, RECEIVE);
signal NEXT_STATE: STATE_TYPE;
...
with NEXT_STATE select
  ZRX <=     B"0001" when APPLY,
             B"0010" when WAITS,
             B"0100" when RESET,
             unaffected when others;
MARK_FLAG <=        BKDET after 5 ns when STROBE= '0' else
                    unaffected;
```

In the first concurrent signal assignment, when NEXT_STATE has the value HOLD or RECEIVE, the value **unaffected** is assigned to signal ZRX. This does not cause any change to the driver, that is, signal ZRX retains its old value. In the second signal assignment, when STROBE has a value other than a '0', MARK_FLAG retains its old value, since an assignment of value **unaffected** does not cause any change to the driver of MARK_FLAG.

It is easier to understand the effects of the **unaffected** value by looking at its equivalent representation in a process statement. An assignment of **unaffected** to a signal translates to a null statement in the equivalent process for the concurrent signal assignment statement. The equivalent process statement for the second example described earlier is:

```
process
begin
  if STROBE = '0' then
    MARK_DET <= BKDET after 5 ns;
  else
    null;                                      -- From the value unaffected.
  end if;
  wait on STROBE, BKDET;
end process;
```

2.22 BLOCK STATEMENT

A block statement is a concurrent statement. It can be used for three major purposes:

1. To disable signal drivers by using guards
2. To limit scope of declarations, including signal declarations
3. To represent a portion of a design

A block statement itself has no execution semantics but provides additional semantics for statements that appear within it. The syntax of a block statement is:

```
block-label: block [ (guard-expression) ] [ is ]
        [ block-header ]
        [ block-declarations ]
begin
    concurrent-statements
end block [ block-label ];
```

The block-header, if present, describes the interface of the block statement to its environment. Any declarations appearing within the block are visible only within the block, that is, between **block** ... **end block.** Any number of concurrent statements can appear within a block, possibly none. Block statements can be nested since a block statement is itself a concurrent statement. The block label present at the beginning of the block statement is necessary; however, the label appearing at the end of the block statement is optional and, if present, must be the same as the one used at the beginning of the block.

If a *guard-expression* appears in a block statement, there is an implicit signal called GUARD of type BOOLEAN declared within the block. The value of the GUARD signal is always updated to reflect the value of the guard expression. The guard expression must be of type BOOLEAN. Signal assignment statements appearing within the block statement can use this GUARD signal to enable Or disable their drivers. Here is an example of a gated inverter.

```
B1: block (STROBE = '1')
begin
    Z <= guarded not A;
end block B1;
```

The signal **GUARD** implicitly declared within block B1 has the value of the expression (STROBE = '1'). The keyword **guarded** can optionally be used with concurrent signal assignment statements within a block statement. This keyword implies that when the value of the GUARD signal is true (that is, the guard expression evaluates to true), the value of the expression **"not A"** is assigned to the target signal Z. If the GUARD is false, events on A do not affect the value of signal Z. That is, the driver to Z for this signal assignment statement is disabled, and signal Z retains its previous value. The block statement is very useful in modeling hardware elements that trigger on certain events, for example, flip-flops and clocked logic.

The only concurrent statements whose semantics are affected by the enclosing block statement are the guarded assignments, that is, the signal assignment statements that use the **guarded** option. The modified meaning is as follows. Whenever an event occurs (an event is a change of value) on any signal used in the expression of a guarded assignment and the signal GUARD has the value TRUE, or if signal GUARD changes from FALSE to TRUE, the signal assignment statement is executed, and the target signal is scheduled to get a new value. If the value of the guard expression is false, the value of the target signal is unchanged.

Every guarded assignment has an equivalent process statement. Here is an example.

```
BG: block (guard-expression)
   signal SIG: BIT;
begin
   SIG <= guarded waveform-elements;
end block BG;

-- The equivalent process statement for the guarded assignment is:
BG: block (guard-expression)
   -- signal GUARD: BOOLEAN;                    --Implicit.
   signal SIG: BIT;
begin
   -- GUARD <= guard-expression;                  --Implicit.
   process
   begin
     if GUARD then
        SIG <= waveform-elements;
     end if;
     wait on signals-in-waveform-elements, GUARD;
   end process;
end block BG;
```

The signal GUARD, even though implicitly declared, can be used explicitly within a block statement. For example,

```
B2: block (CLEAR = '0' and PRESET = '1',)
begin
   Q <= ' 1' when not GUARD else ' 0';
end block B2;
```

In this example, the signal assignment in the block statement is not a guarded assignment; therefore, the driver to signal Q is never disabled. However, the value of Q is determined by the value of the GUARD signal because of its explicit use in the signal assignment statement. The value of the GUARD signal corresponds to the value of the guard expression " CLEAR = '0' **and** PRESET = '1' ". This signal assignment statement is therefore executed any time an event occurs on signal GUARD.

It is also possible to explicitly declare a signal called GUARD, define an expression for it, and then use it within a guarded assignment. Here is an example.

```
B3: block
   signal GUARD: BOOLEAN;
begin
   GUARD <= CLEAR = '0' and PRESET = ' T ';
   Q <= guarded DIN;
end block B3;
```

Here is another example that models a rising-edge triggered D flip-flop.

```
entity D_FLIP_FLOP is
   port (D, CLK: in BIT; Q, QBAR: out BIT);
end D_FLIP_FLOP;
architecture DFF of D_FLIP_FLOP is
begin
   L1: block (CLK = '1' and not CLK'STABLE)
```

```
        signal TEMP: BIT;
    begin
        TEMP <= guarded D;
        Q <=TEMP; QBAR <= not TEMP;
    end block L1;
end DFF;
```

The guard expression uses a predefined attribute called 'STABLE. CLK'STABLE is a new signal of type BOOLEAN, which is true as long as signal CLK has not had any event in the current delta time. This guard expression implies a rising clock edge. The scope of the signal TEMP declared in block L1 is restricted to be within the block. Of the three signal assignments, only the first one is a guarded assignment and, hence, controlled by the guard expression. The other two signal assignments are not controlled by the guard expression and are triggered purely on events occurring on signals in their expressions. When a rising clock edge appears on the signal CLK, say at time T, the value of D is assigned to signal TEMP after delta delay, that is, at time T+Δ. If the value of TEMP is different from its previous value, the assignments to Q and QBAR will be triggered, causing these signals to get new values after another delta delay, that is, at time T+2Δ.

2.23 CONCURRENT ASSERTION STATEMENT

A concurrent assertion statement has exactly the same syntax as a sequential assertion statement. An assertion statement is a concurrent statement by virtue of its place of appearance within the model. If it appears inside a process, it is a sequential assertion statement and is executed sequentially with respect to the other statements in the process; if it appears outside a process, it is a concurrent assertion statement. The semantics of a concurrent assertion statement are as follows. Whenever an event occurs on a signal in the Boolean expression of the assertion statement, the statement is executed.

Here is an example of a concurrent assertion statement used in an SR flip-flop model. It makes a check to ensure that the input signals R and S are never simultaneously zero.

```
library IEEE;
use IEEE.STD_LOGIC_1164.all;
entity SR is
    port (S, R: in STD_LOGIC; Q, NOTQ: out STD_LOGIC);
end SR;
architecture SR_ASSERT of SR is
begin
    assert not (S = '0' and R = '0')
        report "Not valid inputs: R and S are both low"
        severity ERROR;
    -- Rest of model for SR flip-flop here.
end SR_ASSERT;
```

Any time an event occurs on either signal S or R, the assertion statement is executed and the Boolean expression checked. If false, the report message is printed, and the severity level is reported to the simulator for appropriate action.

For every concurrent assertion statement there is an equivalent process statement with the same semantic meaning. The equivalent process statement for the previous example of the concurrent assertion statement is:

```
process
begin
  assert not (S = '0' and R = '0')
    report "Not valid inputs: R and S are both low"
    severity ERROR;
  wait on S, R;
end process;
```

A concurrent assertion statement can be marked as a postponed concurrent assertion statement, such as:

```
postponed assert not (S = '0' and R = '0')
    report "Not valid inputs: R and S are both low"
    severity ERROR;
```

The semantics of such a statement is equivalent to the semantics of its equivalent process statement, which is a postponed process.

2.24 VALUE OF A SIGNAL

A signal gets its value from its drivers. Every concurrent signal assignment statement creates a driver for the target signal. All signal assignments in a process that assign to a particular signal create one driver for that signal.

In a given VHDL description, if a signal has more than one driver, a resolution function is necessary. This function is called with the current value of all the drivers for a signal, and the return value of the function becomes the effective value of the signal.

Here is an example.

```
library IEEE;          -- Library and use clauses can also be associated
                       -- with an architecture body.
use IEEE.STD_LOGIC_1164.all;
architecture DRIVERS of INTERCON is
  signal FRMERR, PARERR: STD_LOGIC := 'Z';
  -- Type STD_LOGIC is defined in package STD_LOGIC_1164 as:
  -- subtype STD_LOGIC is RESOLVED STD_ULOGIC;
signal XFSB: BIT;
begin
  P1: process (...)
    ...
begin
  FRMERR <= ...
  XFSB <= ...
  ...
  XFSB <= ...
end process P1;
P2: process (...)
  ...
begin
  PARERR <= ...
  FRMERR <=...
end process P2;
PARERR <=....
end DRIVERS;
```

In this example, signal FRMERR has two drivers, one from process P1 and one from process P2. Therefore, a resolution function is needed, and this (function RESOLVED) is specified in the signal declaration of FRMERR (STD_LOGIC is a resolved subtype with the resolution function RESOLVED). Similarly, signal PARERR has two drivers and a resolution function is specified in its signal declaration. However, signal XFSB has only one driver from process P1, and therefore needs no resolution function.

The effective value for signals FRMERR and PARERR is the return value of the resolution function, while the effective value of signal XFSB is the current value of the driver itself. Note that a resolution function could be specified for signal XFSB as well, such as in the following declaration.

```
signal XFSB: PULL_UP BIT;
```

In this case, the value of its one driver is the input to the resolution function, and the return value of the function is the effective value of signal XFSB. Clearly, if the function PULL_UP were simply passing on the input value to the output in the case of one driver, then specifying a resolution function simply adds to the simulation overhead since an extra function call is involved.

The recommended practice, therefore, is not to specify a resolution function if a signal has only one driver.

The initial value of a signal, that is, its value at start of simulation, is determined in a similar way from the value of its drivers. The initial value of a driver of a signal is the initial value as specified in the signal declaration, either implicitly, for example, using signal-type 'LEFT for scalars, or explicitly, using ":=" in the signal declaration. If a signal has a resolution function associated with it (it may or may not have more than one driver), the initial value of the signal is computed by passing the initial values of all the drivers to the resolution function, and the return value becomes the initial value of the signal. If a signal does not have a resolution function associated with it (it can only have one driver), the initial value of the signal is the same as the initial value of the driver.

In the architecture body DRIVERS described earlier, the initial value of the two drivers for FRMERR is 'Z' (the explicitly specified initial value). These two values are passed to the resolution function RESOLVED. The return value of the function is the effective initial value of FRMERR. This value may or may not be 'Z', depending on the computation performed by the resolution function. The initial value of signal XFSB is the initial value of the driver (since there is only one driver), which is the implicit initial value of '0' (this is BIT'LEFT).

2.25 PACKAGES

A package provides a convenient mechanism to store and share declarations that are common across many design units. A package is represented by:

1. a package declaration and, optionally,
2. a package body.

2.25.1 Packaage Deularation

A package declaration contains a set of declarations that may possibly be shared by many design units. It defines the interface to the package, that is, it defines items that can be made visible to other design units, for example, a function declaration. A package body, in contrast, contains the hidden details of a package, for example, a function body.

The syntax of a package declaration is:

```
package package-name is
  package-item-declarations --> These may be:
    -- subprogram declarations
    -- type declarations
    -- subtype declarations
    -- constant declarations
    -- signal declarations
    -- variable declarations
    -- file declarations
    -- alias declarations
    -- component declarations
    -- attribute declarations
    -- attribute specifications
    -- disconnection specifications
    -- use clauses
end [ package ] [ package-name ];
```

An example of a package declaration is given next.

```
package SYNTH_PACK is
  constant LOW2HIGH: TIME := 20 ns;
  type ALU_OP is (ADD, SUB, MUL, DIV, EQL);attribute PIPELINE: BOOLEAN;
  type MVL is ('LP,'0', '1','Z');
  type MVL_VECTOR is array (NATURAL range < >) of MVL;
  subtype MY_ALU_OP is ALU_OP range ADD to DIV;
  component NAND2
    port (A, B: in MVL; C: out MVL);
  end component;
end SYNTH_PACK;
```

Items declared in a package declaration can be accessed by other design units by using the library and use clauses. The set of common declarations may also include function and procedure declarations and deferred constant declarations. In this case, the behavior of the subprograms and the values of the deferred constants are specified in a separate design unit called the package body. Since the previous package example does not contain any subprogram declarations or deferred constant declarations, a package body is not necessary.

Consider the following package declaration.

```
use WORK.SYNTH_PACK.all;                  -- Includes all declarations
                                          -- from package SYNTH_PACK.
package PROGRAM_PACK is
  constant PROP_DELAY: TIME;                -- A deferred constant.
  function "and" (L, R: MVL) return MVL;
  procedure LOAD (signal ARRAY_NAME:  inout MVL_VECTOR;
    START_BIT, STOP_BIT, INT_VALUE: in INTEGER);
end package PROGRAM_PACK;                 -- The keyword package after end
                                          -- is optional.
```

In this case, a package body is required because the package declaration contains a deferred constant declaration and two subprogram declarations.

2.25.2 Package Body

A package body primarily contains the behavior of the subprograms and the values of the deferred constants declared in a package declaration. It also may contain other declarations, as shown by the following syntax of a package body.

```
package body package-name is
  package-body-item-declarations --> These are:
    -- subprogram bodies
    -- complete constant declarations
    -- subprogram declarations
    -- type and subtype declarations
    -- file and alias declarations
    -- use clauses
end [ package body ] [ package-name ];
```

The package name must be the same as the name of its corresponding package declaration. A package body is not necessary if its associated package declaration does not have any subprogram or deferred constant declarations. The associated package body for the package declaration PROGRAM_PACK, described in the previous section, is:

```
package body PROGRAM_PACK is
   use WORK .TABLES.all; .
   constant PROP_DELAY: TIME := 15 ns;
   function "and" (L, R: MVL) return MVL is
   begin
     return TABLE_AND(L, R);
     -- TABLE_AND is a 2-D constant defined in another package,
     -- TABLES.
 end "and";

 procedure LOAD (signal ARRAY_NAME: inout MVL_VECTOR;
       START_BIT, STOP_BIT, INT_VALUE: in INTEGER) is
   -- Local declarations here.
 begin
   -- Procedure behavior here.
 end LOAD;
end PROGRAM_PACK;.
```

An item declared inside a package body has its scope restricted to be within the package body, and this item cannot be made visible in other design units. This restriction is in contrast to items declared in a package declaration, which can be accessed by other design units. Therefore, a package body is used to store private declarations that should not be visible, while a package declaration is used to store public declarations, which other design units can access. This is very similar to declarations within an architecture body, which are not visible outside of its scope, while items declared in an entity declaration can be made visible to other design units. An important difference between a package declaration and an entity declaration is that an entity can have multiple architecture bodies with different names, while a package declaration can have at most one package body, the names for both being the same.

2.26 DESIGN FILE

A design file is an ASCII file containing VHDL source. It can contain one or more design units, where a design unit is one of the following:

- ❖ Entity declaration
- ❖ Architecture body
- ❖ Configuration declaration
- ❖ Package declaration
- ❖ Package body

This means that each design unit can also be placed in a separate file.

Figure 2.13 shows the compilation process. The design file is processed by a VHDL analyzer, which, after verifying the syntactic and semantic correctness of the source, compiles each design unit present in the design file into an intermediate form. Each intermediate form is stored in a design library that has been designated as the working library.

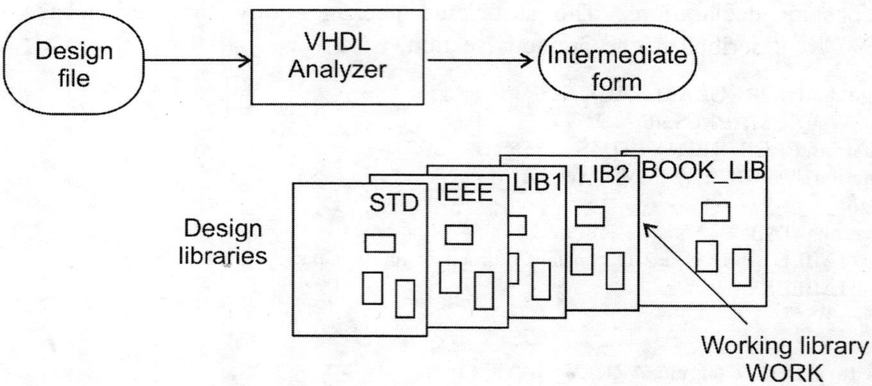

Figure 2.13: A typical compilation process.

2.27 DESIGN LIBRARIES

A compiled design unit is stored in a design library. A design library is an area of storage in the file system of the host environment. The format of this storage is not defined by the language. In one scheme, a design library is implemented on a host system as a file directory, and the compiled design units are stored as files in this directory. The management of the design libraries is also not defined by the language and is again tool-implementation-specific.

An arbitrary number of design libraries may be specified. Each design library has a logical name with which it is referenced inside a VHDL description. The association of the logical names with their physical storage names is maintained by the host environment. There is one design library with the logical name STD predefined by

the language; this library contains the compiled descriptions for the two predefined packages STANDARD and TEXTIO. Exactly one design library must be designated as the working library with the logical name WORK. When a design file is compiled, the intermediate forms of the design units present in the design file are always stored in the working library. Therefore, before compilation begins, the logical name WORK must point to one of the design libraries (see Figure 2.13)

The VHDL source is present in a design file; a design file can contain one or more design units. Design units are further classified as following:

1. *Primary units*: These units are not directly dependenton other design units. They are:
   a. *Entity declarations*: The items declared in an entitydeclaration are implicitly visible within the associated architecture bodies.
   b. *Package declarations*: Items declared within a package declaration can be exported to other designunits using library and use clauses. Items declared within a package declaration are also implicitly visible within the corresponding package body.
   c. *Configuration declarations*.
2. *Secondary units*: These units are always dependentsof a primary unit. These design units do not allow items declared within them to be exported out of the design unit, that is, the items cannot be referenced in other design units. These secondary units are:
   a. *Architecture bodies*: For example, a signal declared in an architecture body cannot be referenced in other design units.
   b. *Package bodies*.

There can be exactly one primary unit with a given name in a single design library. Secondary units associated with different primary units can have identical names in the same design library; also, a secondary unit may have the same name as its associated primary unit. For example, assume there exists an entity called AND_GATE in a design library. It may have an architecture body with the same name, and another entity MY_GATE in the same design library may have an architecture body that also has the name AND_GATE.

Secondary units must coexist with their associated primary units in the same design library; for example, an entity declaration and all of its architecture bodies must reside in the same library. Similarly, a package declaration and its associated package body must reside in a single library.

Even though a configuration declaration is a primary unit, it must reside in the same library as the entity declaration that it configures.

2.28 ORDER OF ANALYSIS

Since it is possible to export items declared in primary units to other design units, a constraint is imposed in the sequence in which design units must be analyzed. A design unit that references an item declared in another primary unit can be analyzed only after that primary unit has been analyzed. For example, if a configuration declaration references an entity COUNTER, the entity declaration for COUNTER must be analyzed before the configuration declaration.

A primary unit must be analyzed before any of its associated secondary units. For example, an entity declaration must be analyzed before its architecture bodies are analyzed.

2.29 IMPLICIT VISIBILITY

An architecture body implicitly inherits all declarations in the entitysince it is tied to that entity by virtue of the statement:

```
architecture architecture-name of entity-name is ...
```

Similarly, a package body implicitly inherits all items declared in the package declaration by virtue of its first statement:

package body *package-name* **is** ...

where the package name is the same as the one in the package declaration.

2.30 EXPLICIT VISIBILITY

Explicit visibility of items declared in other design units can be achieved using the following two clauses:

1. library clause **2.** use clause

A use clause can appear in any declarative part of a design unit. If a library clause or a use clause appears at the beginning of a design unit, it is called a context clause. Figure 2.14 shows an exampie. Items specified in the context clause become visible only to the design unit that follows the context clause; the items are not visible to other primary design units that may be present in the same design file. This means that if a design file contains three primary design units, such as in the example shown in Figure 2.15 context clauses must be specified for each design unit, if needed.for example, the context clauses specified before design unit A are visible to design unit A only; context clauses specified before design unit B are visible to design unit B only.

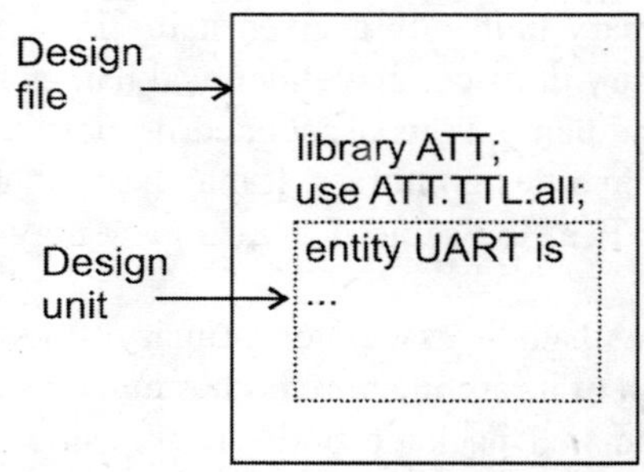

Figure 2.14: Context clause associated with the following design unit.

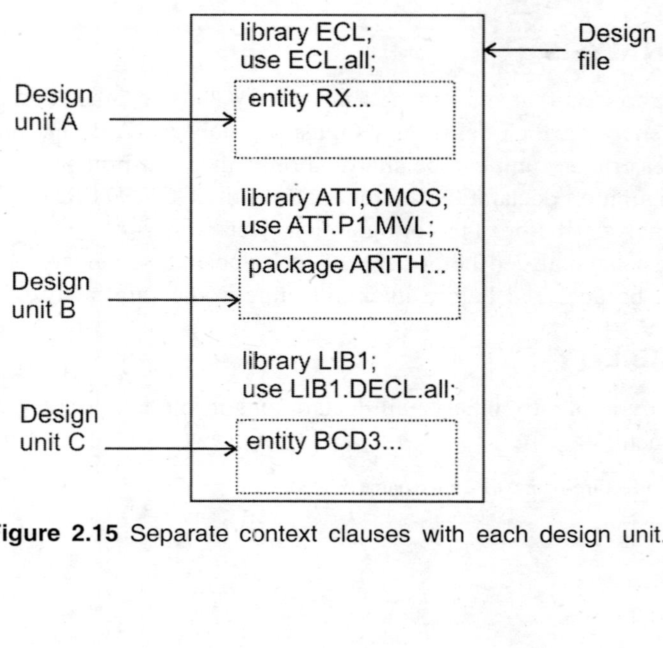

Figure 2.15 Separate context clauses with each design unit.

2.30.1 Library Clause

The library clause makes visible the logical names of design libraries that can be referenced within a design unit. The format of a library clause is:

library *list-of-logical-library-names;*

The library **clause:**

```
library TIL, CMOS;
```

makes the logical names TTL and CMOS visible in the design unit that follows. Note that the library clause does not make design units or items present in the library visible, it makes only the library name visible (it is like a declaration for a library name). For example, it would be illegal to use the name "TTL.SYNTH_PACK.MVL" within a design unit without first declaring the library name using the "**library** TTL;" clause.

The library **clause:**

```
library STD, WORK;
```

is implicitly declared for every design unit.

2.30.2 Use Clause

There are two main forms of the use clause.

use *library-name . primary-unit-name;* -- Form 1.
use *library-name . primary-unit-name. item;* -- Form 2.

The first form of the use clause allows the specified primary unit name from the specified design library to be referenced in a design description. For example,

```
library CMOS;
use CMOS.NOR2;
configuration ... is
    ... use entity NOR2(...);
end;
```

Note that entity NOR2 must be available in compiled form in the design library CMOS before attempting to compile the design unit where it is used.

The second form of the use clause makes the item declared in the primary unit visible; the item can therefore be referenced within the following design unit. For example,

```
library IEEE;
use IEEE.STD_LOGIC_1164.STD_LOGIC;
-- STD_LOGIC is a type declared in STD_LOGIC_1164 package.
--The package STD_LQGIC_1164 is stored in the design library IEEE.
entity NAND2 is
  port (A, B: in STD_LOGIC;...)...
```

If all items within a primary unit are to be made visible, the keyword **all** can be used. For example,

```
use IEEE.STD_LOGIC_1164.all;
```

makes all items declared in package STD_LOGIC_1164 in design library IEEE visible. The use clause:

```
use IEEE.all;
```

makes all names of primary units present in the design library IEEE visible. Items external to a design unit can be accessed by other means as well. One way is to use a selected name. An example of using a selected name is:

```
library IEEE;
use IEEE.STD_LOGIC_1164;
entity NOR2 is
  port (A, B: in STD_LOGIC_1164.STD_LOGIC;...)...
```

Since only the primary unit name was made visible by the use clause, the primary unit name must be used along with the name of the type being referenced, that is; STD_LOGIC_1164.STD_LOGIC must be specified. Another example is shown next. The type VALUE_9 is defined in package SIMPACK, which has been compiled into the CMOS design library.

```
library CMOS;
package P1 is
  procedure LOAD (A, B: CMOS.SIMPACK.VALUE_9;...);
  ...
end P1;
```

In this case, the primary unit name SIMPACK was specified only at the time of usage.

So far, we talked about exporting items across design libraries. What if it is necessary to export items from design units that are in the same library? In this case, there is no need to specify a library clause since every design unit has the following library clause implicitly declared.

```
library STD, WORK;
```

The predefined design library STD contains the packages STANDARD and TEXTIO. The package STANDARD contains declarations for predefined types such as CHARACTER, BOOLEAN, BIT_VECTOR, and INTEGER. The following use clause is also implicitly declared for every design unit.

```
use STD.STANDARD.all;
```

Thus, all items declared within the package STANDARD are available for use in every VHDL description. However, there is no such implicit declaration for the TEXTIO package. If items declared in this package need to be referenced, a use clause of the following form may be used.

```
use STD.TEXTIO.all;
```

2.31 SUBPROGRAMS

A subprogram defines a seqential alongrithm that perfroms a certain computation. There are two Kind ofsubprograms:

1. *Function*: These are usually used for computing a single Value. A fuction executes in Zero simulation time.
2. *Procedures*: These may be used to partition large behavioral descriptions. Procedures can return zero or more values. A procedure may or may not execute in zero simulation time, depending on whether it has a wait statement or not.

A subprogram is defined using a subprogram body. The typical format for a subprogram body is:

> *subprogram-specification* **is**
>   *subprogram-item-declarations*
> **begin**
>   *subprogram-statements* --Sameas sequential-statements.
> **end** [**function** | **procedure**] [*subprogram-name*];

The subprogram-specification specifies the name of a subprogram and defines its interface, that is, it defines the formal parameter names, their class (i.e., **signal, variable, file,** or **constant**), their type, and their mode (whether they are **in, out,** or **inout**). Parameters of mode in are read-only parameters; these cannot be updated within a subprogram body. Parameters of mode out are write-only parameters; their values cannot be used but can only be updated within a subprogram body. Parameters of mode **inout** can be read as well as updated. Files have no mode; they can be read or written to, depending on how the files were opened.

Actuals in a subprogram call are used to pass values to and from a subprogram. Only an actual signal may be associated with a formal parameter of the **signal** class. Only an actual variable may be associated with a parameter of the **variable** class. Only a file may be an actual for a file parameter. An expression may be used to pass a value to a parameter of **constant** class.

When parameters are of a **variable** or **constant** class, values are passed to the subprogram by value. Arrays may or may not be passed by reference. Files are passed by reference. For signals, the reference to the signal, its driver, or both are passed into the subprogram. What this means is that any assignment to a signal in a procedure (signals cannot be assigned values in a function because the parameters are restricted to be of mode **in**) affects the actual signal's driver immediately and is independent of whether the procedure terminates or not. For a signal of any mode, none of the signal-valued attributes, 'STABLE, 'QUIET, 'DELAYED, or TRANSACTION can be used in a subprogram body.

The type of an actual in a subprogram call must match its corresponding formal parameter. If the formal parameter belongs to an unconstrained type, the size of this parameter is determined from the actual that is passed in.

The *subprogram-item-declarations* part contains a set of declarations (e.g., type and object declarations) that are accessible for use within the subprogram. These declarations come into effect every time the subprogram is called. Variables are created and initialized every time the subprogram is called. They remain in existence until the subprogram completes. This is in contrast with variables in a process statement, which get initialized only once at the start of simulation and persist throughout the entire simulation.

The *subprogram-statements* part contains sequential statements that define the computation to be performed by the subprogram. A return statement is a special statement allowed only within subprograms. The format of a return statement is:

> **return** [*expression*];

The return statement causes the subprogram to terminate and control to be returned to the calling program. All functions must have a return statement with an expression; the value of the expression in the return statement is returned to the calling program. For procedures, objects of mode **out** and **inout** return their values to the calling program. Procedures may also affect signals and variables declared globally to it.

The *subprogram-name* appearing at the end of a subprogram body, if present, must be the same as the function or procedure name specified in the subprogram specification part.

2.31.1 Functions

Functions are used to describe frequently used sequential algorithms that return a single value. This value is returned to the calling program using a return statement. Some of their common uses are as resolution functions and type conversion functions. The following is an example of a function body.

```
function LARGEST (TOTAL_NO: INTEGER; SET: PATTERN)
    return REAL is
 -- PATTERN is elsewhere defined to be a type of 1-D array of
 -- non-negative floating-point values.
 variable RETURN_VALUE: REAL := 0.0;
begin
 for K in SET'RANGE loop
   if SET(K) > RETURN_VALUE then
     RETURN_VALUE := SET(K);
   end if;
 end loop;
 return RETURN_VALUE;
end LARGEST;
```

Variable RETURN_VALUE comes into existence with an initial value of 0.0 every time the function is called. It ceases to exist after the function terminates.

The general syntax of a subprogram specification for a function body is:

[pure | impure] function *function-name* (*parameter-list*)
return *return-type*

A pure function is one that returns the same value each time the function is called with the same set of actuals. An impure function is one that potentially returns different values each time it is called with the same set of actuals. For example, NOW is an impure function because it returns different values when called at different times. If neither keyword **pure** nor **impure** is present in the function specification, the function is, by default, a pure function.

The parameter-list describes the list of formal parameters for the function. The only mode allowed for the parameters is mode in. Also, only constants and signal objects can be passed in as parameters. The default object class is **constant.** For example, in function LARGEST, TOTAL_NO is a constant, and its value cannot be modified within the function body. Another example of a function body is shown next. This function returns true if a rising edge has been detected on the input signal. The function has been explicitly marked as a pure function.

```
pure functionVRISE(signalCLOCK-NAME: BIT)
    return BOOLEAN is
begin
  return CLOCK_NAME = '1' and CLOCK_NAME' EVENT;
end VRISE;
```

Here are two examples of impure functions.

```
impure function RANDOM (SEED: REAL) return REAL is
  variable NUM: REAL;
  attribute FOREIGN of RANDOM: function is "NUM = rand(seed)";
begin
  return NUM;
end RANDOM;

impure function USED (TO_ALLOCATE: POSITIVE)
    return POSITIVE is
 -- ALLOCATED is a shared variable declared elsewhere, but
 -- visible to this function.
begin
  ALLOCATED := ALLOCATED + TO_ALLOCATE;
  return ALLOCATED;

end function USED;        -- Keyword function after end is optional.
```

A function call is an expression and can also be used in larger expressions. For example,

```
SUM := SUM + LARGEST(MAX_COINS, COLLECTION);
```

A function call has the form:

function-name (list-of-actuals)

The actuals may be associated by position (the first actual corresponds to the first formal parameter, the second actual corresponds to the second parameter, and so on) or using named association (the association of actuals and formal parameters are explicitly specified). The function call in the last example used positional association. An equivalent function call using named association is:

```
LARGEST (SET => COLLECTION, TOTAL_NO => MAX_COINS)
```

Functions are frequently used to perform type conversion. Here is an example of a function that converts a value from the STD_ULOGIC type to a value from the CHARACTER type.

```
function TO_CHARACTER (ARG: STD_ULOGIC)
    return CHARACTER is
begin
  case ARG is
    when 'U'    => return 'U';
    when 'X'    => return 'X';
    when '0'    => return '0';
    when '1'    => return '1';
    when 'Z'    => return 'Z';
    when 'W'    => return 'W';
    when 'L'    => return 'L';
    when 'H'    => return 'H';
    when '-'    => return '-';
  end case;
end TO_CHARACTER;
```

A function call of the form:

```
TO_CHARACTER (STD_ULOGIC'('U'))
```

returns the character 'U'. However, there is a much more efficient way of performing type conversion, by using look-up tables. Here is the TO_CHARACTER function described using look-up tables.

```
type LOOK_UP is array (STD_ULOGIC) of CHARACTER;
constant TO_CHARACTER: LOOK_UP :=
    ('U' => 'U', 'X' => 'X', '0' => '0',
     1' => '1', 'Z' => 'Z', 'W', => 'W', 'L' => 'L', 'H' => 'H','-' => '-');
```

Then the following expression,

```
TO_CHARACTER (STD_ULOGIC' ('U'))
```

gives the character 'U'. Note that in both schemes, the calling line to perform type conversion does not change. However, in the look-up table scheme, access time is simply the time to look up a table, while in the function scheme, the overhead is that of a function call, a case statement, and a return statement, which is substantially more.

Here is an example of a function that performs byte reversal. The lower-order bits are swapped with the higher-order bits, for example, for a 8-bit vector, the function swaps the 7th bit with the 0th bit, the 6th bit with the 1st bit, and so on.

```
function BYTE_REVERSAL (ARG: STD_LOGIC_VECTOR)
    return STD_LOGIC_VECTOR is
  Variable RESULT:
      STD_LOGIC_VECTOR(ARG'REVERSE_RANGE);
begin
  for J in ARG'RANGE loop
    RESULT(J) := ARG (J);
  end loop;
  return RESULT;
end BYTE_REVERSAL;
```

A function call:

```
BYTE_REVERSAL ("11001110")
```

returns the value:

```
"01110011"
```

2.31.2 Procedures

Procedures allow decomposition of large behaviors into modular sections. In contrast to a function, a procedure can return zero or more values using parameters of mode **out** and **inout.** The syntax for the subprogram specification for a procedure body is:

procedure *procedure-name (parameter-list)*

The *parameter-list* specifies the list of formal parameters for the procedure. Parameters may be constants, variables, or signals, and their modes may be **in, out,** or **inout.** If the class of a parameter is not explicitly specified, the class is by default a **constant** if the parameter is of mode **in;** otherwise, it is a variable if the parameter is of mode **out** or **inout.**

A simple example of a procedure body is shown next. It describes the behavior of an arithmetic logic unit.

```
type OP_CODE is (ADD, SUB, MUL, DIV, LT, LE, EQ);
...
procedure ARITH_UNIT (A, B: in INTEGER; OP: in OP_CODE;
            Z: out INTEGER; ZCOMP: out BOOLEAN) is
begin
  case OP is
    when ADD => Z := A + B;
    when SUB=>Z:=A-B;
    when MUL=>Z:=A*B;
    when DIV=>Z:=A/B;
    when LT => ZCOMP := A < B;
    when LE => ZCOMP := A <= B;
    when EQ => ZCOMP := A = B;
  end case;
end ARITH_UNIT;
```

The following is another example of a procedure body. This procedure rotates the specified signal vector ARRAY_NAME, starting from bit START_BIT to bit STOP_BIT, by the ROTATE_BY value. The object class for parameter ARRAY_NAME is explicitly specified. The variable FILL_VALUE is automatically initialized to '0' every time the procedure is called.

```
procedure ROTATE_LEFT
    (signal ARRAY_NAME: inout BIT_VECTOR;
    START_BH, STOP_BIT: in NATURAL;
    ROTATE_BY: in POSITIVE) is
  variable FILL_VALUE: BIT;    -- Every time the procedure is called,
                               -- initial value of FILL_VALUE is BIT' LEFT, which is '0'.
begin
  assert STOP_BIT >START_BIT.
    report "STOP_BIT is not greater than START_BIT."
    severity NOTE;
  for MACVAR3 in 1 to ROTATE BY loop
    FILL_VALUE := ARRAY_NAME(STOP_BIT);
    for MACVAR1 in STOP_BIT downto (START_B1T+1) loop
      ARRAY_NAME(MACVAR1) <=
                ARRAY_NAME (MACVAR1-1);
    end loop;
    ARRAY_NAME(START_BIT) <= FILL_VALUE;
  end loop;
end procedure ROTATE_LEFT;        -- Keyword procedure after end is
                                  -- optional.
```

Procedures are invoked by using procedure calls. A procedure call can be either a sequential statement or a concurrent statement; this is based on where the actual procedure call statement is present. If the call is inside a process statement or another subprogram, it is a sequential procedure call statement; otherwise, it is a concurrent procedure call statement. The syntax of a procedure call statement is:

[*label*:] *procedure-name* (*list-of-actuals*);

The actuals specify the expressions, signals, **variables, or files,** that are to be passed into the procedure and the names of objects that are to receive the computed values from the procedure. Actuals may be specified using positional association or named association. For example,

```
ARITH_UNIT (Dl, D2, ADD, SUM, COMP);            -- Positional association.
ARITH_UNIT (Z => SUM, B => D2, A => Dl,
        OP => ADD,. ZCOMP => COMP);             -- Named association.
```

A sequential procedure call statement is executed sequentially with respect to the sequential statements surrounding it. A concurrent procedure call statement is executed whenever an event occurs on one of the parameters that is a signal of mode **in** or **inout.** Semanti-cally, a concurrent procedure call is equivalent to a process with a sequential procedure call and a wait statement that waits for an event on the signal parameters of mode **in or inout.** Here is an example of a concurrent procedure call and its equivalent process statement.

```
architecture DUMMY_ARCH of DUMMY is
   -- Following is a procedure body:
  procedure INT_2_VEC (signal D: out BIT_VECTOR;
     START_BIT, STOP_BIT: in NATURAL;
     signal VALUE: in INTEGER) is
  begin
    -- Procedure behavior here.
  end INT_2_VEC;
begin
  -- This is an example of a concurrent procedure call:
  INT_2_VEC (D_ARRAY, START, STOP, SIGNAL_VALUE);
end DUMMY_ARCH;
-- The equivalent process statement for the concurrent
-- procedure call is:
  process
  begin
    INT_2_VEC P_ARRAY, START, STOP, SIGNAL_VALUE);
    -- This is now a sequential procedure call.
    wait on SIGNAL_VALUE;
    -- Since SIGNAL_VALUE is an input signal.
  end process;
```

A procedure can be used as either a concurrent statement or a sequential statement. Concurrent procedure calls are useful in representing frequently used processes.

Concurrent procedures can also be marked as postponed, such as:

```
postponed procedure INT_2_VEC (signal D: out BIT_VECTOR;
     START_BIT, STOP_BIT: in NATURAL;
     signal VALUE: in INTEGER) is
begin
   --Procedure behavior here.
end INT_2_VEC;
```

The semantics of a postponed concurrent procedure call is equivalent to the semantics of its equivalent process statement, which is a postponed process.

A procedure body can have a wait statement, while a function cannot. Functions are used to compute values that are available instantaneously. Therefore, a function cannot be made to wait; for example, it cannot call a procedure with a wait statement in it. A process that calls a procedure with a wait statement cannot have a sensitivity list. This follows from the fact that a process cannot be sensitive to signals and also be made to wait simultaneously. Since a procedure can have a wait statement, any variables and constants

declared in the procedure retain their values through this wait period and cease to exist only when the procedure terminates.

2.31.3 Declarations

A subprogram body may appear in the declarative part of the block in which a call is made. This is not convenient if the subprogram is to be shared by many entities. In such cases, the subprogram body may be described possibly in a package body and then, in the package declaration, the corresponding *subprogram declaration* is specified. If this package declaration is made visible in another design unit using an use clause, the subprogram can be used in this design unit. A subprogram declaration describes the subprogram name and the list of parameters without describing the internal behavior of the subprogram, that is, it describes the interface for the subprogram. The syntax of a subprogram declaration is:

subprogram-specification;

Four examples of procedure and function declarations are shown next.

```
procedure ARITH_UNIT (A, B: in INTEGER; OP: in OP_CODE;
                                  Z: out INTEGER; ZCOMP: out BOOLEAN);
pure functionVRISE (signal CLOCK_NAME: BIT)
      return BOOLEAN;
impure function RANDOM (SEED: REAL) return REAL;
function LARGEST (TOTAL_NO: INTEGER; SET: PATTERN)
   return REAL;                              --This is a pure function.
```

Another reason subprogram declarations are necessary is to allow two subprograms to call each other recursively; for example,

```
procedure PROC_P (...)...
   variable LOCAL:...
begin
   ...
   LOCAL := FUNC_Q (B);                      -- illegal function call.
end PROC.P;
function FUNC_Q (...)...
begin
   ...
   PROC_P(...);
end FUNC.Q;
```

The call to function FUNC_Q in procedure PROC_P is illegal, since FUNC_Q has not yet been declared. This can be corrected by writing the function declaration for FUNG_Q before procedure PROC_P.

2.32 SUBPROGRAM OVERLOADING

Sometimes it is convenient to have two or more subprograms with the same name. In such a case, the subprogram name is said to be *overloaded;* also, the corresponding subprograms are said to be overloaded. For example, consider the following two declarations.

```
function COUNT (ORANGES: INTEGER) return INTEGER;
function COUNT (APPLES: BIT) return INTEGER;
```

Both functions are overloaded since they have the same name, COUNT. When a call to either function is made, it is possible to identify the exact function to which the call is made from the type of actuals passed because they have different parameter types. For example, the function call

```
COUNT(20)
```

refers to the first function, since 20 is of type INTEGER, while the function call

```
COUNT('1')
```

refers to the second function, since the type of actual is BIT.

Here is another example.

```
function TO_CHARACTER (ARG: STD_LOGIC
  return CHARACTER;
function TO_CHARACTER (ARG: STD_LOGIC_VECTOR)
  return CHARACTER;
```

Both functions are overloaded. The function call:

```
TO_CHARACTER ('0')
```

will call the first function since the argument is of type STD_LOGIC, while the function call:

```
TO_CHARACTER ("10010")
```

will call the second function since the argument is of type STD_LOGIC_VECTOR.

If two overloaded subprograms have the same parameter types and result types, it is possible for one subprogram to hide the other subprogram. This can happen if a subprogram is declared within another subprogram's scope. Here is an example.

```
architecture HIDING of  DUMMY_ENTITY is
  function ADD (A, B: BIT_VECTOR) return BIT_VECTOR is
  begin
    -- Body of function here.
  end ADD;
begin
  SUM_IN_ARCH <= ADD (IN1, IN2);
process
  function ADD (C, D: BIT_VECTOR) return BIT_VECTOR is
  begin
    -- Body of function here.
  end ADD;
begin
   SUMJN_PROCESS <= ADD (IN1, IN2);
   SUM_IN_ARCH <= HIDINGADD(IN1, IN2);
  end process;
end HIDING;
```

The function ADD declared in the architecture body is hidden within the process because of the second function, ADD, declared within the declarative part of the process. This function can still be accessed by qualifying the function name with the architecture name, as shown in the second statement in the process.

It is also possible for two overloaded subprograms to be directly visible within a region, by using use clauses. In such a case, a subprogram call may be ambiguous, and hence an error, if it is not possible to determine which of the overloaded subprograms is being called. Here is an example.

```
package P1 is
  function ADD (A, B: BIT_VECTOR) return BIT_VECTOR;
end P1;
package P2 is
  function ADD (X,Y: BIT_VECTOR) return BIT_VECTOR;
end P2;
use WORK.P1.all, WORK.P2.all;
architecture OVERLOADED of DUMMY_ENTITY is
begin
  SUM_CORRECT <= ADD (X => IN1,Y => IN2);
  SUM_ERROR <= ADD (IN1, IN2);                    -- An error.
end OVERLOADED;
```

The function call in the first signal assignment statement is not an error, since it refers to the function declared in package P2 (the formal parameters are explicitly specified in the association), while the call in the second signal assignment statement is ambiguous and hence an error.

Two overloaded subprograms can also have a different number of parameters but have the same parameter types and result types. In this case, the number of actuals supplied in the subprogram call identifies the correct subprogram. Here is an example of such a set of functions that determine the smallest value from a set of 2, 4, or 8 integers.

```
function SMALLEST (Al, A2: INTEGER) return INTEGER;
function SMALLEST (Al, A2, A3, A4: INTEGER) return INTEGER;
function SMALLEST (Al, A2, A3, A4, A5, A6, A7, A8: INTEGER)
  return INTEGER;
```

A call such as:

```
... SMALLEST (4, 5) ...
```

refers to the first function, while the function call:

```
... SMALLEST (20, 45, 52,1, 89, 67, 91, 22) ...
```

refers to the third function. This flexibility helps in writing code that is easy to decipher since the same subprogram name can be made to serve differently when used with a different set of inputs.

A call to an overloaded subprogram is ambiguous, and hence an error, if it is not possible to identify the exact subprogram being called using the following information:

i. Subprogram name
ii. Number of actuals
iii. Types and order of actuals
iv. Names of formal parameters (if named association is used)
v. Result type (for functions)

Notice that overloading does not distinguish by subtypes. Here is an example to illustrate this point.

```
type LOGIC5 is ('X', '0', 1', 'D', 'Z');
subtype LOGIC5_01 is LOGIC5 range '0' to '1';

function RISING _EDGE (signal CLOCK: LOGIC5)
  return BOOLEAN;
function RISING_EDGE (signal CLOCK: LOGIC5_01)
  return BOOLEAN;
```

The second function declaration hides the first function, because both have the same parameter base types and the same return types. Thus, a call such as:

```
signal CK: LOGIC5_01;
...
...RISING_EDGE (CK)
```

is an error.

Overloading also does not distinguish by the parameter modes.

```
procedure UNSIGNED_PLUS (A, B: in STD_LOGIC_VECTOR;
                                C: out STD_LOGIC_VECTOR);
procedure UNSIGNED_PLUS (A, B: in STD_LOGIC_VECTOR;
                                C: inout STD_LOGIC_VECTOR);
```

The second procedure hides the first procedure because both the procedures have the same parameter base types. The mode of the third parameter is different, but the mode is irrelevant when checking if two subprograms hide each other.

2.33 OPERATOR OVERLOADING

Operator overloading is one of the most useful features in the language. When a standard operator symbol is made to behave differently based on the type of its operands, the operator is said to be overloaded. The need for operator overloading arises from the fact that the predefined operators in the language are defined only for operands of certain predefined types. For example, the and operation is defined for arguments of type BIT and BOOLEAN, and for one-dimensional arrays of BIT and BOOLEAN only. What if the arguments were of type MVL (where MVL is a user-defined enumeration type with values 'U', '0', '1', and 'Z')? In such a case, it is possible to argument the and operation as a function that operates on arguments of type MVL. The **and** operator is then said to be overloaded. The operator in the expression:

```
S1 and S2
```

where S1 and S2 are of type MVL, would then refer to the and operation that was defined by the model writer as a function. The operator in the expression:

```
CLKl and CLK2
```

where CLK1 and CLK2 are of type BIT, would refer to the predefined **and** operator.

Function bodies are written to define the behavior of overloaded operators. The number of parameters in such a function must match in cardinality with that of the predefined operator. The function thus has, at most, two parameters; the first one refers to the left, or the only, operand of the operator, and the second parameter, if present, refers to the second operand. Here are some examples of function declarations for such function bodies.

```
type MVL is ('U', '0', '1', 'Z');
function "and" (L, R: MVL) return MVL;
function "or" (L, R: MVL) return MVL;
function "not" (R: MVL) return MVL;
```

Since the **and,** or, and **not** operators are predefined operator symbols, they have to be enclosed within double quotes when used as overloaded operator function names. Having declared the overloaded functions, the operators can now be called using any of the following two different types of notations:

1. Standard operator notation
2. Standard function call notation

Here are some examples of these two types of notations based on the overloaded operator function declarations that appeared earlier.

```
signal A, B, C: MVL;
signal XX Z: BIT;

A <='Z' or '1';                -- #1: standard operator notation.
B <= "or" ('0', 'Z');          -- #2: function call notation.
X <= not Y;                    -- #3
Z <= X andY;                   -- #4
C <= (A or B) and (not C);     -- #5
Z <= (X and Y) or A;           -- #6
```

The or operator in the first statement refers to the overloaded operator because the type of the left operand is MVL. This is the standard operator notation since the overloaded operator symbol appears just like the standard operator symbol. An example of the function call notation is shown in the second statement in which the overloaded function, **or**, is explicitly called. The operators in the third and fourth statements refer to the predefined operators since their operands are of type BIT. The sixth statement would be an error, assuming that there are no overloaded **or** operators defined with the first parameter type of BIT and the second parameter type of MVL.

The last example brings up a very interesting point. In overloaded operator functions, it is not necessary for both operands to have the same type. In the previous case, if another or overloaded function with a declaration such as:

```
function "or" (L: BIT; R: MVL) return BIT;
```

were defined, the sixth assignment statement would not be an error.

It is often the case that a user deals mostly with vectors; therefore, it is necessary to provide a capability to perform arithmetic operations on vectors. Arithmetic operations are not predefined in the language to work on vectors. Overloaded functions again need to be defined. Here are some examples of overloaded operator declarations.

```
function "+" (OPD1, OPD2: BIT_VECTOR) return BIT_VECTOR;
function "–" (OPD1, OPD2: BIT_VECTOR) return BIT_VECTOR;

function "+" (OPD1, OPD2: STD_LOGIC_VECTOR)
  return STD_LOGIC_VECTOR;
function "–" (OPD1, OPD2: STD_LOGIC_VECTOR)
  return STD_LOGIC_VECTOR;
```

Examples of calls to these overloaded functions are given next.

```
variable A, B, C, CAB: BIT_VECTOR (3 downto 0);
variable D, E, F, FED: STD_LOGIC_VECTOR (0 to 7);
...
CAB:= A + B – C;
FED:= D – F + E;
```

In the first variable assignment statement, the arguments for the + and – operators are BIT vectors; therefore, these operations call the corresponding "+" and "–" overloaded functions for evaluation, respectively. Similarly, in the second variable assignment, the + and -operators operate on STD_LOGIC vectors; therefore, for evaluation, their corresponding overloaded operator functions for "+" and "–" are called.

2.34 SIGNATURES

In earlier versions of the language, it was not possible to uniquely identify an overloaded subprogram or an overloaded enumeration literal. For example,

```
-- Two enumeration type declarations
type MYBIT is ('0', 1');
type FOUR_VALUE is ('U', '0', 1', 'Z');
-- Two overloaded operator function declarations:
function "+" (A, B: BIT_VECTOR) return BIT_VECTOR;
function "+" (A, B: STD_LOGIC_VECTOR)
    return STD_LOGIC_VECTOR;
```

In the enumeration type declarations, '0' and '1' are overloaded enumeration literals, while in the two function declarations, "+" is an overloaded operator function. If there was a need to either define an alias for the first "+" or to assign an attribute (aliases and attributes are discussed in Chapter 10), it was not possible to uniquely distinguish either one of the "+" functions in earlier version of the language. Similarly, it was not possible to uniquely identify the '0' or '1' enumeration literal so that an attribute could be associated with or an alias created for either one of them.

Signatures have been provided as a means of distinguishing overloaded subprograms and overloaded enumeration literals. A *signature* shows the parameter types and result type of an overloaded subprogram or an overloaded enumeration literal. A signature is of the form:

[*first-parameter-type, second-parameter-type,...*
  [**return** *function-return-type*]]

Note that the outer square brackets ([ ]) are part of the syntax and do not represent optional items. Here are the signatures for the previously declared "+" overloaded functions.

```
[ BIT_VECTOR, BIT_VECTOR return BIT_VECTOR ]
[ STD_ LOGIC_VECTOR, STD_LOGIC_VECTOR
    return STD_LOGIC_VECTOR ]
```

To determine the parameter and result types of an enumeration literal, an enumeration literal is treated as equivalent to a function with no parameters, with the function name being the same as the enumeration literal and the return type of the function being the enumeration type. For example, the equivalent functions for the '0' enumeration literals declared previously in type MYBIT and FOUR_VALUE are:

```
function '0' return MYBIT;
function '0' return FOUR_VALUE;
```

Therefore, the signatures for each of these enumeration literals are:

```
[ return MYBIT ]
[ return. FOUR_VALUE ]
```

Signatures can be used in alias declarations, attribute specifications, and attribute names to uniquely identify an overloaded subprogram or an overloaded enumeration literal.

2.35 DEFAULT VALUES FOR PARAMETERS

It is possible to assign default values to input parameters in a subprogram declaration. In such a case, if no actual is specified for a formal parameter that has a default value, then the default value is used as a value for the formal.

Here is an example of such a specification.

```
function MY_AND (A: in BIT := '0', B: in BIT := '1') return BIT;
```

When a default value is specified for a parameter, then it is optional to pass an actual value in a subprogram call. When no actual is passed in for such a parameter (or the keyword **open** is used as an actual), the parameter assumes the default value. The default value is ignored if an actual value (other than keyword **open**) is explicitly specified. Here are some examples of function calls.

```
...MY_AND()...
 -- Formal parameters A and B have the default values
 -- of '0' and '1' respectively.
... MY_AND (B => SIG_P)...
 -- SIG_P is passed in as the value for B. Since no actual is
 -- specified for A, A has its default value '0'.
... MY_AND (B => SIG_P, A => SIG_Q)...
 -- Both the default values are ignored since actuals are
 -- explicitly specified.
... MY_AND (A => SIG_Q, B => open) ...
 -- Using keyword open as an actual is equivalent to not specifying
 -- a value; thus, B has a default value of '1'.
```

Here is another example. This function can convert an integer value to a value of type TIME.

```
function TO_TIME (ARG: INTEGER; SCALE: TIME := ns)
    return TIME is
begin
    return ARG* SCALE;
end TO_TIME;
```

Notice that a default value has been specified for the formal parameter SCALE of 1 ns. A function call such as:

```
TO_TIME (56)
```

uses a SCALE of 1 ns, while a function call:

```
TO_TIME (56,10 ms)
```

uses a SCALE of 10 ms; the default value is ignored in this case because an actual value has been explicitly specified.

An input signal parameter in a subprogram cannot have a default value associated with it. For example,

```
procedure TO_VECTOR (signal VALUE: in INTEGER;...);
```

VALUE is a signal parameter and cannot have a default value associated with it.

2.36 STRUCTRAL MODELING

An entity is modeled as a set of components connected by signals, that is, as a netlist. The behavior of the entity is not explicitly apparent from its model. The component instantiation statement is the primary mechanism used for describing such a model of an entity.

2.36.1 An Example

Consider the circuit shown in Figure 2.16 and its VHDL structural model.

```
entity GATING is
  port (A, CK, MR, DIN: in BIT; RDY, CTRLA: out BIT);
end GATING;

architecture STRUCTURE_VIEW of GATING is
  component AND2
    port (X,Y: in BIT; Z: out BIT);
  end component;
  component DFF
    port (D, CLOCK: in BIT; Q, QBAR: out BIT);
  end component;
  component NOR2
    port (DA, DB: in BIT; DZ: out BIT);
  end component;
  signal S1, S2: BIT;
begin
  Dl: DFF port map (A, CK, S1, S2);
  Al: AND2 port map (S2, DIN, CTRLA);
  Nl: NOR2 port map (S1, MR, RDY);
end STRUCTURE_VTEW;
```

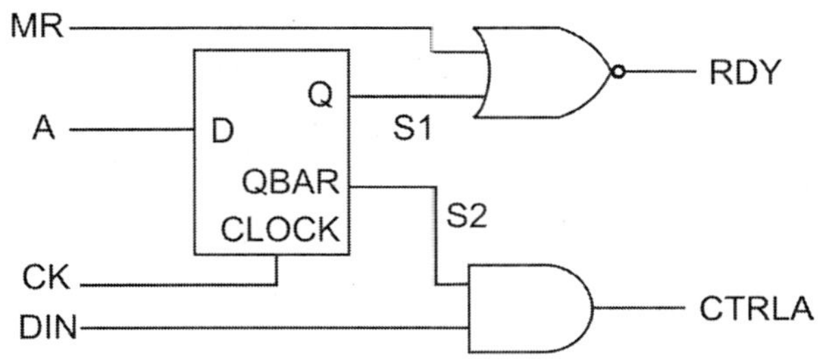

Figure 2.16: A circuit generating control signals.

Three components, AND2, DFF, and NOR2, are declared. These components are instantiated in the architecture body via three component instantiation statements, and the instantiated components are connected to each other via signals S1 and S2. The component instantiation statements are concurrent statements, and their order of appearance in the architecture body is therefore not important. A component can, in general, be instantiated any number of times. However, each instantiation must have a unique component label; as an example, A1 is the component label for the AND2 component instantiation.

2.37 COMPONENT DECLARATION

A component instantiated in a structural description must first be declared using a component declaration. A component declaration declares the name and the interface of a component. The interface specifies the mode and the type of ports. The syntax of a simple form of component declaration is:

```
component component-name [ is ]
  [ port (list-of-interface-ports); ]
end component [ component-name ];
```

The *component-name* may or may not refer to the name of an entity already existing in a library. If it does not, it must be explicitly bound to an entity; otherwise, the model cannot be simulated (the model may still be analyzed). The binding information can be specified using a configuration. Configurations are discussed in the next chapter.

The *list-of-interface-ports* specifies the name, mode, and type for each port of the component in a manner similar to that specified in an entity declaration. The names of the ports may also be different from the names of the ports in the entity to which it may be bound (different port names can be mapped in a configuration). In this chapter, we will assume that an entity of the same name as that of the component already exists and that the name, mode, and type of each port matches the corresponding ones in the component. Some examples of component declarations are:

```
component NAND2
  port (A, B: in STD_LOGIC; Z: out STD_LOGIC); end component;
component MP
  port (CK, RESET, RDN, WRN: in BIT;
          DATA_BUS: inout INTEGER range 0 to 255;
          ADDR_BUS: in BIT_VECTOR(15 downto 0));
end component MP;
```

```
component RX
  port (CK, RESET, ENABLE, DATAIN, RD: in BIT;
          DATA_OUT: out INTEGER range 0 to (2**8 –1);
          PARITY_ERROR, FRAME_ERROR,
          OVERRUN_ERROR: out BOOLEAN);
 end component;
```

Component declarations appear in the declarations part of an architecture body. Alternately, they may also appear in a package declaration. Items declared in this package can then be made visible within any architecture body by using the library and use clauses. For example, consider the entity GATING described in the previous section. A package such as the one shown next may be created to hold the component declarations.

```
package COMP_LIST is
  component AND2
    port (X,Y: in BIT; Z: out BIT);
  end component;
  component DFF
    port (D, CLOCK: in BIT; Q, QBAR: out BIT);
end component;
component NOR2
  port (DA, DB: in BIT; DZ: out BIT);
end component;
end COMP_LIST;
```

Assuming that this package has been compiled into design library DES_LIB, the architecture body can be rewritten as:

```
library DES_LIB;
use DES_LIB.COMP_LIST.all;
architecture STRUCTUREJ_VIEW of GATING is
  signal S1, S2: BIT;
  -- No need for specifying component declarations here, since they
  -- are made visible to architecture body
  -- using library and use clauses.
begin
  -- The component instantiations here.
end STRUCTURE_VIEW;
```

The advantage of this approach is that the package can now be shared by other design units, and the component declarations need not be specified inside every design unit.

2.38 COMPONENT INSTANTIATION

A component instantiation statement defines a subcomponent of the entity in which it appears. It associates the signals in the entity with the ports of that subcomponent. A format of a component instantiation statement is:

component-label: component-name [**port map** (*association-list)*];

The *component-label* can be any legal identifier and can be considered as the name of the instance. The *component-name* must be the name of a component declared earlier using a component declaration.

The *association-list* associates signals in the entity. called *actuals,* with the ports of a component, called *formals.* An actual may be a signal. An actual for an input port may also be an expression. An actual may also be the keyword **open** to indicate a port that is not connected.

There are two ways to perform the association of formals with actuals:

1. Positional association
2. Named association

In positional association, an *association-list* is of the form:

$actual_1$, $actual_2$, $actual_3$,..., $actual_n$

Each actual in the component instantiation is mapped by position with each port in the component declaration. That is, the first port in the component declaration corresponds to the first actual in the component instantiation, the second with the second, and so on. Consider an instance of a NAND2 component.

```
-- Component declaration:
component NAND2
  port (A, B: in STD.LOGIC; Z: out STD_LOGIC);
end component;
-- Component instantiation:
Nl: NAND2 port map (S1, S2, S3);
```

N1 is the component label for the current instantiation of the NAND2 component. Signal S1 (which is an actual) is associated with port A (which is a formal) of the NAND2 component, S2 is associated with port B of the NAND2 component, and S3 is associated with port Z. Signals S1 and S2 thus provide the two input values to the NAND2 component, and signal S3 receives the output value from the component. The ordering of the actuals is therefore important.

If a port in a component instantiation is not connected to any signal, the keyword **open** can be used to signify that the port is not connected. For example,

```
N3: NAND2 port map (S1, open, S3);
```

The second input port of the NAND2 component is not connected to any signal. An input port may be left open only if its declaration specifies an initial value. For the previous component instantiation statement to be legal, port B of the component declaration for NAND2 must have an initial value expression.

```
component NAND2
  port (A, B: in STD_LOGIC := '0'; Z: out STD_LOGIC);
  -- Both A and B have an initial value of '0'; however, only
  -- the initial value of B is necessary in this case.
end component;
```

A port of any other mode may be left unconnected as long as it is not an unconstrained array.

In named association, an *association-list* is of the form:

$formal_1$ => $actual_1$, $formal_2$ => $actual_2$,..., $formal_n$ => $actual_n$

For example, consider the component NOR2 in the entity GATING described in the first section. The instantiation using named association may be written as:

```
Nl: NOR2 port map (DB => MR, DZ => RDY, DA => S1);
```

In this case, the signal MR (an actual), which is declared in the entity port list, is associated with the second port (port DB, a formal) of the NOR2 gate, signal RDY is associated with the third port (port DZ), and signal S1 is associated with the first port (port DA) of the NOR2 gate. In named association, the ordering of the associations is not important since the mapping between the actuals and formals is explicitly specified. An important point to note is that the scope of the formals is restricted to be within the port map part of the instantiation for that component; for example, the formals DA, DB, and DZ of component NOR2 are relevant only within the port map of instantiation of component NOR2.

For either type of association, there are certain rules imposed by the language. First, the types of the formal and actual being associated must be the same. Second, the modes of the ports must conform to the rule that if the formal is readable, so must the actual be; and if the formal is writable, so must the actual be. Since a locally declared signal is considered to be both readable and writable, such a signal may be associated with a formal of any mode. If an actual is a port of mode **in,** it may not be associated with a formal of mode **out** or **inout;** if the actual is a port of mode **out,** it may not be associated with a formal of mode **in** or **inout;** if the actual is a port of mode **inout,** it may be associated with a formal of mode **in, out,** or **inout.**

It is important to note that an actual of mode **out** or **inout** indicates the presence of a source for that signal, and therefore, it must be resolved if that signal is multiply driven. A buffer port can never have more than one source; therefore, the only kind of actual that can be associated with a buffer port is another buffer port or a signal that has at most one source.

Another example of a component instantiation is:

```
Ml: MICRO port map (UDIN(3 downto 0), WRN, RDN, STATUS(0),
                    STATUS(l), UDOUT(0 to 7), TXDATA);
```

The first actual of the port map refers to a slice of the vectored UDIN signal, WRN and RDN are 1-bit signals, STATUS (0) and STATUS (1) refer to the 0 th and 1st element of the STATUS array, UDOUT(0 to 7) refers to a slice of the UDOUT vector, and TXDATA refers to an entire vector signal. This example shows that the signals used to interconnect components can also be any of the following:

- Slices
- Vectors
- Array elements

Not only can an actual be any of the above form, even a formal can be a slice, a vector or an array element. This is shown using an example.

```
component CHILD6
  port (INP: in BIT_VECTOR(0 to 4);
        OUTP: out BIT_VECTOR(3 downto 0));
end component;
signal QB: BIT__VECTOR(4 downto 0);
signal RESET: BIT;
signal SET: BIT_VECTOR(0 to 3);
...
INST6B: CHILD6 port map (
                    INP(0 to 2) => QB(4 downto 2),
                    INP (3) => RESET,
                    INP(4) => QB(0),
                    OUTP => SET);
```

The first formal, INP (0 **to** 2), is an example of a slice, the second and third formals are examples of array elements and the last formal, OUTP, is the entire vector.

Constraints are passed to a formal port from an actual if the formal port is unconstrained.

```
signal QBUS: BIT_VECTOR (3 downto 0);
signal PAND, QAND: BIT;
signal PBUS: BIT_VECTOR (0 to 8);
component REDUCE_AND
   port (D: in BIT_VECTOR; AND_OUT: out BIT);
end component;
...
 P1: REDUCE_AND port map (QBUS, QAND);
 P2 : REDUCE_AND port map (PBUS, PAND);
```

The size of the formal port D has not been explicitly specified in the component declaration for REDUCE_AND. In this case, the size of the formal comes from the actual during instantiation. From the first instantiation, port D has the constraint "3 **downto** 0"; from the second instantiation, formal port D has the constraint "0 **to** 8". The constraints for the formal port are on a "per-instance basis", that is, based on the actual in an instantiation, the formal port gets that constraint which could be different for different instances.

2.39 OTHER EXAMPLES

A complete structural model for a 9-bit parity generator circuit is shown in Figure 2.17

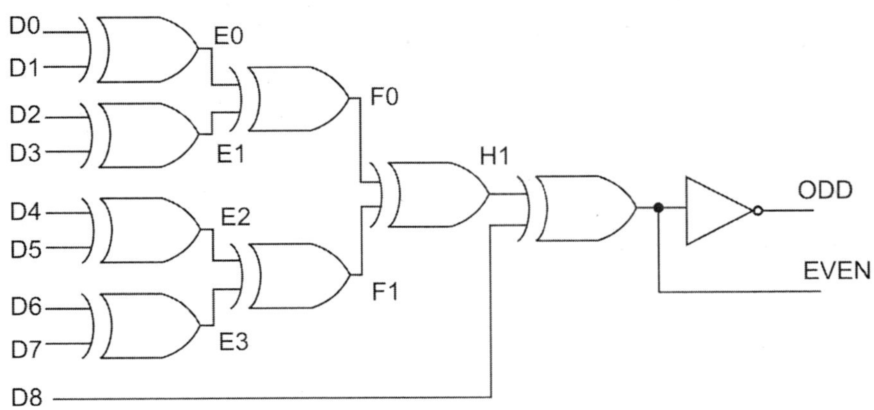

Figure 2.17: A 9-bit parity generator circuit.

```
library IEEE;
use IEEE.STD_LOGIC_1164.all;
entity PARITY_9_BIT is
  port ( D: in STD_LOGIC_VECTOR(8 downto 0);
           ODD: out STD_LOGIC
     EVEN: buffer STD_LOGIC);
end PARITY_9_BIT;
```

```
architecture PARITY_STR of PARITY_9_BIT is
  component XOR2
    port (A, B: in STD_LOGIC; Z: out STD_LOGIC);
  end component;
  component INV2
    port (A: in STD_LOGIC; Z: out STD_LOGIC);
  end component;
  signal E0, E1, E2, E3, F0, F1, H0: STD_LOGIC;
begin
  XE0: XOR2 port map (D(0), D(1), E0);
  XE1: X0R2 port map (D(2), D(3), E1);
  XE2:XOR2 port map (D(4),D(5),E2);
  XE3: XOR2 port map (D(6), D(7), E3);
  XF0: XOR2 port map (E0, E1, F0);
  XF1: XOR2 port map (E2, E3, F1);
  XH0: XOR2 port map (F0, F1, H0);
  XEVEN: XOR2 port map (H0, D(8), EVEN);
  XODD: INV2 port map (EVEN, ODD);
end PARITY_STR;
```

In this example, port EVEN is of mode **buffer** since the value of this port is being read as well as written to inside the architecture. If this port were declared to be of mode **inout,** external signals defined outside the PARITY_9_BIT design would be able to drive this port, which may not be desired.

An example of a decade counter using J-K flip-flops is shown in Figure 2.18 Its structural model is described next.

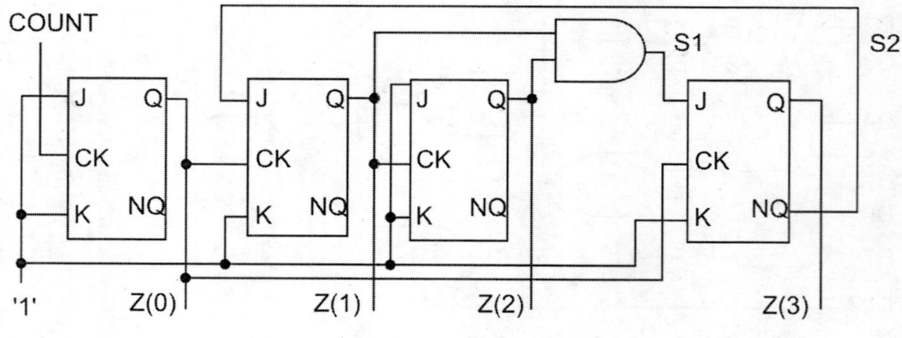

Figure 2.18 decade counter.

```
entity DECADE_CTR is
  port (COUNT: in BIT; Z: buffer BIT_VECTOR (0 to 3));
end DECADE_CTR;

architecture NET_LIST of DECADE_CTR is
  component JK_FF
    port (J,K, CK: in BIT; Q, NQ: buffer BIT);
  end component;
```

```
    component AND_GATE
      port (A, B: in BIT; C: out BIT);
    end component;
    signal S1, S2: BIT;
begin
    A1: AND_GATE port map (Z(2), Z(l), S1);
    JK1: JK_FF port map ('1', '1', COUNT, Z(0), open);
    JK2: JK_FF port map (S2, '1', Z(0), Z(l), open);
    JK3: JK_FF port map ('1', '1', Z(l), Z(2), open);
    JK4: JK_FF port map (S1, '1', Z(0), Z(3), S2);
end NET_LIST;
```

This example illustrates the point that a constant value or, in general, a constant expression can be used as an actual in a port map.

Structural models can be simulated only after the entities that the components represent are modeled and placed in a design library. The lowest-level entities must be behavioral models. The simulation semantics of a component instantiation are best understood by an example. Consider the component instantiation A1 in the previous example. Assume that this instance is bound to an entity with the same name and identical port names. Its equivalent behavioral representation is:

```
Al: block                                  -- A block for each instantiation.
    port (A, B: in BIT; C: out BIT);       -- Ports in component
                                           -- declaration.
    port map (C => S1, A => Z(2), B => Z(l)); -- Association list.
begin
  AND_GATE: block                          -- The entity block.
    port (A, B: in BIT; C: out BIT);       -- Ports of entity.
    port map (A => A, B => B, C => C);  -- Association of
                  -- component ports with entity ports.
    -- Declarations that occur in entity declaration and
    -- architecture body of AND_GATE entity appear here.
begin
    -- Behavior in architecture body for AND_GATE entity.
    -- For example,
    -- C <= A and B after 10 ns;
  end block AND_GATE;
end block Al;
```

As shown, a block statement can also have a port list and a port map. The port list specifies the ports through which the block communicates with its external environment. The port map specifies the mapping between the ports and the signals in the block's external environment to which these ports connect.

As a final example, consider the three-bit up-down counter circuit shown in Figure 2.11 It's structural model follows.

```
entity UP_DOWN is
  port (CLK, CNT_UP, CNT_DOWN: in BIT;
        Q0,Q1, Q2: buffer BIT);
end UP_DOWN;
architecture COUNTER of UP_DOWN is
```

```
    component JK_FF
      port (J, K, CK: in BIT; Q, QN: buffer BIT);
    end component;
    component AND2
      port (A, B: in BIT; C: out BIT);
    end component;
    component OR2
     port (A, B: in BIT; C: out BIT);
    end component;
    signal S1, S2, S3, S4, S5, S6, S7, S8: BIT;
  begin
    JK1: JK_FF port map ('1', '1', CLK, Q0, S1);
    Al: AND2 port map (CNT_UP, Q0, S2);
    A2: AND2 port map (S1, CNT_DOWN, S3);
    Ol: OR2 port map (S2, S3, S4);
    JK2: JK_FF port map ('1', '1', S4, Q l, S5);
    A3: AND2 port map (Ql, CNT_UP, S7);
    A4: AND2 port map (S5, CNT_DOWN, S6);
    O2: OR2 port map (S7, S6, S8);
    JK3: JK_FF port map ('1', '1', S8, Q2, open);
  end COUNTER;
```

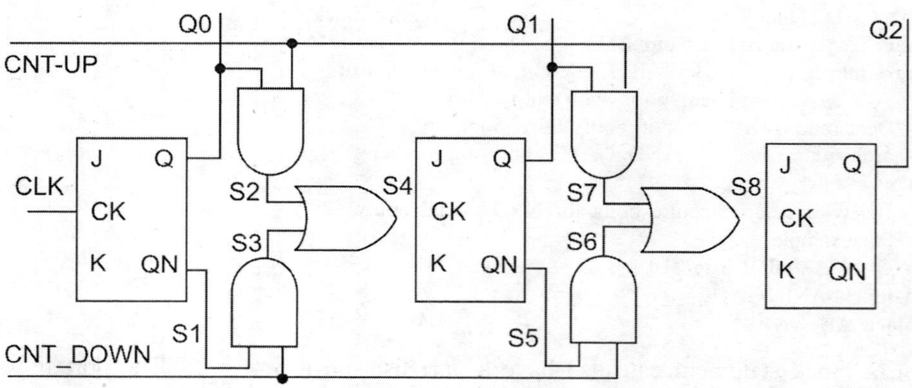

(J,K inputs of all flip-flops connected to'1')

Figure2.19: A 3-bit up-down counter.

2.40 RESOLVING SIGNAL VALUES

If outputs of two components drive a common signal, the value of the signal must be resolved using a resolution function. This is similar to the case of a signal being assigned using more than one concurrent signal assignment statement. For example, consider the circuit shown in Figure 2.20 which shows two and gates driving a common signal RS1, which is inverted to produce the result in Z.

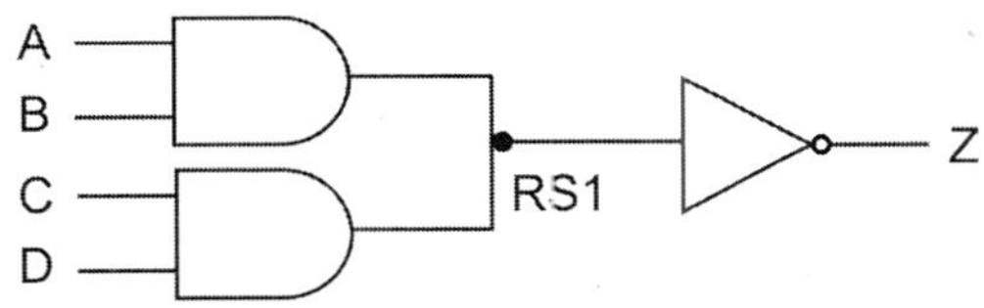

Figure 2.20: Two components driving a common signal.

```
entity DRIVING_SIGNAL is
  port (A, B, C, D: in BIT; Z: out BIT);
end DRIVING_SIGNAL;

    -- Assume PULL_UP is the name of a function defined in package
    -- RF_PACK that has been compiled into the working library.
    use WORK.RF_PACK.PULL_UP;
    architecture RESOLVED of DRIVING_SIGNAL is
      signal RS1: PULL_UP BIT;
      component AND2
        port (IN1, IN2: in BIT; OUT1: out BIT);
      end component;
      component INV
        port (X: in BIT;Y: out BIT);
      end component;
    begin
      Al: AND2 port map (A, B, RS1);
      A2: AND2 port map (C, D, RS1);
      I1: INV port map (RS1, Z);
    end RESOLVED;
```

The key point here is that even though an assignment to signal RS1 is not being made explicitly using signal assignment statements, the signal RS1 is being driven by two output ports, and therefore must be resolved using a resolution function. In the previous example, the PULL_UP resolution function is associated with signal RS1. This implies that the values of the outputs of the and gates are passed through the resolution function before a value is assigned to signal RS1. In general, each **out, inout,** and **buffer** port of a compo hent creates a driver for the signal with which it is associated.

2.41 GENERICS

It is often useful to pass certain types of information into a design description from its environment. Examples of such information are rise and fall delays and the size of interface ports. This is accomplished by using generics. Generics of an entity are declared along with its ports in the entity declaration. An example of a generic N-input and gate is shown next.

```
entity AND_GATE is
  generic (N: NATURAL);
  port (A: in BIT_VECTOR (l to N); Z: out BIT);
end AND_GATE;

architecture GENERIC_EX of AND_GATE is
begin
```

```
    process(A)
      variable AND_OUT: BIT;
  begin
    AND_OUT := '1';
    for K in 1 to N loop
      AND_OUT := AND_OUT and A(K);
      exit when AND-OUT = '0';
    end loop;
    Z <= AND_OUT;
  end process;
end GENERIC_EX;
```

In this example, the size of the input port has been modeled as a generic. By doing this, we have modeled an entire class of and gates with a variable number of inputs using a single behavioral description. The AND_GATE entity may now be used with a different number of input ports in different instantiations.

A generic declares a constant object of mode in (that is, the value can only be read) and can be used in the entity declaration and its Corresponding architecture bodies. The value of this constant can be specified as a globally static expression in one of the following:

1. Entity declaration
2. Component declaration
3. Component instantiation
4. Configuration specification
5. Configuration declaration

The value of a generic must be determinable at elaboration time, that is, a value for a generic must be explicitly specified at least once using any of the ones mentioned.

The value for a generic may be specified in the entity declaration for an entity, as shown in this example. This is the default value for the generic. It can be overridden by others.

```
entity NAND_GATE is
  generic (M: INTEGER := 2);                    -- M models the number of inputs.
  port (A: in BIT_VECTOR(M downto 1); Z: out BIT);
end NAND_GATE;
```

Two other ways of specifying the value of a generic are in a component declaration and in a component instantiation. The following examples demonstrates these.

```
entity ANOTHER_GEN_EX is
end;
architecture GEN_IN_COMP of ANOTHER_GEN_EX is
  -- Component declaration for NAND_GATE:
  component NAND_GATE
    generic (M: INTEGER);
    port (A: in BIT_VECTOR (M downto 1); Z: out BIT);
  end component;
  -- Component declaration for AND_GATE:
  component AND_GATE
    generic (N: NATURAL := 5);
    port (A: in BIT_VECTOR(1 to N); Z: out BIT);
end component;
signal S1, S2, S3, S4: BIT;
```

```
    signal SA: BIT_VECTOR (1 to 5);
    signal SB: BIT_VECTOR (2 downto 1);
    signal SC: BIT_VECTOR (1 to 10);
    signal SD: BIT_VECTOR (5 downto 0);
  begin
    -- Component instantiations:
    Nl: NAND_GATE generic map (6) port map (SD, S1);
    Al: AND_GATE generic map (N => 10) port map (SC, S3);
    A2: AND_GATE port map (SA, S4);
    -- N2: NAND_GATE port map (SB, S2);
  end GEN_IN_COMP;
```

For the purposes of this discussion, we shall assume that the components NAND_GATE and AND_GATE are bound to the entities NAND_GATE and AND_GATE described earlier. The component declaration for AND_GATE specifies a value for the generic. When this component is instantiated and a new generic value is assigned using a generic map as in instance A1, the new value, 10, overrides the value specified in the component declaration, 5. When the AND_GATE component is instantiated and no generic map is specified, as in instance A2, the value of the generic specified in the component declaration, 5, is used. In the case of instance N1, again the value supplied by the generic map (i.e., 6) overrides the value assigned to the generic in the entity declaration for NAND_GATE (i.e.,2). The instance N2, shown as a comment, is illegal, since neither the component instantiation nor the component declaration supply a value for the generic.

The architecture body described earlier uses the more general form of syntax for component declaration and component instantiation statement, which is:

```
-- Component declaration:
component component-name [ is ]
  [ generic (list-of-generics); ]
  [ port (list-of-interface-ports); ]
end component [ component-name ];
-- Component instantiation statement:
component-label: component-name
          [ generic map (generic-association-list) ]
          [ port map (port-association-list) ];
```

Values for generics may also be specified in a configuration specification or in a configuration declaration. We shall see this later in the section on configurations. A model of a nor gate with generic rise and fall delays is shown next.

```
entity NOR2 is
  generic (PT_HL, PT_LH: TIME range 0 ns to TIME' HIGH);
  port (DA, DB: in BIT; DZ: out BIT);
end NOR2;

architecture NOR2_DELAYS of NOR2 is
signal TEMP: BIT;
begin
  TEMP <= not (DA or DB);
  DZ <= TEMP after PT_HL when TEMP = '0' else
      TEMP after PT_LH;
end NOR2_DELAYS;
```

Since no default values were provided for the generics, the values must be provided when this entity is instantiated or configured.

Consider an or gate constructed using two nor gates; each nor gate has the behavior described previously. The rise and fall delays are specified when the NOR2 component is instantiated. In the following example, different propagation delays are specified in each component instantiation statement.

```
entity OR2 is
  port (A, B: in BIT; C: out BIT);
end OR2;
architecture OR2_NOR2 of OR2 is
  component NOR2
    generic (PT_HL, PT_LH:TIME);
    port (A, B: in BIT; Z: out BIT);
  end component;
  signal S1: BIT;
begin
  Nl: NOR2 generic map (5 ns, 3 ns) port map (A, B, S1);
  N2: NOR2 generic map (6 ns, 5 ns) port map (S1, S1, C);
end;
```

Other uses of generics include modeling ranges of subtypes; for example,

```
subtype ALUBUS is INTEGER range TOP downto 0;
—TOP is a generic.
```

Generics can also be used to control the number of instantiations of a component in a generate statement.

2.42 WHY CONFIGURATIONS?

A question that is often asked is why are configurations needed? There are two main reasons.

1. Sometimes it may be convenient to specify multiple views for a single entity and use any one of these for simulation. This can be easily done by specifying one architecture body for each view and using a configuration to bind the desired architecture body. For example, there may be three architecture bodies, called FA_BEH, FA_STR, and FA_MIXED, corresponding to an entity FULL_ADDER. Any one of these can be selected for simulation by specifying an appropriate configuration.
2. Similar to the previous case, it may be desirable to associate a component with any one of a set of entities. The component declaration may have its name and the names, types, and number of ports and generics different from those of its entities. For example, a declaration for a component used in a design may be:

```
component OR2
  port (A, B: in BIT; Z: out BIT);
end component;
```

and the entities that the above component may possibly be bound to are:

```
entity OR_GENERIC is
  port (N: out BIT; L, M: in BIT);
end OR_GENERIC;
entity OR_HS is
  port (X,Y: in BIT; Z: out BIT);
end OR_HS;
```

The component names and the entity names, as well as the port names and their order, are different. In one case we may be interested in using the OR_HS entity for the OR2 component, and in another case, the OR_GENERIC entity. This can be achieved by appropriately specifying a configuration for the component. The advantage is that when components are used in a design, arbitrary names for components and their interface ports can be used, and these can later be bound to specific entities prior to simulation.

A configuration is therefore used to bind the following pairs:

1. An architecture body to its entity declaration
2. A component with an entity

Note that a configuration does not have any simulation semantics associated with it; it only specifies how a top-level entity is organized in terms of lower-level entities by specifying the bindings between the entities. The language provides two ways of performing this binding:

1. By using a configuration specification
2. By using a configuration declaration

2.43 CONFIGURATION SPECIFICATION

A configuration specification is used to bind component instantiations to specific entities stored in design libraries. The specification appears in the declarations part of the architecture or block in which the components are instantiated. Binding of a component to an entity can be done on a per-instance basis, for all instantiations of a component, or for a selected set of instantiations of a component. Instantiations of different components can also be bound to the same entity

Figure 2.21 shows a logic diagram for a one-bit full-adder. Its structural model is described next.

```
library HS_LIB, CMOS_LIB;
entity FULL_ADDER is
  port (A, B, CIN: in BIT; SUM, COUT: out BIT);
end;
architecture FA_STR of  FULL_ADDER is
  component XOR2
    port (Dl, D2: in BIT; DZ: out BIT);
  end component;
  component AND2
    port (Z: out BIT; A0, Al: in BIT);
  end component;
  component OR2
    port (Nl, N2: in BIT; Z: out BIT);
  end component;

-- Following are four configuration specifications:
for Xl,X2:XOR2
  use entity WORK.XOR2(XOR2BEH);     -- Binding the entity with
      -- more than one instantiation of a component.
for A3:AND2
  use entity HS_LIB.AND2HS(AND2STR)
    port map (HS_B=>A1, HS_Z=>Z, HS_A=>A0);        -- Binding
      -- the entity with a single instantiation of a component.
for all: OR2
 use entity CMOS_LIB.OR2CMOS(OR2STR);          -- Binding the
```

```
        -- entity with all instantiations of OR2 component.
   for others: AND2
      use entity WORK. A_GATE(A_GATE_BODY)
         port map (A0, Al, Z);          -- Binding the entity with all unbound
                                        -- instantiations of AND2 component.
   signal S1, S2, S3, S4, S5: BIT;
begin
   X1: XOR2 port map (A, B, S1);
   X2: XOR2 port map (S1, CIN, SUM);
   Al: AND2 port map (S2, A, B);
   A2: AND2 port map (S3, B, CIN);
   A3: AND2 port map (S4, A, CIN);
   Ol: OR2 port map (S2, S3, S5);
   Ol: OR2 port map (S4, S5, COUNT);
end FA_STR;
```

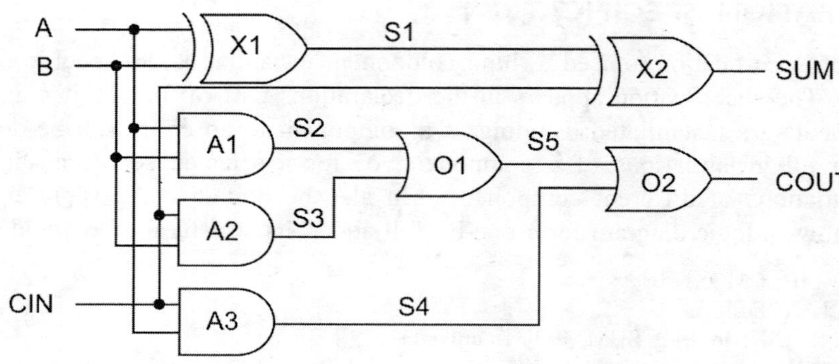

Figure 2.21: A1-bit full-adder circuit.

There are four configuration specifications appearing in the declarative part of the architecture body. The first specification indicates that instances X1 and X2 of component XOR2 are bound to the entity represented by the entity-architecture pair XOR2 and XOR2BEH, which reside in library WORK. The second specification binds the AND2 component with instantiation label A3 to the entity represented by the entity-architecture pair AND2HS and AND2STR, which is present in design library HS_LIB. The mapping of the component (AND2) ports and the entity (AND2HS) ports is specified using named association; for example, port HS_A of the AND2HS entity is mapped to port A0 of the AND2 component. The third specification implies that all instances of component OR2 use the entity represented by the specified entity-architecture pair present in the design library CMOS_LIB. The last specification implies that all unbound instances of component AND2, that is, instances A1 and A2, are bound to the entity A_GATE using the architecture A_GATE_BODY, which resides in library WORK.

The previous example showed that different instances of thesame component can be bound to different entities. Figure 2.22 picts this binding.

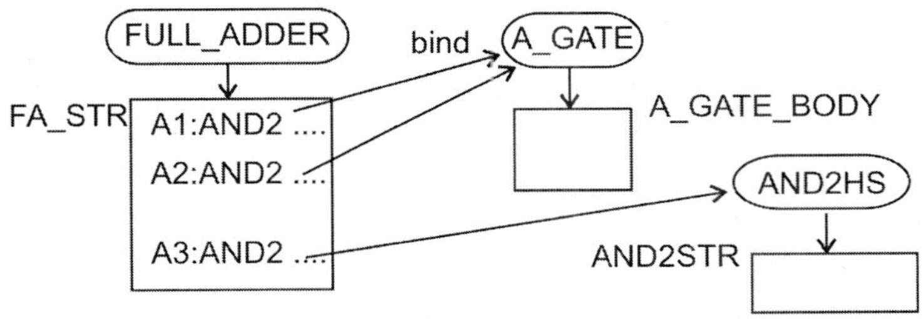

Figure 2.22: Different instances bound to different entities.

Similarly, it is also possible to bind different components to the same entity. An example is shown in Figure 2.23 This figure shows that there is nothing special about the component name AND2. By binding an instance of component AND2 to an entity called OR_GATE, this instance is made to behave as specified in the architecture of entity OR_GATE. Using such a binding may cause confusion to the reader, even though it is correct, and it may be what was intended. Such bindings may sometimes be necessary, for example, while debugging a model we may want to see the effect of specifying an and gate to behave like an or gate without changing the rest of the description.

This flexibility of being allowed to bind a component instance to any entity may result in a complex maze of bindings. An example is shown in Figure 2.24

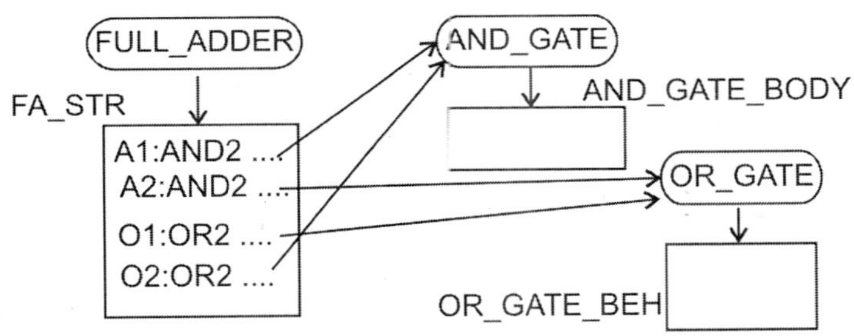

Figure 2.23: Different components bound to same entity.

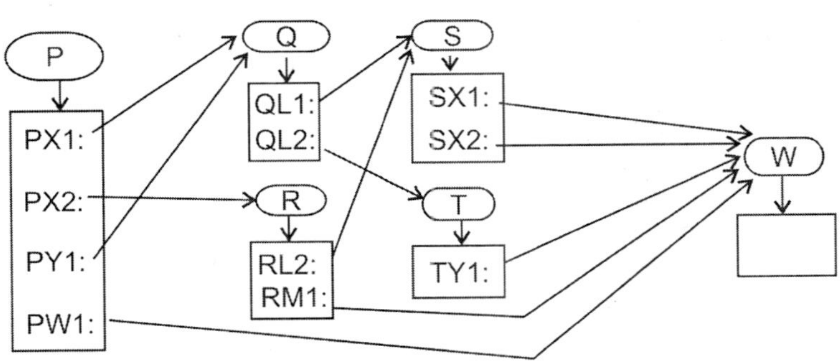

Figure 2.24: complex maze of bindings.

Entity P has four component instances, PX1 and PX2 of component type PX, PY1 of component type PY, and PW1 of component type PW. Component instances PX1 and PY1 are bound to entity

Q, while PX2 is bound to R. Component instances QL1 and QL2 (of component type QL) in entity Q are bound to entities S and T, respectively. Component instances RL1 and RM1 in entity R are bound to entities S and W, respectively. All component instances in S and T, and component PW1 in entity P, are bound to a single entity W. In other words, all the entities P, Q, R, S, and T have been built hierarchically using a single primitive component, W.

The syntax of a configuration specification is:

for *list-of-comp-labels* : *component-name binding-indication;*

The *binding-indication* specifies the entity represented by the entity-architecture pair, and the generic and port bindings, and one of its forms is:

```
use entity entity-name [ (architecture-name ) ]
  [ generic map (generic-association-list) ]
  [ port map (port-association-list) ]                    -- Form 1
```

The list of component labels may be replaced with the keyword all to denote all instances of a component; it may also be the keyword others to specify all as yet unbound instances of a component. The generic map is used to specify the values for the generics or provide the mapping between the generic parameters of the component and the entity to which it is bound. The port map is used to specify the port bindings between the component and the bound entity. Additional examples of configuration specifications appear in the following architecture body.

```
architecture DUMMY of DUMMY is
  component NOR_GATE
    generic (RISE_TIME, FALL_TIME: TIME);
    port (SO, S1: in BIT; Q: out BIT);
end component;
component AND2_GATE
  port (DOUT: out BIT; DIN: in BIT_VECTOR);
end component;
for Nl, N2: NOR_GATE
 use entity WORK.NOR2(NOR2_DELAYS)
   generic map (PT_HL => FALL_TIME,
                PT_LH => RISE_TIME)
   port map (S0, S1, Q);
for all: AND2_GATE
use entity WORK.AND2 (GENERIC_EX)
  generic map (10)
  port map (A => DIN, Z => DOUT);
signal WR, RD, RW, S1, S2: BIT;
signal SA: BIT_VECTOR (1 to 10);
begin
  Nl: NOR_GATE generic map (2 ns, 3 ns) port map (WR, RD, RW);
  Al: AND2_GATE port map (S1, SA);
  N2: NOR_GATE generic map (4 ns, 6 ns)
               port map (S1, SA(2), S3);
end DUMMY;
```

The entity declarations for the entities that are bound to components NOR_GATE and AND2_GATE are:

```
entity NOR2 is
  generic (PT_HL, FT. LH:TTME);
  port (DA, DB: in BIT; DZ: out BIT);
```

```
end NOR2;

entity AND2 is
   generic (N: NATURAL := 5);
   port (A: in BIT_VECTOR(l to N); Z: out BIT);
end AND2;
```

In the binding for N1 and N2, the generic map specifies the mapping of generic names from the entity NOR2 to the component NOR_GATE using named association. The generic values supplied in the instantiations are therefore passed to the NOR2 entity through this mapping. The port binding is specified using positional association; that is, ports SO, S1, and Q of NOR_GATE component map to ports DA, DB, and DZ, respectively, of the NOR2 entity. The configuration specification for the AND2_GATE specifies the value 10 for the generic explicitly (using positional association), which overrides the default value 5 specified in the entity declaration for the AND2 entity. The component declaration for AND2_GATE should not specify any generics, since the values are passed directly to the actual generics of the entity. The port mapping for the AND2_GATE is specified using named association.

How are the generic map and port map values in a component instantiation passed into its bound entity via the configuration specification? A look at the elaboration of a component instantiation helps us to understand this. Let us take the N1 instantiation in the previous architecture body as an example. Elaboration transforms this component instantiation into the following block statement.

```
Nl: block                         -- A block for the component instantiation.
  -- Generics of the component:
  generic (RISE_TIME, FALL_TIME: TIME);
  -- Generic map in instantiation:
  generic map (RISE_TIME => 2 ns, FALL_TIME => 3 ns);
  port (SO, S1: in BIT; Q: out BIT);          -- Ports of component.
  port map (SO => WR, S1 => RD, Q => RW);              -- Port map in
                                                       -- instantiation.
begin
NOR2: block                                     -- A block for the bound entity.
  generic (PT_HL, PT_LH: TIME);                             -- Its generics.
  generic map (PT_TL => FALL_TIME, PT_LH => RISE_TIME);
   -- Generic map in configuration specification.
  port (DA, DB: in BIT; DZ: out BIT);                  -- Its ports.
  port map (DA => SO, DB => S1, DZ => Q);              -- Port map in
                                                       -- specification.
  -- Other declarations in the architecture
  -- body NOR2_DELAYS appear here.
 begin
  -- Statements in architecture body NOR2_DELAYS appear here.
 end block NOR2;
end block Nl;
```

The block N1 is created from the component instantiation of NOR_GATE. The generic map and port map for this block are the generic map and port map specified in the component instantiation for N1; that is, they specify the mapping between the values and signals in the component instantiation statement with the generics and ports of component NOR_GATE. The inner block with label NOR2 represents the entity NOR2 that the component instantiation N1 is bound to in the configuration specification. The generic map and port map of this block specify the generic map and port map that appear in the configuration specification; that is, they specify the mapping between the component NOR_GATE and the entity NOR2.

2.44 CONFIGURATION DECLARATION

Configuration specifications have to appear in an architecture body. Therefore, to change a binding it is necessary to change the architecture body and re-analyze it. This may be cumbersome and time-consuming. To avoid this, a configuration declaration may be used to specify a binding.

A configuration declaration is a separate design unit. It therefore allows for late binding of components, that is, the bindings can be performed after the architecture body has been written. It is also possible to have more than one configuration declaration for an entity, each of which defines a different set of bindings for components in a single architecture body, or possibly specifies a unique entity-architecture pair.

The typical format of a configuration declaration is:

configuration *configuration-name* of *entity-name* is
  block-configuration
end [**configuration**] [*configuration-name*];

It declares a configuration with the name *configuration-name,* which applies to the entity *entity-name.* A *block-configuration* defines the binding of components in a block, where a block may be an architecture body, a block statement, or a generate statement. A block configuration is a recursive structure of the form:

for *block-name*
  component-configurations
  block-configurations
end for;

The *block-name* is the name of an architecture body, a block statement label, or a generate statement label. The top-level block is always an architecture body. A *component-configuration* binds components that appear in a block to entities, and is of the form:

for *list-of-comp-labels* : *comp-name [binding-indication*;]
  [*block-configuration*]
end for;

The block configuration that appears within a component configuration defines the bindings of components at the next level of hierarchy in the entity-architecture pair specified by the binding indication.

There are two other forms of binding indication in addition to the one shown in the previous section. These are:

use configuration *configuration-name* -- Form 2
use open -- Form 3

In form 2, the binding indication specifies that the component instances are to be bound to a configuration of a lower-level entity as specified by the configuration name. This implies that a configuration declaration with such a name must exist. In form 3, the binding indication indicates that the binding is not yet specified and is to be deferred. Both these forms of binding indication may also be used in a configuration specification.

Here is an example of a configurationdeclaration that specifies the component configurations for all component instances in architecture FA_STR of entity FULL_ADDER described in the previous section.

```
library CMOS_LIB;
configuration FA_CON of  FULL_ADDER is
  for FA_STR
    use  WORK.all;
    for Al, A2, A3 : AND2
      use entity CMOS_LIB.BIGAND2 (AND2STR);
    end for;
    for others: OR2          -- use defaults, i.e. use OR2 from
                             -- library WORK.
    end for;
    for all: XOR2
      use configuration WORK.XOR2CON;
    end for;
  end for;
end FA_CON;
```

The configuration with name FA_CON binds architecture FA_STR with the FULL_ADDER entity. The components within this architecture body, instances A1, A2, and A3, are bound to the entity BIGAND2 that exists in the design library CMOS_LIB. For all instances of component OR2, the default bindings are used ; these are the entities in the working library with the same names as the component names. The last component configuration shows a different type of binding indication. In this case, all component instances are bound to a configuration instead of an entity-architecture pair. All instances of component XOR2 are bound to a configuration with name XOR2CON, which exists in the working library. This type of binding may also be specified in a configuration specification.

The power of the configuration declaration lies in the fact that the sub-components in an entire hierarchy of a design can be bound using a single configuration declaration. For example, consider a full-adder circuit composed of two half-adders and an or gate. The half-adder circuit is in turn composed of xor and and gates. The hierarchy for this full-adder is shown in Figure 2.25

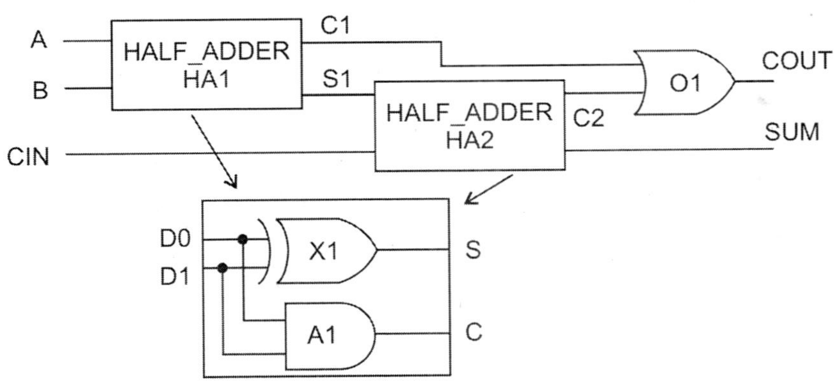

Figure 2.25: A hierarchical 1-bit full-adder.

The structural models for the full-adder and half-adder circuits are described next. A configuration declaration that specifies the bindings for components used in the entire hierarchy of the full-adder is also shown.

```
entity FULL_ADDER is
   port (A, B, CIN: in BIT; COUT, SUM: out BIT);
end FULL ADDER;
architecture FA_WITH_HA of FULL_ADDER is
   component HALF_ADDER
      port (HA, HB: in BIT; HS, HC: out BIT);
   end component;
   component OR2
      port (A, B: in BIT; Z: out BIT);
   end component;
   signal S1, C1, C2: BIT;
begin
   HA1: HALF_ADDER port map (A, B, S1, C1);
   HA2: HALF_ADDER port map (S1, CIN, SUM, C2);
   Ol: OR2 port map (C1, C2, COUT);
end FA_WITH_HA;
entity HA is
   port (DO, Dl: in BIT; S, C: out BIT);
end HA;

architecture HA_STR of HA is
   component XOR2
      port (X,Y: in BIT; Z: out BIT);
   end component;
   component AND2
      port (L, M: in BIT; N: out BIT);
   end component;
begin
   X1: XOR2 port map (DO, Dl, S);
   Al: AND2 port map (DO, Dl, C);
end HA_STR;
-- A configuration for the FULL_ADDER that is built
-- using half-adders.
library ECL;
configuration FA_HA_CON of FULL_ADDER is
   for FA_WITH_HA                         -- Top-level block configuration.
      for HA1, HA2: HALF_ADDER
         use entity WORK.HA(HA_STR)
            port map (DO => HA, Dl => HB, S => HS, C => HC);
         for HA_STR                       -- Nested block configuration.
            for all:XOR2
               use entity WORK.XOR2(XOR2);
            end for; for Al:AND2
               use configuration ECL.AND2CON;
            end for;
         end for;
      end for;
for O1: OR2
```

```
      use configuration WORK.OR2CON;
    end for;
  end for;
end FA_HA_CON;
```

The top-level block configuration specifies the bindings of component instances present in the architecture body FA_WITH_HA. Instances HA1 and HA2 are bound to an entity specified by the entity-architecture pair, entity HA and architecture HA_STR. The nested block configuration specifies the binding of component instances present in the architecture body HA_STR. In this way, a configuration can be nested to any arbitrary depth and can be used to bind all components in a hierarchy.

The previous example shows that, when components in a hierarchy are bound, a single configuration declaration may be used to replace a set of configuration specifications. If configuration specifications were used in the earlier example, they would have to be included separately in the architecture bodies FA_WITH_HA and HA_STR, and then these bodies would have to be recompiled every time a binding is changed. Note that a component instance must not be bound in both a configuration specification and a configuration declaration.

Therefore, configurations provide the mechanism by which architecture bodies may contain technology-independent components. Technology-specific mappings can be specified separately using configuration declarations.

2.45 DEFAULT RULES

A large amount of overhead is introduced if the bindings for every component in a design must be individually specified. Fortunately, the language provides default binding rules. For each unbound component instance:

- An entity that is visible and has the same name as that of the component is used to bind to the instance; if no such entity exists, a default binding indication of **"use open"** is used.
- The most recently analyzed architecture for that entity is used; it is an error if no architecture body exists.
- For each port or generic in the component instance, there must exist a corresponding port or generic in the entity that matches by name, type, and mode; any ports or generics in the entity that are unassociated are treated as open; if any match fails, it is an error.

These default rules help to avoid writing specific bindings in cases where the component names are the same as the entity names. Another advantage of these default rules is that it allows for the usage of standard component names such as SN7400 and SN7402. In cases such as these, no bindings are necessary. However, a minimum configuration may sometimes be necessary. This could be of the form:

```
library TTL_LIB;
configuration TLC_CON of TLC is
  for TLC_STRUCTURE
    use TTL_LIB.all;
  end for;
end TLC_CON;
```

TLC is an entity that has some component instantiations defined within its architecture body. The use clause "use TTL_LIB.all" makes all entities in library TTL_LIB visible. Therefore, these entities will get bound to the component instances in the architecture body by virtue of the default rules.

2.46 CONVERSION FUNCTIONS

It is possible for the port types of an entity to differ from the port types of the component being bound to. For example, given the component declaration:

```
component COUNTER
  port (CTR: inout BIT_VECTOR;
        CLK, RST: in BIT;
        PAR: out BIT);
  end component;
```

the component COUNTER is bound to an entity with the declaration:

```
entity COUNTER is
  port (Q: inout STD_LOGIC_VECTOR(3 downto 0);
  CLOCK, RESET: in STD_LOGIC;
  PARITY: out STD_LOGIC);
end COUNTER;
```

This scenario occurs quite often in practice, for example, when a vendor supplies a library of entity declarations using the vendor's favorite type, while the user models a design using a different data type. In such a case, *conversion functions* can be used to convert the values from one type to another by specifying this function in the association list of the binding.

In the binding infromation of a configuration specification, when data flows from a component to an entity, that is, for an input port, the conversion function is specified with the component port name; this means that the conversion function is applied on the value of the component port before assigning it to the entity port. For an output port, the conversion function is specified with the entity port name, since data flows from entity to component, while for an inout port, the conversion function is specified on both the component port and the entity port.

Here is an example.

```
for all: COUNTER use entity WORK.COUNTER
  port map (
     TO_BITVECTOR (Q) => TO_STDLOGICVTICTOR (CTR),
     CLOCK =>TO_STDLOGIC(CLK),
     RESET =>TO_STDLOGIC(RST),
     TO_BIT(PARTY) => PAR);
```

CLOCK and RESET are input ports of the entity. Therefore, data flows from the component ports (which are CLK and RST) to the entity ports. In this case, the conversion function TO_STDLOGIC appears on the component ports. The value of the component port is the parameter to the conversion function. The value returned by the conversion function is then passed to the entity port.

For the output port PARITY, the conversion function is specified with the entity port. The entity port value is passed as a parameter to the conversion function TO_BIT, and its return value is passed on to the component port PAR.

For the output port Q, conversion functions appear on both the component ports and the entity ports, since data travels both ways. When data is passing from component to entity, the conversion function TO__STDLOGICVECTOR is called, and the return value is passed to the entity port. If data is flowing from the entity to the component, the conversion function TO_BITVECTOR is called, and its return value is assigned to the component port CTR.

Conversion functions can be used wherever there is an association of generics, ports, or parameters, such as in parameter associations in a function call or a procedure call, in port map and generic map of a block statement, or a component instantiation statement.

2.47 DIRECT INSTANTIATION

A component declaration declares the interface of a component. This component can then be instantiated using a component instantiation statement. However, before the entity that contains the component can be simulated, the component instance needs to be bound or linked to an entity architecture pair or to a configuration. This binding is specified using additional constructs, that is, by using a configuration specification or a configuration declaration.

However, it is possible to directly instantiate the entity-architecture pair or a configuration in a component instantiation statement. This saves the additional binding step necessary when using components. Here are the two additional forms of the component instantiation statement that can be used to directly instantiate an entity or a configuration.

```
component-label: entity entity-name [ (architecture-name) ]
                        [ generic map (generic-association-list) ]
                        [ port map (port-association-list) ];
component-label: configuration configuration-name
                        [ generic map (generic-association-list) ]
                        [ port map (port-association-list)];
```

Here is the description of the hierarchical 1-bit full-adder using direct instantiation. Note that no configuration declaration is necessary or possible in this case, since the component instantiations directly instantiate the appropriate entity-architecture pairs or configurations. Also, no component declarations are necessary or possible. The description is therefore much more compact.

```
entity FULL_ADDER is
  port (A, B, ON: in BIT; COUT SUM: out BIT);
end FULL_ADDER;

architecture FA_WITH_HA of FULL_ADDER is
  signal S1, C1, C2: BIT;
begin
  HA1: entity WORK.HA (HA_STR) port map (A, B, S1, C1);
  HA2: entity WORK.HA (HA_STR) port map (S1, CIN, SUM, C2);
  Ol: configuration WORK.OR2CON port map (C1, C2, COUT);
end FA_WTTH_HA;

entity HA is
  port (D0, Dl: in BIT; S, C: out BIT);
end HA;

library ECL;
architecture HA_STR of HA is
begin
  X1: entity WORK.XOR2(XOR2) port map (D0, Dl, S);
  Al: configuration ECL.AND2CON port map (D0, Dl, C);
end HA_STR;
```

Here is another example. This is the full-adder example with architecture FA_STR, described earlier in Section 2.43 Note that, again in this case, no configuration specifications and component declarations are necessary, thereby making the description very compact.

```
library SLAB, CMOS_LIB;
entity FULL_ADDER is
  port (A, B, CIN: in BIT; SUM, COUT: out BIT);
end FULL_ADDER;

architecture FA_STR of FULL_ADDER is
  signal S1, S2, S3, S4, S5: BIT;
begin
  X1: entity WORK.XOR2 (XOR2BEH) port map (A, B, S1);
  X2: entity WORK.XOR2 (XOR2BEH) port map (S1, CIN, SUM);
  Al:  entity WORK.A_GATE (A_GATE_BODY) port map (S2, A, B);
  A2: entity WORK.A_GATE (A_GATE_BODY)
       port map (S3, B, CIN);
  A3: entity HS_LIB.AND2HS (AND2STR)
       port map (S4, A, CIN);
  Ol: entity CMOS_LIB.OR2CMOS (OR2STR)
       port map (S2, S3, S5);
  02: entity CMOS_LIB.OR2CMOS (OR2STR)
       port map (S4, S5, COUT);
 end FA_STR;
```

2.48 INCREMENTAL BINDING

It is possible to delay the bindings of ports and generics and to override pre-specified generics using the incremental binding mechanism. Bindings present in a configuration specification represent the primary bindings. However, this may not contain the complete binding information. Missing information could be any of the following:

- Unassociated ports and generics
- Open ports and generics

Here is an example that contains incomplete binding information.

```
entity FULL_ADDER is
  port (A, B, CIN: in BIT; SUM, COUT: out BIT);
end FULL_ADDER;
architecture FA_STR_INCR of FULL_ADDER is
  component XOR2
    port (Dl, D2: in BIT; DZ: out BIT);
  end component;
  component AND2
    port (Z: out BIT; A0, Al: in BIT);
  end component;
  component OR2
    port (Nl, N2: in BIT; Z: out BIT);
end component;
```

```
for X1, X2: XOR2
  use entity WORK.MY_XOR2;          -- No generic map and port map
    -- specified. Since architecture name is not specified,
    -- default is the most recently analyzed architecture.
for others: AND2 use entity WORK.MY_AND2(ARCH_BODY)
  port map (HS_B => Al, HS_A => open);          -- Port HS_Z is not
    -- associated and port HS_A is open.
for all: OR2 use entity WORK.MY_OR2
    generic map (TPHL => 2 ns, TPLH => 3 ns);       -- These generic
      -- values will later be overridden by the ones specified in the
      -- configuration declaration.
begin
  ...
end FA_STR_INCR;
```

The three entities to which the above components are bound are:

```
entity MY_XOR2 is
  generic (TPHL, TPLH: TIME);
  port (XA, XB: in BIT; XZ: out BIT);
end MY_XOR2;
entity MY_AND2 is
  port (HS_A, HS_B: in BIT; HS_Z: out BIT);
end MY_AND2;

entity MY_OR2 is
  generic (TPHL, TPLH: TIME);
  port (N1, N2: in BIT; Z: out BIT);
end MY_OR2;
```

In the bindings for the XOR2 components, the port and generic maps are missing. In the AND2 component bindings, port HS_Z is not associated, and in the bindings for the OR2 component, generic values are specified. However, generic values specified in a configuration declaration can override the generic values specified in a configuration specification. All the extra and missing information can later be supplied in a configuration declaration, which may exist in a different file. Here is a configuration declaration with the missing information.

```
configuration FA_NCREMENTAL of  FULL_ADDER is
  for FA_STR_INCR
   -- First component configuration: for Xl, X2: XOR2
   port map (A0, Al, Z)
   generic map (TPHL => 2 ns, TPLH => 5 ns);
end for;
-- Second component configuration:
for all: AND2
  port map (HS_A => '1', HS_Z => Z);
end for;
-- Third component configuration:
for all: OR2
  generic map (TPHL => 4 ns, TPLH => 6 ns);
end for;
 end for;
end FA_NCREMENTAL;
```

In the configuration declaration, there is no need to specify the entity name to which the components are bound (by using **"use entity . . ."**), since this information is already provided in the configuration specifications that are present inside the architecture body. In the first component configuration, the missing port map and generic map is provided. In the second component configuration, the un-associated port and the open port are bound. In the third component configuration, the new set of generic values specified override the generic values "TPHL => 2 ns, TPLH => 3 ns" specified in the configuration specification.

REVIEW QUESTIONS

2.1. Explain the difference between concurrent and sequential statements.

2.2. Explain conditional statement in brief.

2.3. Write short note on resolution function.

2.4. Define packages and libraries in brief.

2.5. What is the difference between function and procedures ?

2.6. What is the package body and package declaration ? Explain with the example.

2.7. Explain component declaration and component instantiation in brief.

2.8. Explain the use of generics.

2.9. Differentiate between a process and wait statement. Can they be used simultaneously in a program.

2.10. In behavioural modelling, explain
(a) Architecture Body (b) Process Statement
(c) Signal Assignment Statement (d) Wait Statement
(e) Case Statement

2.11. What are EXIT and ASSETRT statements in sequential processing?

2.12. How case statement is different from loop statement?

2.13. Differentiate ARRAYS and LOOPS. Explain with the help of examples.

2.14. Write short notes on:
(a) Package
(b) Difference between If and Case statements
(c) LOOP statement
(d) Next statement

2.15. What is library clause and why it is used?

2.16. What are technology libraries? How delay effects are used in delay models?

2.17. What is the difference between sequential and concurrent statements?

2.18. Explain the significance of conditional signal assignment statements and selected signal assignment statements.

2.19. Write a short note on Functions.

2.20. How will you compare component declaration and component installation?

2.21. Write a short note on:
(a) Generics (b) Libraries
(c) Component Declaration

2.22. What are generics and why they are being used?

2.23. What is the significance of "Wait for One" statement? Explain by an example.

UNIT 3

COMBINATIONAL CIRCUIT DESIGN

3.1 MULTIPLEXER

Multiplexer mean "many into one". It is a device that allows digital information from several sources to be routed on to a single line for transmission over that line to a common destination.

Multiplexer (MUX) sometimes referred to as DATA SELECTOR is a combinational circuit that is designed to gate one out of several inputs to output. The input selected for connection to the output is controlled by a set of SELECT INPUTS. Generally in an *N*: 1 MUX. There are

- *N* input lines = 2^n where *n* are the select lines.
- 1 output line.

Example:

In a 4: 1 M U X there are 2 select lines, 4 input lines and 1 output line. The circuit diagram and functional block diagram is shown in Fig. 3.1. There are 4 inputs I_0, I_1, I_2, I_3 Which are selectively transmitted to the output Y depending on select input combinations. In this case, two selects inputs S_0, S_1 are required as 2^2= 4 The truth table shows various combinations of S_0 and S and selected output. The Boolean expression for output

$$Y = S'_1S_0'I_0 + S'_1S_0I_2 + S_1S'_0I_2 + S_1S_0I_3$$

The truth table and logic diagram is shown in Fig. 3.1.

Behavioural Model for 4:1 MUX (Using IF Statement)

```
Library ieee;
use ieee. std_ logic_1164.all;
use ieee.std_logic_arith.all;
enity MUX1 is
    port (I0, I1, I2, I3, S0, S1 : in bit;
    Y : OUT BIT);
end MUX1;
Architecture MUX_beh of MUX1 is
Begin
  Process (I0, I1, I2, I3, S0, S1)
```

S_1	S_0	Y
0	0	I_0
0	1	I_1
1	0	I_2
1	1	I_3

(a)

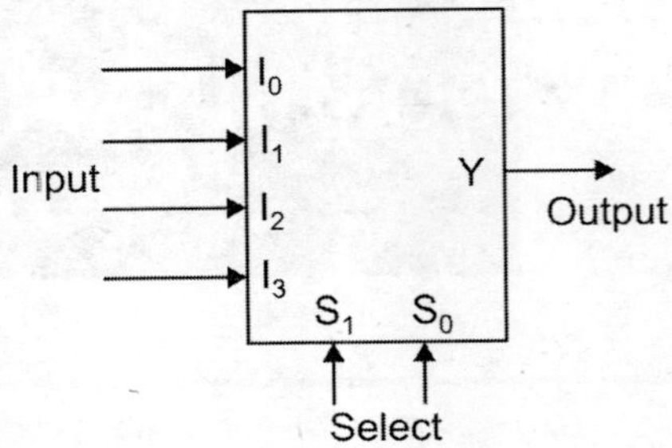

(b)

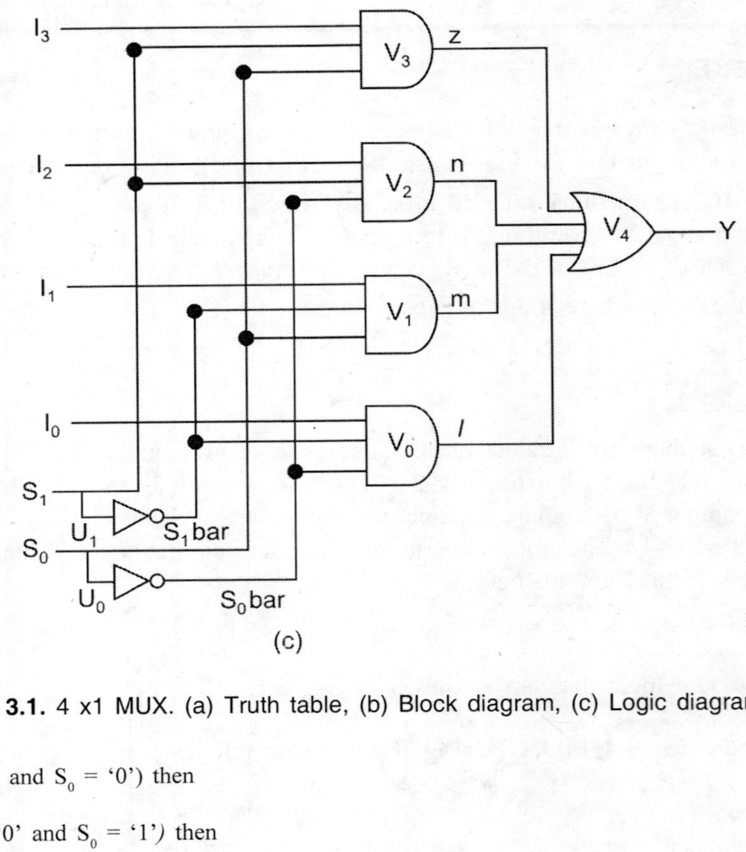

(c)

Fig. 3.1. 4 x1 MUX. (a) Truth table, (b) Block diagram, (c) Logic diagram.

```
Begin
If (S1 = '0' and S0 = '0') then
   Y<= I0;
elsif (S1 = '0' and S0 = '1') then
   Y<= I1;
elsif (S1 = '1' and S0 = '0') then
   Y<= I2;
elsif (S1 = '1' and S0 = '1') then
   Y< = I3;
end if;
   end process;
end MUX_beh;
```

Behavioural Model for 4:1 MUX (Using Case Statement)

```
Library ieee
   Use ieee.std_logic_1164.all;
   Use ieee. std_logic_arith.all;
   entity MUX1 is
     port (I0, I1, I3 :in bit
      S : in bit_vector (0 to 1);
      Y : out bit);
   end MUX1;
   Architecture MUX_beh of MUX1 is
   begin
     process (I0, I1, I2, I3, S) begin
case S is
   when "00" => Y <= I0 ;
   when "01" => Y <= I1 ;
   when "10" => Y <= I2 ;
   when "11" => Y <= I3 ;
 end case;
 end process;
end MUX_beh;
```

Data Flow Model for 4 : 1 MUX

```
Library ieee;
Use ieee.std_logic_1164.all;
Use ieee.std_logic_arith : all;
entity MUX1 is
   port (I0, I1, I2, I3, S0 , S1 : in bit;
      Y : out bit);
end MUX1;
Architecture MUX_DF of MUXl is
   begin
      Y <= I0 when S1 = '0' and S0 = '0' else
           I1 when S1 = '0' and S1 = '1' else
           I2 when S1 = '1' and S0 = '0' else
           I3 when S1 = '1' and S0 = '1';
end MUX_DF;
```

Structural Model for 4:1 MUX

```
Library ieee;
Use ieee.std_logic_1164.all;
Use ieee.std_logic_arith.all;
entity MUX1 is
   port (I 0, I1, I2, I3, S0 , S1 : in bit;
                          Y : out bit);
end MUX1;
Architecture mux_str of MUX1 is
   Component and_g
      Port (a, b, c : in bit;
```

```
        01: out bit);
end component;
component or_g
   port (a1, b1, c1, dl : in bit;
        02: out bit);
end component;
component inv_g
   port (a2 : in bit;
        03: out bit) ;
end component;
signal S0 bar, S1 bar, 1, m, n, z : bit;
   begin
   U0 : inv_g port map (S0 , S0 bar) ;
   U1 : inv_g port map (S1 , S1bar) ;
   V0 : and_g port map (I0 , S1bar, S0bar, 1);
   V1 : and_g port map (I1, S1bar, S0, m) ;
   V2 : and_g port map (I2, S1, S0bar, n) ;
   V3 : and_g port map (I3, S1, S0, z);
   V4 : or_g port map (1, m, n, z, Y) ;
end mux_str;
```

3.2 DEMULTIPLEXER

A demultiplexer is a circuit that receives infromation on signal line and transmits this infromation on one of 2^n possible output lines.

Demultiplexer (DEMUX), performs the reverse operation of a MUX, is a digital circuit having one input and many outputs. It receives information from a single line and directs it to one of the many output lines *i.e.* DEMUX takes one input data source and selectively distributes it to one of *N* output channels. Generally in an *N*: 1 MUX, there are

- 1 input line
- *N* output line = 2^n where *n* are the select lines.

Example : In a 1 : 4 DEMUX there are 2 select lines, 1 input line and 4 output lines. The circuit diagram and functional block diagram is shown in Fig. 3.2. There is 1 input I which is selectively transmitted to one of the outputs Y_0, Y_1, Y_2, Y_3 depending on select input combinations. In this case, two select inputs S_0, S_1 are required as 2 = 4. The truth table shows various combinations of S_0 and S_1 and selected output.

The truth table and logic diagram is shown in Fig.3.2

Behavioural Model for 1 : 4 DEMUX (Using IF Statement)

```
Library ieee;
Use ieee.std_logic_1164.all; Use
ieee.std_logic_arith.all; entity DEMUX1 is
   port (I, S0, S1 : in bit;
      y : out bit_vector (0 to 3));
   end DEMUXl;
Architecture DEMUX__beh of DEMUXl is
begin
   Process (I, S0, S1)
```

```
        begin
          If (S1= '0' and S0 = '0') then
            Y <= '1000';
          elsif (S1 = '0' and S0 = '1') then
            Y<= '0100';
          elsif (S1 = '1' and S0 = '0') then
            Y <= '0010';
          elsif (S1 = '1' and S0 = '1') then
            Y <= '0001';
          end if;
        end process;
      end DEMUX_beh;
```

S_1	S_0	Y_3	Y_2	Y_1	Y_0
0	0	0	0	0	1
0	1	0	0	1	0
1	0	0	1	0	0
1	1	1	0	0	0

(a)

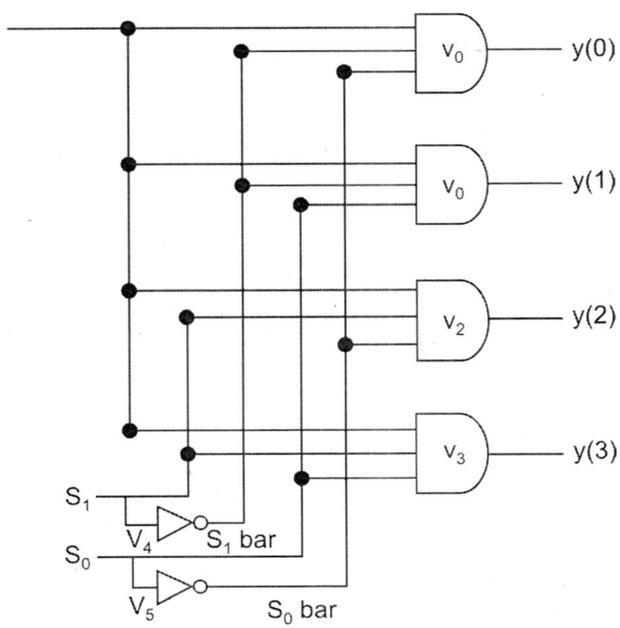

(b)

Fig. 3.2. 1 : 4 Demux (a) Truth Table, (b) Logic diagram.

Behavioural Model for 1 : 4 DEMUX (Using Case Statement)

```
library ieee;
Use ieee.std_logic_1164.all;
Use ieee.std_logic_arith.all;
entity DEMUX1 is
     port (I : in bit;
           S : in bit_vector (0 to 1);
           Y : out bit_vector (0 to 3));
end DEMUX1;
Architecture DEMUX_beh of DEMUXl is
begin
     Process (I, S)
begin
     cases is
        When "00" => Y <= '1000';
        When "01" => Y <= '0100';
        When "10" => Y <= '0010';
        When "11" => Y <= '0001';
     end case;
  end process;
end DEMUX_beh;
```

Data Flow Model for 1 : 4 DEMUX

```
library ieee;
use ieee.std_logic_1164.all;
use ieee.std_logic_arith.all;
entity DEMUX1 is
   port (I, S1, S0; in bit;
      y : out bit_vector (0 to 3));
end DEMUX1;
Architecture DEMUX_DF of DEMUX1 is
begin
  Y <= 'l000;when S1 = '0' and S0 = '0 ' else
      '0100' When S1 = '0' and S0 = '1' else
      '0010' when S1 = '1' and S0 = '0' else
      '0001' when S1 = '1' and S0 = '1';
end DEMUX_DF;
```

Strctural Model For 1 : 4 DEMUX

```
Library ieee;
Use ieee.std_logic_1164.all;
Use ieee. std_logic_arith.all;
entity DEMUX 1 is
   port (I, S1, S0 : in bit;
      y : out bit_vector (0 to 3));
end DEMUX1;
Architecture DEMUX_str of DEMUX 1 is
   Component and_g
        Port (a, b, c : in bit; 0 : out bit);
```

```
    end component;
    component inv_g
        port (d : in bit;
        Z : out bit);
    end component;
    signal S0bar, S1bar, Z : bit;
     begin
      V0 : and_g port map (I, S1bar, S0bar, y(0));
      V1 : and_g port map (I, S1bar, S0, y (l));
      V2 : and_g port map (I, S1, S0 bar, y (2));
      V3 : and_g port map (I, S1, S0, y (3));
      V4 : inv_g port map (S1, S1bar);
      V5 : inv_g port map (S0, S0bar);
end DEMUX_str;
```

3.3. ENCODER

An encoder is a digital circuit that performs the inverse operation of a decoder. An encoder has 2^n (or fever) input lines and n output lines. The output lines generate the binary code corresponding to the input value. A priority encoder is an encoder circuit that includes the priority function. The operation of the priority encoder is such that if two or more inputs are equal to flat the same time, the input having the highest priority will take precedence. in addition to the outputs, the priority encoder circuit has a third output designated by V; this is a valid bit indicator that is set to 1 when 1 or more inputs are equal to 1(If all inputs are 0, there is no valid input and $V = 0$. The other two outputs are not inspected, when $V = 0$ and are specified as Don't care condition.)

Behavioural Model For Priority Encoder

```
library ieee;
use ieee.std_logic_1164.all;
use ieee.std_logic_arith.all;
entity Pr_En is
   port (I : in bit_vector (0 to 3);
      Y : out bit_vector (1 downto 0);
      V : out bit);
end Pr_En;
Architecture Pr_Enbeh of Pr_En is
begin
   Process (I)
   begin
```

I(0)	I(1)	I(2)	I(3)	y(1)	y(0)	V
0	0	0	0	x	x	0
1	0	0	0	0	0	1
x	1	0	0	0	1	1
x	x	1	0	1	0	1
x	x	x	1	1	1	1

(a)

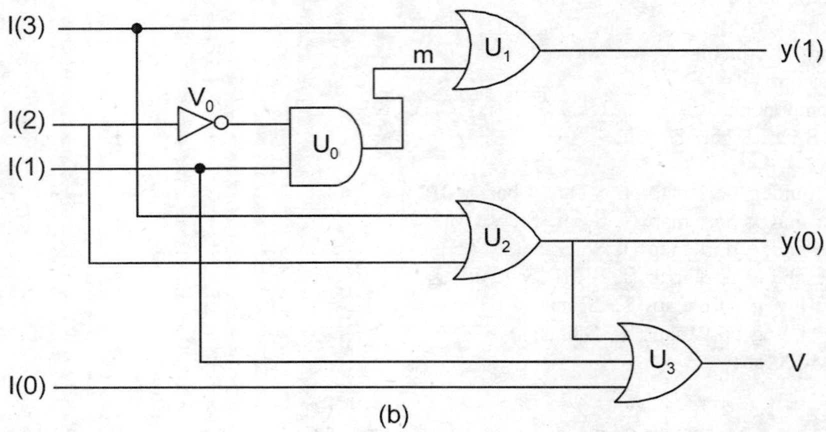

Fig. 3.3: Priority Encoder (a) Truth table, (b) Logic diagram.

```
If I = '0000' then
  V <= '0';
else
  {
    If I (3) = '1' then
      Y<= '11';
    elsif I(2) = 'l' then
      Y<= '10';
    elsif I(1) = '1' then
      Y<= '01';
    elsif I(0) = '1' then
      Y<='00';
    end if;
    }
  end if;
  end process;
end Pr_EnBeh;
```

Data Flow Model for Priority Encoder

```
library ieee;
use ieee.std_logic_1164.all;
use ieee.std_logic_arith.all;
entity Pr_En is
  port (I : in bit_vector (0 to 3);
      y : out bit_vector (1 downto 0);
      V:out bit);
end Pr_En;
Architecture Pr_En DF of Pr_En is
begin
  Y <= '11' when I(3) = '1' else
      '10' when I (2) = '1' else
      '01' when I (1) = '1' else
```

```
'00';
V <= '0' when I ='0000' else '1';
end Pr_En DF;
```

Structural Model for Priority Encoder

```
library ieee;
use ieee.std_logic_1164.all;
use ieee.std_logic_arith.all;
entity Pr_En is
  port (I0, I1, I2, I3 : in bit;
     Y : out bit_vector (1 down to 0);
     V : out bit);
end Pr_En;
Architecture Pr_Enstr of Pr_En is
  Component and_g
      port (a, b : in bit;
        01: out bit);
  end component;
  component or_g
      port (a1, b1 : in bit;
        02: out bit);
  end component;
  component orl_g
      port (a2, b2, c2 : in bit;
        02 : out bit);
  end component;
  Component inv_g
      port (d : in bit;
        03 : out bit);
  end component;
  signal I2 bar, m : bit;
  begin
    V0 : and_g portmap (I1, I2 bar, m);
    V1 : Or_g portmap (I3, m, y(1));
    V2 : Or_g portmap (I3, I2, y(0));
    V3 : Orl_g portmap (I0, I1, y(0),V) ;
    U0 : inv_g portmap (I2, I2 bar);
  end Pr_Enstr;
```

3.4. DECODERSICODE CONVERTER)

A decoder is a combination circuit that converts binary information form n input lines to a minimum of 2^n output lines. The decoders are referred to as n to m line decoders, where $m <= 2^n$ The name decoder is also used in conjunction with code ccnverter.

EXAMPLE :

In a 2 : 4 decoder, the two inputs are decoded into 4 outputs. The two inverters provide complement to the inputs. The circuit is enabled by using a enable input (En). When the value of En input is 0 none of the outputs is available i.e. the circuit operates only when the value of En, input is 1. The truth table or functional block diagram is shown in Fig. 3.4.

BEHAVIOURAL MODEL FOR 2:4 DECODER (USING IF STATEMENT)

```
library ieee;
use ieee.std_logic_1164.all;
use ieee.std_logic_arith.all;
entity Dec is
    port (I0, I1, En : in bit;
        y : out bit_vector (0 to 3));
end Dec;
  Architecture Dec_Beh of Dec is
  begin
    process (I0, I1, En)
    Variable I0bar, I1 bar : bit;
```

En	I_0	I_1	y(0)	y(1)	y(2)	y(3)
0	x	x	0	0	0	0
1	0	0	1	0	0	0
1	0	1	0	1	0	0
1	1	0	0	0	1	0
1	1	1	0	0	0	1

(a)

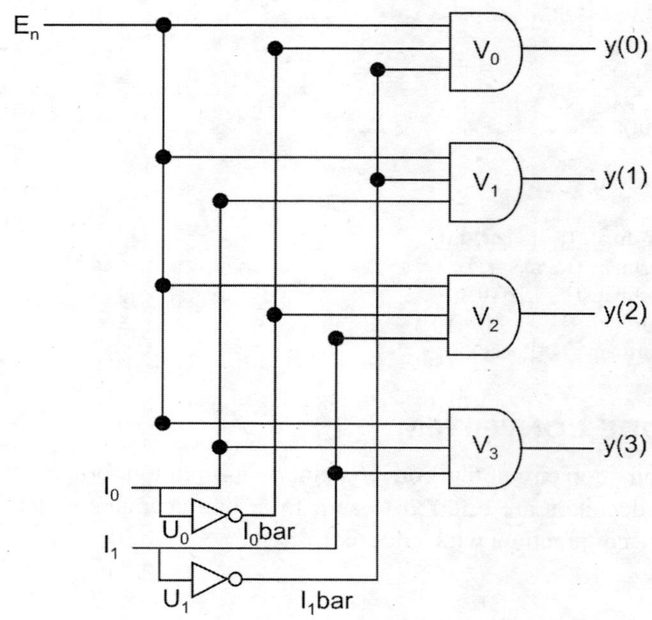

(b)

Fig. 3.4. 2 : 4 Decoder (a) Truth table, (b) Logic diagram

```
begin
  I0bar := not I0;
  I1bar := not I1;
  if (En = '1') then
  Y(0) <= I0bar and I1bar;
  Y(l) <= I0bar and I1 ;
  Y(2) <= I0 and I1bar;
  Y(3) <= I0 and X1;
else
  Y <= "0000";
end if;
end process;
end Dec_Beh;
```

Data Flow Model for 2:4 Decoder

```
library ieee;
use ieee.std_logic_1164.all;
use ieee.std_logic_arith.all;
entity Dec is
  port (I0, I1, En : in bit ;
    y : out bit_vector (0 to 3));
end Dec;
Architecture Dec_df of Dec is
    Signal I0bar, I1bar : bit;
      begin
      I0bar <= not I0 ;
      I1bar <= not I1;
  y(0) <= I0 bar and I1bar and En;
  y(l) <= I0 and I1bar and En;
  y(2) <= I0 bar and I1 and En;
  y(3) <= I0 and I1 and En;
end Dec_df;
```

Structural Model for 2:4 Decoder

```
library ieee;
Use ieee. std_logic_1164. all;
Use ieee. std_logic:_arith.all;
entity Dec is
      port (I0, I1, En : in bit ;
      y : out bit_vector (0 to 3));
end Dec;
Architecture Deic_str of Dec is
      Component and_g
      port (a, b, c; in bit;
        01 : out bit);
end component;
component i:nv_g
port (d : in bit;
      02 : out bit);
end component ;
```

```
signal I0bar: I1bar : bit;
    begin
    V0 : and_g portmap (I0bar, I1bar, En, y(0));
    V1 : and_g portmap (I0bar, I1, En, y(1));
    V2 : and_g portmap (I0, I1bar, En, y(2));
    V3 : and_g portmap (I0, I1, En, y(3));
    U0 : inv_g portmap (I0, Z0bar);
    U1 : inv_g portmap (I1, I1bar) ;
end Dec_str;
```

Example: BCD to Seven Segment Decoder or Code Converter

I_0	I_1	I_2	I_3	a	b	c	d	e	f	g
0	0	0	0	1	1	1	1	1	1	0
0	0	0	1	0	1	1	0	0	0	0
0	0	1	0	1	1	0	1	1	0	1
0	0	1	1	1	1	1	1	0	0	1
0	1	0	0	0	1	1	0	0	1	1
0	1	0	1	1	0	1	1	0	1	1
0	1	1	0	1	0	1	1	1	1	0
0	1	1	1	1	1	1	0	0	0	0
1	0	0	0	1	1	1	1	1	1	1
1	0	0	1	1	1	1	1	0	1	1

(a)

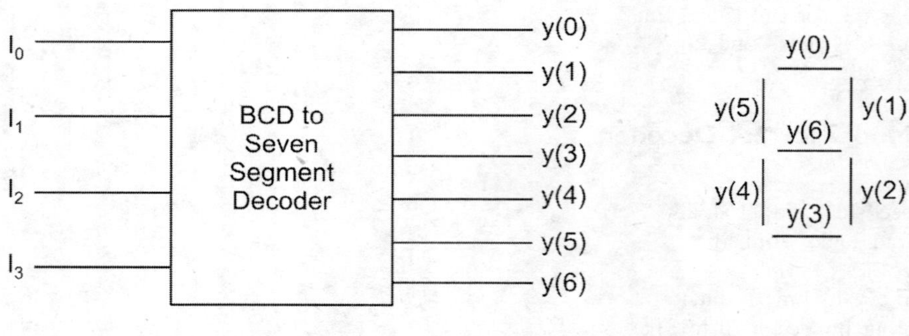

Fig. 3.5: (a) Truth Table, (b) BCD to seven segment decoder

Behavioral Model

```
Library ieee;
  Use _ ieee.std_logic_1164.all;
  Use ieee.std_logic_arith.all;
  entity BCD is
    port (J : in bit_vector (0 to 3);
```

```
        y : out bit_vector (0 to 6));
      end BCD;
      Architecture BCD_beh of BCD is
     begin
      Process (I)
      begin
        Case I is
   When "0000"  => Y < = "1111110";
   When "0001" => Y < = "0110000";
   When "0010" => Y <= "1101101";
   When "0011" => Y < = "1111001";
   When "0100" => Y < = "0110011";
   When "0101" => Y < = "1011011";
   When "0110" => Y < = "1011110";
   When "0111" => Y < = "1110000";
   When "1000" => Y <= "1111111";
   When "1001" => Y <= "1111011";
   When others  => Y < = "1111111";
      end case;
   end process;
end BCD_beh;
```

Dataflow Model

```
   Library ieee;
   Use ieee.std_logic_1164.all;
   Use ieee.std_logic_arith.all;
   entity BCD is
      port (I : in bit_vector (0 to 3);
      y : Out bit_vector (0 to 6));
   end BCD;
   Architecture BCD_DF of BCD is
      begin
y <= "1111110" when X = "0000" else
   "0110000" when I = "0001" else
   "1101101" when I = "0010" else
   "1111001" when I = "0011" else
   "0110011" when I = "0100" else
   "1011011" when I = "0101" else
   "1011110" when I = "0110" else
   "1110000" when I = "0111" else
   "1111111" when I = "1000" else
   "1111011" when I = "1001"
end BCD_df;
```

Structural Model

```
   Library ieee;
   Use ieee.std_logic_1164.all;
   Use ieee.std_logic_arith.all;
   entity BCD is
     Port (I : in bit_vector (0 to 3);
```

```
y : out bit_vector (0 to 6));
end BCD;
Architecture BCD_stn of BCD is
   Component decod_g
       Port (I : in bit_vector (0 to 3);
       01 : out bit_vector (0 to 6));
    end component;
    begin
     V0 : decod_g portmap (I, y);
    end BCD_str;
```

Example : 3:8 Decoder or Binary to Octal Converter

Behavioral Model

```
Library ieee;
Use ieee.std_logic_1164.all;
Use ieee.std_logic_arith.all;
entity Dec is
   port (I0, I1, I2, En : in bit;
      y : out bit_vector (0 to 7));
end Dec;
Architecture Dec_beh of Dec is
begin
   process (I0, I1, I2)
   Variable X0bar, I1bar, I2bar : bit;
   begin
      If (En = '1')
y(0) <= I0bar and I1bar and I2bar;
y(l) <= I0bar and I1bar and I2;
y(2) <= I0bar and I1 and I2bar;
y(3) <= I0bar and I1 and I2;
y(4) <= I0 and l1bar and I2bar;
```

E_n	I_0	I_1	I_2	y_0	y_1	y_2	y_3	y_4	y_5	y_6	y_7
0	x	x	x	0	0	0	0	0	0	0	0
1	0	0	0	1	0	0	0	0	0	0	0
1	0	0	1	0	1	0	0	0	0	0	0
1	0	1	0	0	0	1	0	0	0	0	0
1	0	1	1	0	0	0	1	0	0	0	0
1	1	0	0	0	0	0	0	1	0	0	0
1	1	0	1	0	0	0	0	0	1	0	0
1	1	1	0	0	0	0	0	0	0	1	0
1	1	1	1	0	0	0	0	0	0	0	1

(a)

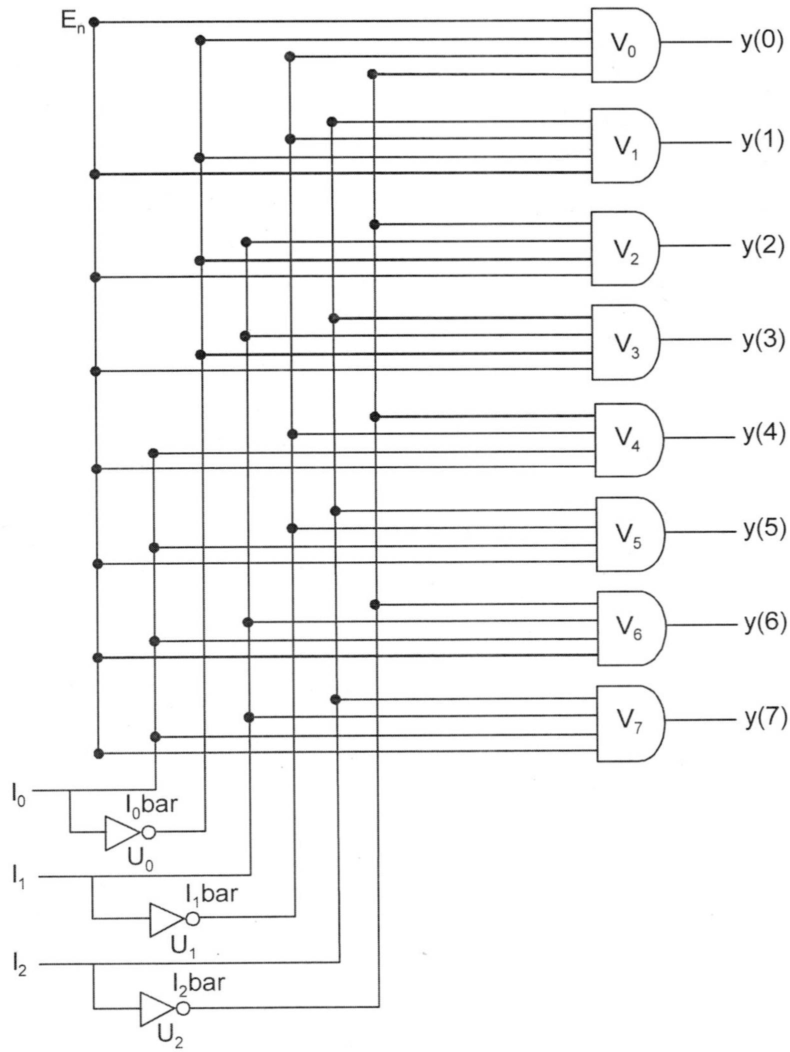

Fig. 3.6. 3:8 Decoder *(a)* Truth table, *(b)* Logic diagram.

```
    Y (5) <= I0 and I1bar and I2;
    y (6) <= I0 and I1 and I2bar;
    y (7) <= I0 and I1 and I2 ;
    else
      y <= "00000000"
    end if;
  end process;
end Dec_beh;
```

Dataflow Model

```
Library ieee;
Use ieee.std _logic__1164.all;
Use ieee.std_logic_arith.all;
entity Dec is
   port (I0, I1, I2, En : in bit;
     y : out bit_vector (0 to 7));
    end Dec;
Architecture Dec_df of Dec is
    begin
    I0bar <= not I0;
    I1bar <= not I1;
    I2bar <= not I2;
y (0) <= I0bar and I1bar and I2bar and En;
y (l) <= I0bar and I1bar and I2 and En;
y (2) <= I0bar and I1 and I2bar and En;
y (3) <= I0bar and I1 and I2 and En;
y (4) <= I0 and I1bar and I2bar and En;
y (5) <= I0 and I1bar and I2 and En;
y (6) <= I0 and I1and I2bar and En;
y (7) <= I0 and I1and I2 and En;
end Dec_df;
```

Structural Model

```
Library ieee;
Use ieee.std_logic_1164.all;
Use ieee.std_logic_arith.all;
entity Dec is
   port (I0, I1, I2, En : in bit;
    y : out bit_vector (0 to 7));
    end Dec;
Architecture Dec_str of Dec is
    Component and_g
          port (a, b, c, d : in bit;
          01 : out bit);
    end component;
    component inv_g
          port (d : in bit;
          02 : out bit);
    end component;
    signal I0 bar, I1 bar, I2 bar : bit;
    begin
      V0 : and_ g portmap (I0bar, I2bar, I2bar, En, y(0));
      V1 : and_ g portma (I0bar, I1bar, I2, En, y(l));
      V2 and_ g portmap (I0bar, I1, I2bar, En, y(2));
      V3 : and_g portmap (I0bar, I1, I2, En, y(3));
      V4 : and_ g portmap (I0, I1bar, I2bar, En, y(4));
      V5 : and_g portmap (I0, I1bar, I2, En, y(5));
      V6 : and_g portmap (I0, I1, I2bar, En, y(6));
      V7 : and_g portmap (I0, I1, I2, En, y(7));
      U0 : inv_g portmap (I0, I0bar) ;
```

```
        U1 : inv_g portmap (I1, I1,bar) ;
        U2 : inv_g portmap (I2, I2bar);
    end Dec_str;
```

3.5. COMPARATOR

The comparator is a combinational circuit that compares two numbers A and B and determines their relative magnitudes. The outcome of comparison is specified by three binary variables that indicate whether A > B, A = B or A < B.

Dataflow Model for 4 Bit Comparator

```
Library ieee;
Use ieee.std_logic_1164.all;
Use ieee.std_logic_arith.all;
entity comp is
    port (A, B : in bit_vector (0 to 3);
      AgtB, AeqB, AltB; out bit);
end comp;
Architecture comp_df of comp is
    begin
    AeqB <="l" when A = B else '0';
    AgtB < = "1" when A > B else '0';
    AltB <="1" when A < B else '0';
end comp_df
```

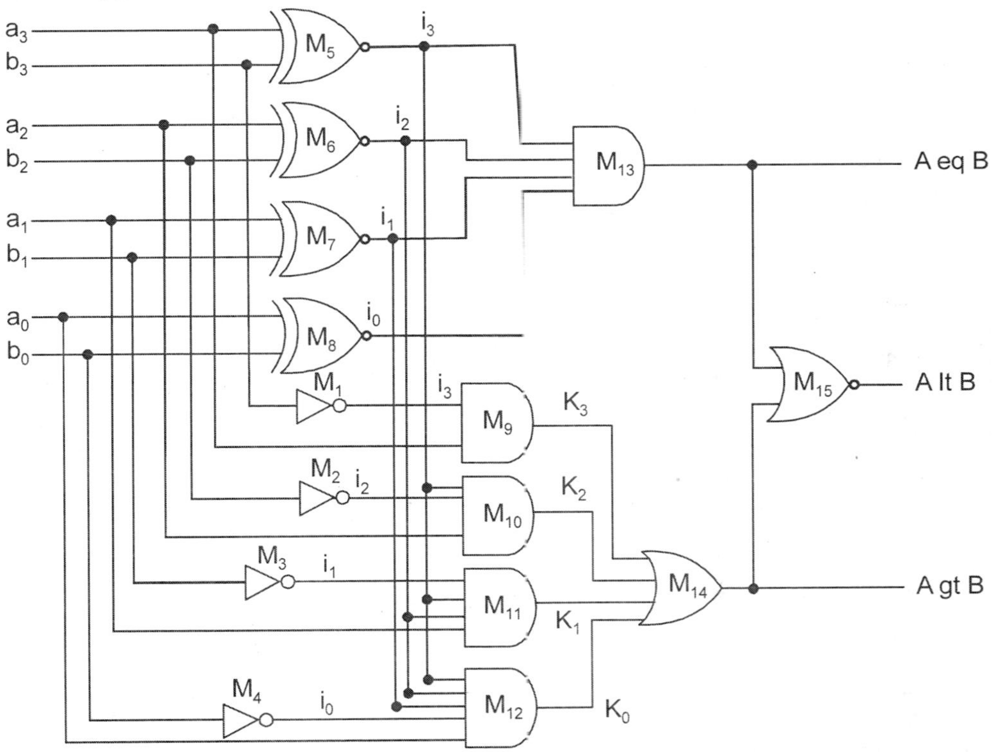

Fig. 3.7: Comparator

Behavioural Model for 4-Bit Comparator

```
library ieee;
use ieee.std_logic_1164.all;
use ieee.std_logic_arith.all;
entity comp is
    port (A, B : in vector (0 to 3); AgtB, AeqB
                                    AltB : out bit);
end comp;
Architecture eomp_beh of comp is
   begin
      process (A, B)
   begin
     if (A > B) then
     AgtB <= '1';
     AeqB <= '0';
     AltB <= '0';
     elsif (A < B) then
       AltB <= '1';
       AeqB <= '0';
       AgtB <= ' 0';
   else - - A = B
     AeqB <= '1';
     AgtB <= '0';
     AltB <= '0';
   end if;
   end process;
end comp_beh;
```

Structural Model for 4 Bit Comparator

```
Library ieee;
Use ieee.std_logic_1164.all;
Use ieee.std_logic_arith.all;
entity comp is
   port (A, B : in bit_vector (0 to 3);
     AgtB, AeqB, AltB : out bit);
   end comp;
Architecture comp_str of comp is
   Component notl
        port (X1 : in bit;
               R1 : out bit);
   end component;
   component and2
       port (X2, Y2 : in bit;
              R2 : out bit);
       end component;
       Component and3
       port (X3, Y3, Z3 : in bit;
```

```
            R3 : out bit);
        end component;
    Component and4
        port (X4, Y4, Z4, W4 : in bit;
            R4 : out bit);
        end component;
        component and5 ;
            port (X5, Y5, Z5, W5, U5 : in bit; R5 : out bit);
        end component;
        component or6
            port (X6, Y6, Z6, W6 : in bit;
              R6 : out bit);
        end component;
        component nor7
            port (X7, Y7 : in bit;
              R7 : out bit);
        end component;
        component Xnor8
            port (X8, Y8 : in bit;
              R8 : in bit);
        end component;
        signal i0, i1, i2, i3, j0, jl, j2, j3, k0, kl, k2, k3 : bit;
        begin
          Ml : not 1 portmap (b3, i3);
          M2 : not 1 partmap (b2, i2);
          M3 : not 1 portmap (b1, i1);
          M4 : not 1 portmap (b0, i0);
          M5 : Xnor 8 portmap (a3, b3, j3);
          M6 : Xnor 8 portmap (a2, b2, j2);
          M7 : Xnor 8 portmap (a1, b1, J1);
          M8 : Xnor 8 portmap (a0, b0, j0);
          M9 : and 2 portmap (a3, i3, k3);
          M10 : and 3 portmap (a2, i2, j3, k2);
          M11 : and 4 portmap (a1, i1, j2, j3, k1);
          M12 : and 5 port map (a0, i0, j1, j2, j3, k0);
          M13 : and 4 portmap (j0, j1, j2, j3, AeqB);
          M14 : Or 6 portmap (k0, k1, k2, k3, AgtB);
          M15 : nor 7 portmap (AeqB, AgtB, AltB);
    End comp_str;
```

3.6. BOOLEAN FUNCTION

The three models for the given Boolean function. $Z = (A.B + C.\ D)$

Input				*Output*
A	B	C	D	Z
0	0	0	0	0
0	0	0	1	0
0	0	1	0	0
0	0	1	1	1
0	1	0	0	0
0	1	0	1	0
0	1	1	0	0
0	1	1	1	1
1	0	0	0	0
1	0	0	1	0
1	0	1	0	0
1	0	1	1	1
1	1	0	0	1
1	1	0	1	1
1	1	1	0	1
1	1	1	1	1

(a)

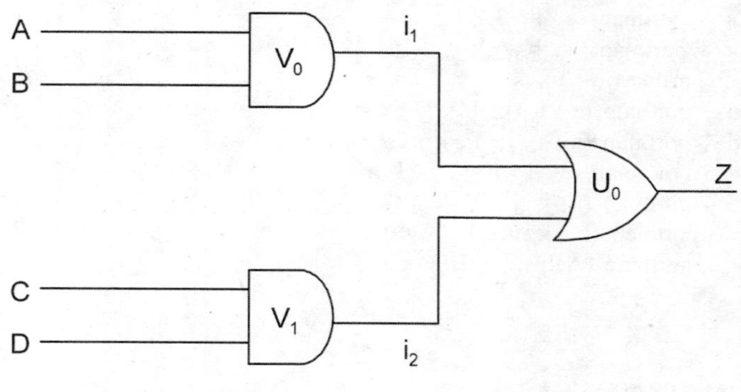

(b)

Fig. 3.8: (a) Truth table, (b) Logic diagram.

Behavioural Model for Given Boolean Function

```
Library ieee;
Use ieee.std_logic_1164.all;
```

```
Use ieee.std_logic_arith.all;
entity Bool_Func is
   port (A : in bit_vector (0 to 3);
      Z : out bit);
end Bool_Func;
Architecture Bool_Funcbeh of Bool_Func is
begin
   process (A, Z)
      begin
      Case A is
         When "0000" => Z <= 0;
         When "0001" => Z <= 0;
         When "0010" => Z <= 0;
         When "0011" => Z <= 1;
         When "0100" => Z <= 0;
         When "0101" => Z <= 0;
         When "0110" => Z <= 0;
         When "0111" => Z <= 1;
         When "1000" => Z <= 0;
         When "1001" => Z <= 0;
         When "1010" => Z <= 0;
         When "1011" => Z <= 1;
         When "1100" => Z <= 1;
         When "1101" => Z <= 1;
         When "1110" => Z <= 1;
         When "1111" => Z <= 1;
      end case;
   end process;
end Bool_Funcbeh;
```

Dataflow Model for Given Boolean Function

```
Library ieee;
Use ieee.std_logic_1164.all;
Use ieee.std_logic_arith_all;
entity Bool_Func is
     port (A : in bit_vector (0 to 3);
         Z : out bit);
     end Bool_Func;
Architecture Bool_Funcdf of Bool_Func is
begin
   Z <= 0 when A = '0000' else
      0 when A = '0001' else
      0 when A = '0010' else
      1 when A = '0011' else
      0 when A - '0100' else
      0 when A = '0101' else
      0 when A = '0110' else
      1 when A = '0111' else
      0 when A = '1000' else
      0 when A = '1001' else
      0 when A = '1010' else
```

```
        1 when A = '1011' else
        1 when A - '1100'  else
        1 when A = '1101' else
        1 when A = '1110' else
        1 when A = '1111' else
end Bool_FuncdF;
```

Structural Model for Given Boolean Function

```
library ieee;
Use ieee.std_logic_1164.all;
Use ieee.std_logic_arith_all;
entity Bool_Func is
    port (A, B, C, D : in bit;
      Z : out bit;
end Bool_Func;
Architecture Bool_Funcstr of Bool_Func is
Component and_g
     port (x, y : in bit;
       01 : out bit);
     end component;
     component or_g
         port (x1, y1 : in bit;
         02 : out bit);
     end component;
     signal i1, i2 : bit;
       begin
       V0 : and_g portmap (A, B, i1);
       V1 : and_g portmap (C. D, i2);
       U0 : or_g portmap (i1, i2, z);
   end Bool_Funcstr;
```

Example:

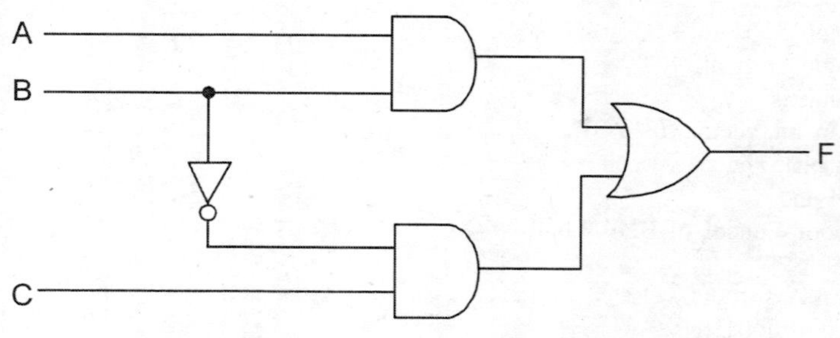

Fig. 3.9: VHDL code for simple logic function.

```
entity SLF is
   port (A, B, C : in bit;
          F : out bit);
   end SLF;
   architecture SLF1 of SLF is
begin
   F <= (A and B) OR (NOT B AND C);
end SLF1;
```

PROBLEMES

3.1. Write VHDL code for 3-bit binary to gray code converter in structural modeling.

3.2. Write VHDL code for full adder using two half adder in behavioural modeling.

3.3. Write VHDL code for 3 bit binary to excess-3 code converter in dataflow modeling.

3.4. Write VHDL code for boolean function F = AB + CDE in dataflow modeling.

3.5. Write VHDL code, for boolean function F = (AB' + A'B) (C + D')E in structural modeling.

3.6. Using process statement implement four input MUX symbol and its functions.

3.7. Design a 2:4 decoder ciruit. Give its entity declaration and structuralmodel and behavioural model. Also draw the waveform giving relationbetween its inputs and outputs.

3.8. What is the difference between

(i) Decoder and Demultiplexer

(ii) Encoder and Multiplexer.

3.9. Draw a complete structural model for a 9-bit parity generator circuit in structural modelling.

3.10. Write down truth table, working behavioural/data flow architecture for the following.

(a) Code converter

(b) 4-bit comparator.

Also draw the circuit and output waveform.

UNIT 4

SEQUENTIAL CIRCUIT DESIGN

4.1 SHIFT REGISTER

A register is a group of flip-flops. Each flip-flop is capable of storing 1 bit of information. It is a sequential circuit. A register capable of shifting its binary information in one or both directions is called a shift register. The logical configuration of a shift register consists of a chain of flip-flops in cascade with the output of one flip-flop connected to the input of the next flip-flop.

4.2 BIT REGISTER

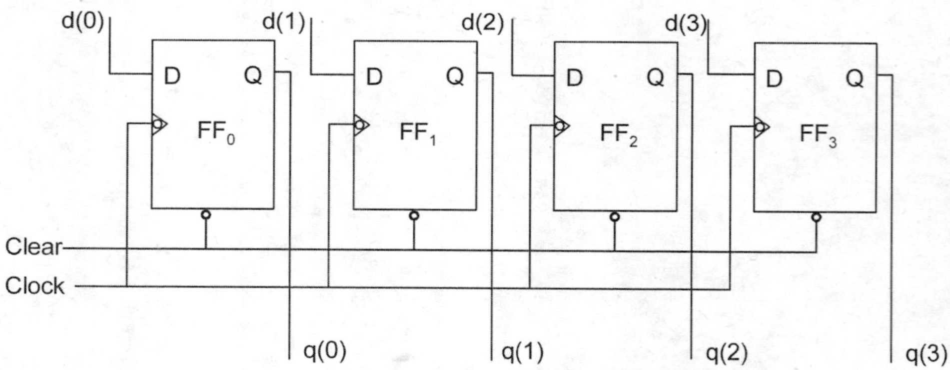

Fig. 4.1: 4-bit register.

Behavioural Model

```
Library ieee;
Use ieee. std_logic_1164 .all;
Use ieee. std logic_arith.all;
entity reg is
   port (D : in bit_vector (0 to 3);
      clock : in bit;
      clear : in bit;
      Q : out bit_vector (0 to 3));
end reg;
```

```
architecture reg_beh of reg is
   signal Q_tmp : bit_vector (0 to 3);
begin
if clear = '0' then
   Q_tmp <= "0000";
elsif (clock = '1' and clock' event) then
   Q_tmp <= D;
end if;
end process;
   Q < = Q_tmp;
end reg_beh;
```

Data Flow Model

```
Library ieee;
Use ieee.std_logic_1164.all;
Use ieee.std_logic_arith.all;
entity reg is
   port (D : in bit_vector (0 to 3);
   clock : in bit;
   clear : in bit;
   Q : out bit_vector (0 to 3);
end reg;
architecture reg_df of reg is
   signal Q_tmp : bit_vector (0 to 3);
   begin
    Q_tmp <= "0000"
    when clear = '0'
    else
       Q_tmp <= D when clock = '1'
     and clock'event;
   Q <= Q_tmp;
end reg_df;
```

Structural Model

```
Library ieee;
Use ieee.std_logic_1164.all;
Use ieee.std_logic_arith.all;
entity reg is
port (D : in bit_vector (0 to 3);
   clock : in bit;
   clear : in bit;
   Q : out bit_vector (0 to 3));
end reg;
architecture reg_str of reg is
   component dff
       port (a, b, c : in bit; 0 : out bit);
   end component;
begin
   FIF0 : d_ff portmap (d(0), clock, clear, q(0));
   FF1 : d_ff portmap (d(l), clock, clear, q(l));
```

```
    FF2 : d_ff portmap (d(2), clock, clear, q(2));
    FF3 : d_ff portmap (d(3), clock, clear, q(3));
end reg_str;
```

4.3 4-BIT SERIAL IN AND SERIAL OUT RIGHT SHIFT REGISTER

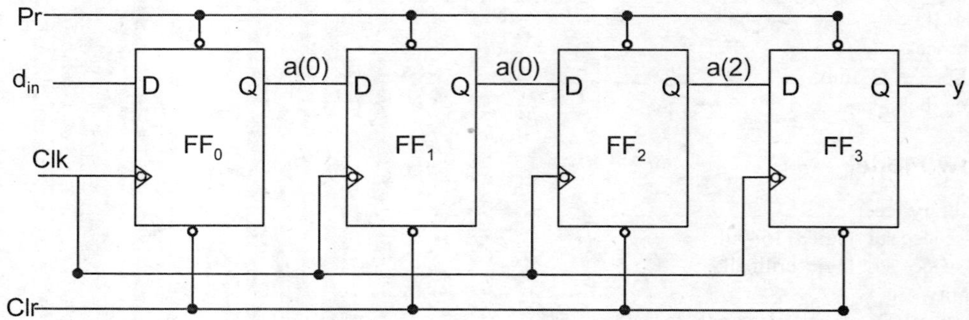

Fig. 4.2: Serial in serial out right shift register

Behavioural Model

```
Library ieee;
Use ieee.std_logic,,1164.all;
Use ieee. std__logic_arith.all;
entity SSft_reg is
   port (din, clk, clr, pr : in bit;
      y : out bit);
   end SSft_reg;
   Architecture SSft_regbeh of SSft_reg is
      signal a(0), a(l), a(2) : bit;
   begin
    process (clk, clr, pr)
    begin
       if (clr = '1' and pr = 'l' and clk = '0' and clk'event) then
            a (0) <= din;
            a (l) <= a (0);
            a (2) <= a (l);
            y <= a (2);
         elsif (clr = '1' and pr = '0') then
            a(0) <= '1';
            a(l) <= '1';
            a(2) <= '1';
               y <= 'l';
         elsif (clr = '0' and pr = '1') then
            a(0) <= '0';
            a(l) <= '0';
            a(2) <= '0';
```

```
            y <= '0';
        end if;
      end process;
    end SSft_ regbeh;
```

Dataflow Model

```
Library ieee;
Use ieee.std_logic_1164.all;
Use ieee.std_logic_arith.all;
entity SSft_reg is
   port (din, clk, clr, pr : in bit;
         y : out bit);
end SSft_reg;
Architecture SSft_regdf of SSft_reg is
   signal a : bit_vector (0 to 3);
begin
  a : <= din & a (0) & a (1) & a (2) when
  clr = '1' and pr = '1' and clk = '0' and
clk'event
   else
  '1111' when clr = ' 1' and pr = '0' else
  '0000' when clr = '0' and pr = '1' else
 end SSft_reg df;
```

Structural Model

```
Library ieee;
Use ieee.std_logic_1164.all;
Use ieee.std_logic_arith.all;
entity SSft_reg is
   port (din; clk, crl, pr : in bit;
        y : out bit);
end ssft_regstr of ssft_reg is
Architecture SSft_regstr of ssft_reg is
   Component d_ff
        port (a, b, c, d : in bit;
           01 : out bit);
        end component;
   signal a : bit_vector ( 0 to 2 ) ;
begin
  FF0 : d_ff portmap (din, clk, clr, pr, a(0));
  FF1 : d_ff portmap (a(0), clk, clr, pr, a(l));
  FF2 : d_ff portmap (a(l), clk, clr, pr, a(2));
  FF3 : d_ff portmap (a(2), clk, clr, pr, y);
end SSft_regstr;
```

4.4 BIT SERIAL IN AND PARALLEL OUT SHIFT REGISTER

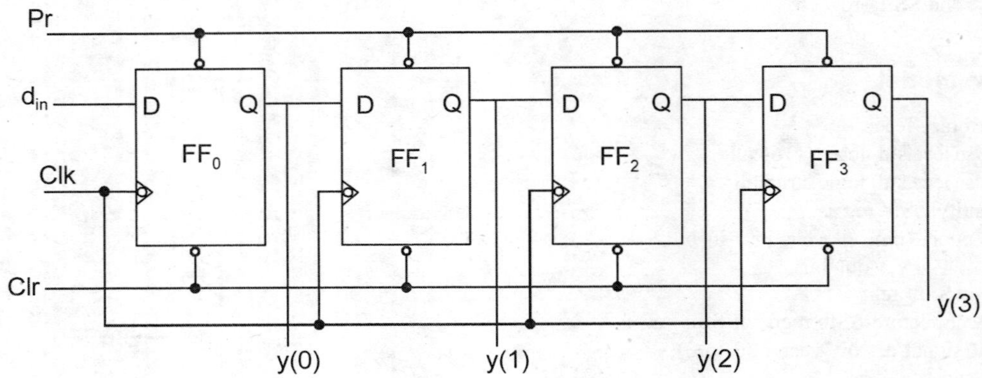

Fig. 4.3: Serial in parallel out shift register.

Behavioural Model

```
Library ieee;
Use ieee.std_logic_1164.all;
Use ieee.std_logic_arith.all;
entity Spsft_reg is
   port (din, clk, clr, pr : in bit;
      y : out bit_vector (0 to 3));
   end Spsft_reg;
   architecture Spsft_regbeh of Spsft_reg is
      begin
         process (clk, clr, pr)
         begin
if (clr = '1' and pr = '1' and clk = '0' and clk' event)
   then
     y (0) < = din;
     y (1) < = y (0);
     y (2) < = y (2);
elsif (clr = '1' and pr = '0') then
     y (0) <= '1';
     y (1) < = '1';
     y (2) < = '1';
     y (3) < = '1';
elsif (clr = '0' and pr = '1')
   then
   y (0) < = '0';
   y (1) < = '0';
   y (2) < = '0';
```

```
        y (3) < = '0';
    end if;
  end process;
  end psft_regbeh;
```

Dataflow Model

```
Library ieee;
Use ieee.std_logic_1164.all;
Use ieee.std_logic_arith.all;
entity Spsft_reg is
   port(din, clk, clr, pr : in bit;
      y : out bit_vector (0 to 3));
   end Spsft_reg;
   Architecture Spsft_regdf of Spsft_reg is
begin
   y <= din & y(0) & y(1) & y(2)
   when clr = '1' and pr = '1' and clk = '0'
      and clk'event
else
   '1111', when clr = '1' and pr = '0' else
   '0000' when clr = '0' and pr = '1' else
end Spsft_regdf;
```

Structural Model

```
Library ieee;
Use ieee.std_logic_1164.all;
Use ieee.std_logic_arith.all;
entity Spsft_reg is
   port (din,clk, clr, pr : in bit;
      y : out bit_vector (0 to 3));
   end Spsft_reg;
   Architecture Spsft_regstr of Spsft_reg is
      Component d_ff
         Port (a, b, c, d : in bit;
           01 : out bit);
   end component;
begin
   FF0 : d_ff portmap (din, clk, clr, pr, y(0));
   FF1 : d_ff portmap (din(0), clk, clr, pr, y(0));
   FF2 : d_ff portmap (din(1), clk, clr, pr, y(1));
   FF3 : d_ff portmap (din(2), clk, clr, pr, y(2));
end Spsft_regstr;
```

4.5 4-BIT PARALLEL IN AND PARALLEL OUT SHIFT REGISTER

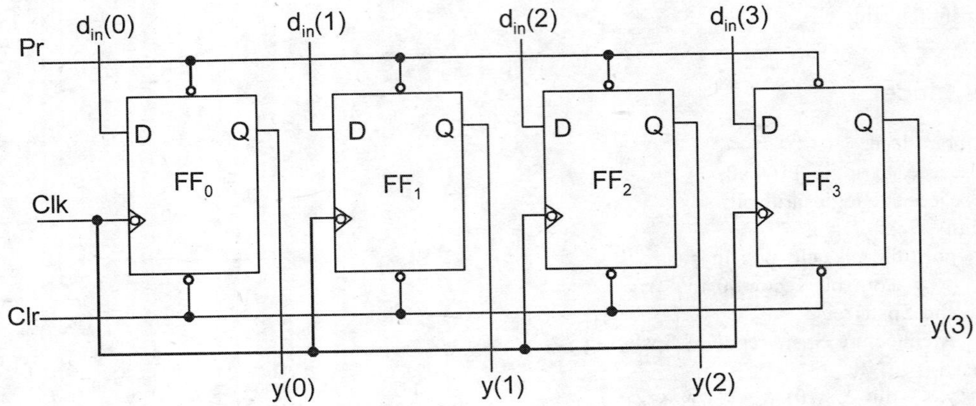

Fig. 4.4: Parallel in parallel out shift register.

Behavioural Model

```
Library ieee;
Use ieee.std_logic_1164.all;
Use ieee.std_logic_arith.all;
entity psft_reg is
  port (din : in bit_vector (0 to 3);
     clk, clr, pr : in bit;
        y: out bit_vector (0 to 3));
end psft_reg;
Architecture psft_regbeh of Psft_reg is
begin
   process (clk, clr, pr)
    begin
       if (clr = '1' and pr = '1' and clk = '0'
          and clk'event) then
          y <= din;
      elsif (clr = '1' and pr = '0') then
        y <= '1111';
      elsif (clr = '0' and pr = '1') then
        y <= '0000';
   end if;
 end process;
end psft_regbeh;
```

Dataflow Model

```
Library ieee;
Use ieee.std_logic_1164.all;
Use ieee.std_logic_arith.all;
entity psft_reg is
```

```
    port (din: in bit_vector (0 to 3);
       clk, clr, pr : in bit;
       y : out bit_vector (0 to 3));
     end psft_reg;
Architecture psft_regdf of psft_reg is
   begin
       y <= din when clr = 'l' and pr = '1' and
          clk = "0' and clk'event else
     'llll' when clr = '1' and pr = '0' else
     '0000' when clr = '0' and pr = '1' else
    end psft_regdf;
```

Structural Model

```
Library ieee;
Use ieee.std_logic_1164.all;
Use ieee.std_logic_arith.all;
entity psft_reg is
   port (din : in bit_vector (0 to 3);
      clk, clr, pr : in bit;
        y : out bit_vector (0 to 3));
    end psft_reg;
  Architecture psft_regstr of psft_reg is
        Component d_ff
           port (a, b, c, d : in bit; 01 : out bit);
       end component;
    begin
       FF0: d_ff portmap (din(0), Clk, clr, pr, y(0));
       FF1: d_ff portmap (din(l), clk, clr, pr, y(l));
       FF2: d_ff portmap (din(2), clk, clr, pr, y(2));
       FF3: d_ff portmap (din(3), clk, clr, pr, y(3));
end psft_regstr;
```

4.6 4-BIT UP COUNTER (DECADE COUNTER)

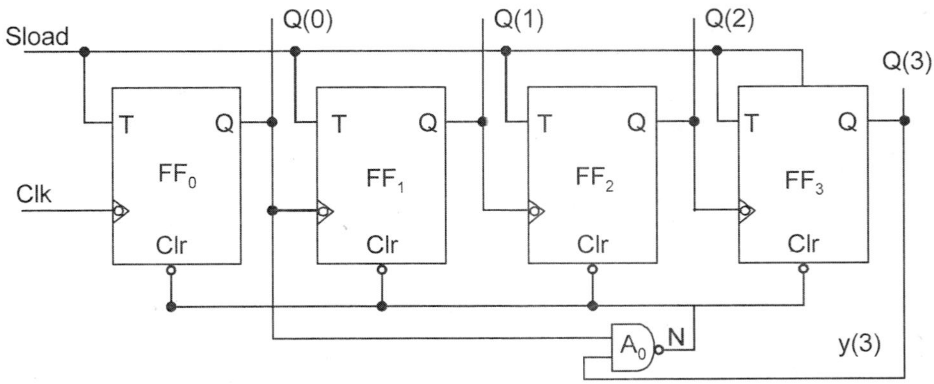

Fig. 4.5: Decade Counter.

Behavioural Model

```
Library ieee;
Use ieee. std_logic_1164.all;
Use ieee.std_logic_arith.all;
entity upcount is
   port (clk, Sload, clr : in bit;
          Q : out bit_vector (3 downto 0));
    end upcount;
    architecture upcount_beh of upcount is
    signal tmp : bit_vector (3 downto 0);
      begin
        process (clk)
        begin
    if (clk' event and clk = '1') then
        if clr = 1 then
          tmp = "0000";
    elseif Sload = 1 then
      if tmp = "1010" then
        tmp <= "0000";
    else tmp <= tmp + 1;
      endif;
      endif;
      endif;
    end process;
    Q <= tmp;
    end upcount_beh;
```

Dataflow Model

```
Library ieee;
Use "ieee.std_logic_1164.all;
Use ieee.std_logic_arith.all;
entity up count is
  port (clk, Solad, clr : in bit;
    Q : out bit_vector (3 downto 0));
  end upcount;
      architecture upcount_df of upcount is
        signal tmp : bit_vector (3 downto 0);
begin
  tmp <= '1010' when Sload = '1' and clk = '1'
      and clk'event else
  tmp <= tmp + 1;
Q <= tmp;
end upcount_df;
```

Structural Model

```
Library ieee;
Use ieee.std_logic_1164.all;
Use ieee.std_logic_arith.all;
entity upcount is
```

```
port (clk, Sload, clr : in bit;
     Q : out bit_vector (0 to 3));
 end upcount;
 architecture upcount_str of upcount is
   Component t_ff
        port (a, b, c : in bit;
           01  :out bit);
   end component;
   Component nand_gate
        port (al, bl : in bit;
           02  :out bit);
   end component;
   signal N : bit;
     begin
     A0 : nand_gate portmap (Q(l), Q(3), N);
     FF0 : t_ff portmap (clk, Sload, N, Q(0));
     FF1 : t_ff portmap (Q(0), Sload, N, Q(l));
     FF2 : t_ff portmap (Q(l), Sload, N, Q(2));
     FF3 : t_ff portmap (Q(2), Sload, N, Q(3));

 end upcount_str;
```

4.7 BIT DOWN COUNTER

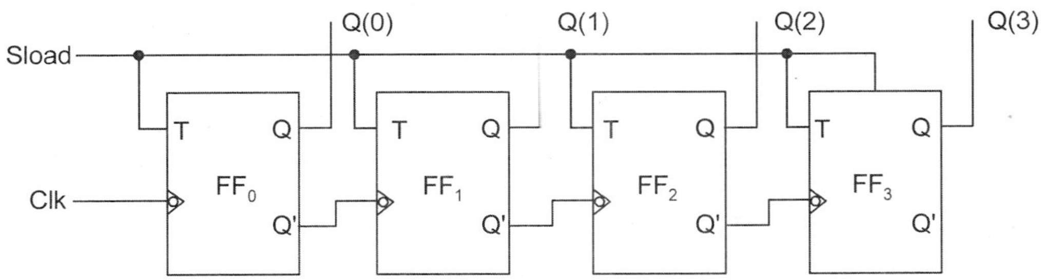

Fig. 4.6: Down counter.

Behavioural Model

```
Library ieee;
Use ieee.std_logic_1164.all;
Use ieee.std_logic_arith.all;
entity dwncount is
   port (clk, Sload : in bit;
      Q : out bit_vector (3 downto 0));
end dwncount;
architecture dwncount_beh of dwncount is
```

```
signal tmp : bit_vector (3 downto 0);
begin
process (clk)
  begin
  if (clk'event and Clk = '1') then
  if (Sload = '0') then
    tmp <= "1111";
  else
    tmp <= tmp-1;
  end if;
end if;
end process;
Q <= tmp;
end dwncount_beh;
```

Dataflow Model

```
Library ieee;
Use ieee.std_logic_1164.all;
Use ieee.std_logic_arith.all;
Entity dwncount is
   port (clk, Sload : in bit;
      Q : out bit_vector (3 downto 0))
   end dwncount;
   architecture dwncount_df of dwncount is
     signal tmp : bit_vector (3 downto 0);
      begin
      tmp <= 'llll' when Sload = '0' and clk =
'1' and clk'event else
            tmp <= tmp - 1;
      Q <= tmp;
      end dwncount_df;
```

Structura Model

```
Library ieee;
Use ieee.std_logic_1164.all;
Use ieee.std_logic_1164.all;
entity dwncount is
   port (clk, Sload : in bit;
      Q: out bit_vector (3 downto 0)
   end dwncount;
   architecture dwncount_str of dwncount is
      component t_ff
         port (a, b : in bit;
            01 : out bit);
      end component;
      signal Q(0)bar, Q(l)bar, Q(2)bar; bit
```

```
        begin
            FFO : t_ff portmap (Sload, clk, Q(0));
        FF1 : t_ff portmap (Sload, Q(0)bar, Q(1));
        FF2 : t_ff portmap (Sload, Q(l)bar, Q(2));
        FF3 : t_ff portmap (Sload, Q(2)bar, Q(3));
end dwncount_str;
```

4.8 BIT UP/DOWN COUNTER

Behavioural Model

```
Library ieee;
Use ieee.std_logic_1164.all;
Use ieee.std_logic_arith.all;
entity UDCount is
    port (clk, clr, Up_Dwn, Sload; in bit;
            Q : out bit_vector (3 downto 0)
        end UDCount;
        Architecture UDCount_beh of UD Count is
            signal tmp : bit_vector (3 downto 0);
            begin
              process (clk, clr, sload)
              begin
              if (Sload = '1') then
              if (clr = '1') then
              tmp <= "0000";
              elsif (clk'event and Clk = '1') then
                if (Up_Dwn = 'l') then
                tmp <= tmp + 1;
              else
```

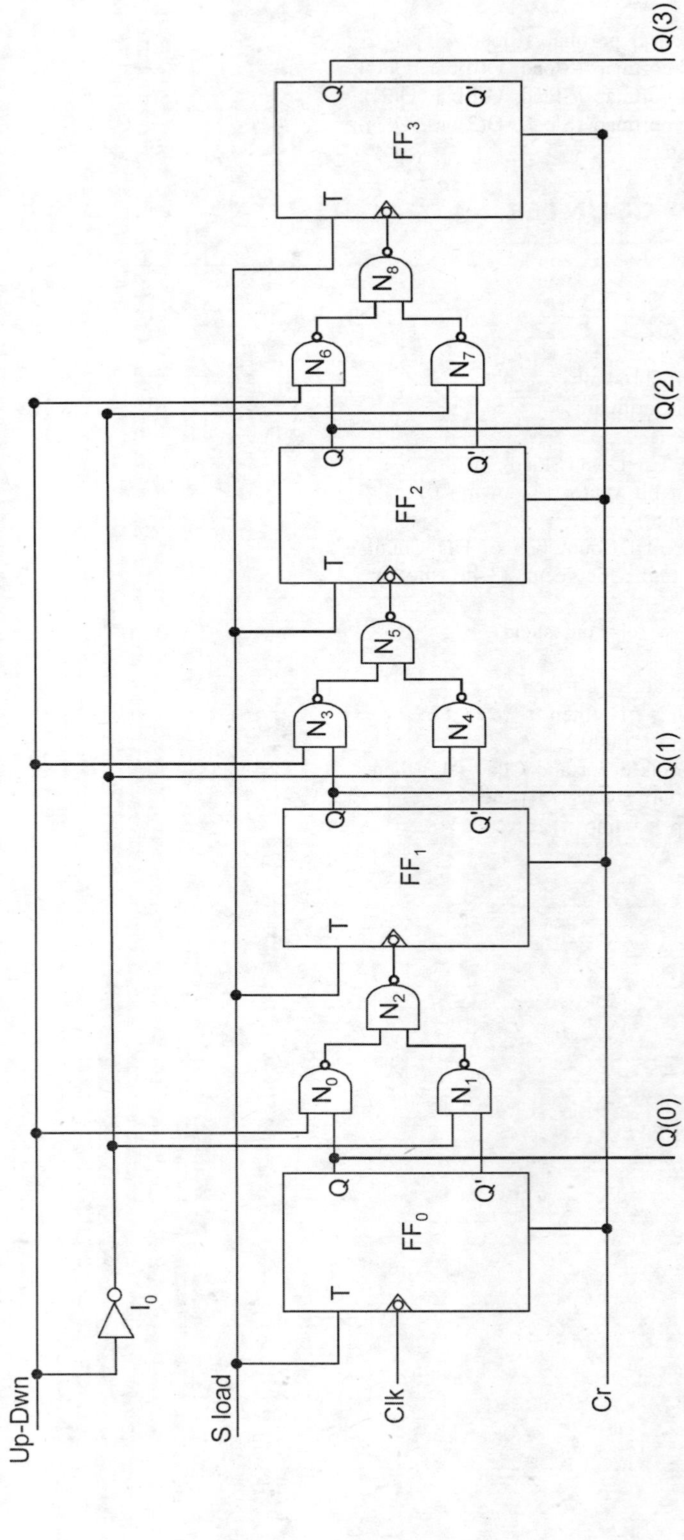

Fig: 4.7: Up/Down Counter.

```
            tmp <= tmp - 1;
          end if;
       end if;
       end if;
       end process;
    Q <= tmp;
    end UDCount_beh;
```

Dataflow Model

```
    Library ieee;
    Use ieee.std_logic_1164.all;
    Use ieee.std_logic_arith.all;
    entity UDCount is
       port (clk, clr, Up_Dwn, Sload : in bit,
          Q : out bit_vector (3 downto 0));
        end UD Count;
    architecture UDcount_beh of UDcount is
       signal tmp : bit_vector (3 downto 0);
         begin
            tmp <= '0000' when Sload = '1' and
                   clr = '1' else
            tmp <= tmp + 1 when Sload = '1' and
                 clk = '1' and clk'event and
                 up_dwn = '1' else
            tmp <= tmp - 1;
        Q <= tmp;
      end UDcount_df;
```

Structural Model

```
    Library ieee;
    Use ieee.std__logic_1164.all;
    Use ieee.std_logic_arith.all;
    entity UDcount is
       port (clk, clr,Sload, Up_Dwn : in bit;
          Q : out bit_vector (3 downto 0));
    end UDcount;
    architecture UDcount_str of UDcount is
       component t_ff
            port (A, B, C : in bit; X, Y : out bit);
        end component;
        component nandl
            port ( Al, Bl : in bit;
            X1 : out bit);
        end component;
        component in2
            port (A2 : in bit;
              X2 : out bit);
        end component;
```

```
        signal Up_Dwnbar, Q(0)bar, Q(l)bar, Q(2)bar,
S1, Q(3)bar, S2, Si, S4, S5, S6, S7, S8, S9 : bit
    begin
        I0 : inv2 portmap (Up_Dwn, Up_Dwnbar);
        FFO : d_ff, portmap (Sload, clk, clr, Q(0),
Q(0) bar);
        FF1 : t_ff portmap (Sload, S3, clr, Q(l), Q(l)bar);
        FF2 : t_ff portmap (Sload, S6, clr, Q(2), Q(2)bar);
        FF3 : t_ff portmap (Sload, S9, clr, Q(3), Q(3)bar);
        NO : nandl portmap (Up_Dwn, Q(0), S1);
        Nl : nandl portmap (Up_Dwnbar, Q(0)bar, S2);
        N2 : nandl portmap (S1, S2, S3);
        N3 : nandl portmap (Up_Dwn, Q(l), S4);
        N4 : nandl portmap (Up_Dwnbar, Q(l)bar, S5);
        N5 : nandl portmap (S4, S5, S6);
        N6 : nandl portmap (Up_Dwn, Q(2), S7);
        N7 : nandl portmap (Up_Dwnbar, Q(2)bar, S8);
        N8 : nandl portmap (S7, S8, S9);
    end UDCount_str;
```

4.9 DIFFERENCE BETWEEN COMBINATIONAL AND SEQUENTIAL CIRCUIT

There are many applications in which digital outputs are required to be generated in accordance with the sequence in which the input signals are received. This requirement cannot be satisfied using a combinational logic system. These application require outputs to be generated that are not only dependent on the present input conditions but they also depend upon the past history of these inputs. The past history is provided by feedback from the output back to the input.

Fig.4.8.shows the block diagram of sequential circuit. As shown in the Fig. 9.1, memory elements are connected to the combinational circuit as a feedback path.

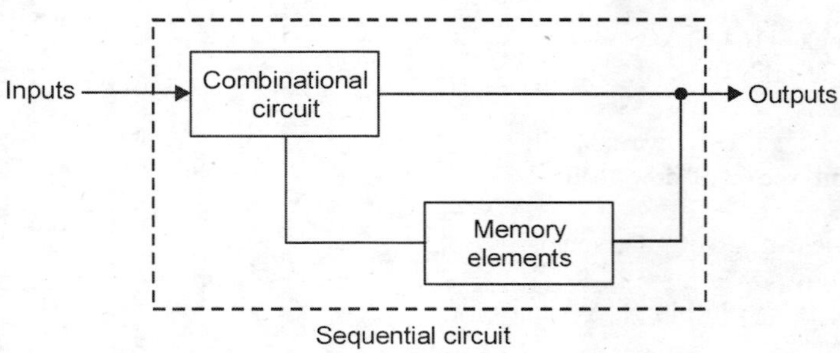

Fig. 4.8: Block diagram of sequential circuit

The information stored in the memory elements at any given time defines the present state of the sequential circuit. The present state and the external inputs determine the outputs and the next state of the sequential circuit. Thus we can specify the sequential circuit by a time sequence of external inputs, internal states (present states and next states), and outputs.

4.1(a) Table: Comparison between combinational and sequential circuits.

Sr.No.	Combinational circuits	Sequential circuits
1.	In combinational circuits, the output variables are at all times dependent on the combination of input variables.	In sequential circuits, the output variables dependent not only on the present input variables but they also depend up on the **past** history of input variables in the sequential circuit
2.	Memory unit is not required in combinational circuits	Memory unit is required to store the past history of input variables in the sequential circuit
3.	Combinational circuits are faster in speed because the delay between input and output is due to propagation delay of gates.	Sequential circuits are slower than the combinational circuits.
4.	Combinational circuits are easy to design.	Sequential circuits are comparatively harder to design.
5.	Parallel adder is a combinational circuit	Serial adder is a sequential circuit.
6.	More expensive circuit	Cheaper circuit.

The sequential circuits can be classified depending on the timing of their signals: **Synchronous sequential circuits and Asynchronous sequential circuits:** In synchronous sequential circuits, signals can affect the memory elements only at discrete instants of time. In asynchronous sequential circuits change in input signals can affect memory element at any instant of time. The memory elements used in both circuits are flip-flops which are capable of storing 1-bit binary information.

Table 4.1(b): Comparison between Synchronous and as asynchronous sequential circuits.

Sr.No.	Synchronous sequential circuits	Asynchronous sequential circuits
1.	In synchronous circuits, memory elements are clocked flip flops	In asynchronous circuits, memory elements are either unlocked flip-flops or time delay elements
2.	In synchronous circuits, the change in input signals can affect memory element upon activation of clock signal	In asynchronous circuits change in input signals can affect memory element at any instant of time.
3.	The maximum operating speed of clock depends on time delays involved.	Because of absence of clock, asynchronous circuits can operate faster than synchronous circuits.
4.	Easier to design.	More difficult to design.
5.	A synchronous sequential circuit is a system whose behaviour is defined from the knowledge of its signals at discrete constants of times i.e. the input signals can affect the memory elements upon the activation of clock	The behaviour of asynchronous sequential circuit depends upon the order in which the inputs change and the state of the circuits can be affected at any constant of line i.e, the change in input signals can affect the memory element at any instant of time.

4.10 LATCHES AND FLIP-FLOPS

Latches and flip-flops both are bistable elements. These are the basic building blocks of most sequential circuits. The main difference between latches and flip-flops is in the method used for changing their state. We use the name flip-flop for a sequential device that normally samples its inputs and changes its outputs only at times determined by clocking signal. On the other hand, we use the name latch for a sequential device that checks all of its inputs continuously and changes its outputs accordingly at any time independent of a clocking signal. Many times enable signal is provided with the latch. When enable signal is active output changes occur as input changes. But when enable signal is not activated input changes do not affect output.

4.11 RING COUNTER

Fig 4.9 shows the logic digrams for ten-bit ring counter. As shown in the Fig.4.9, the Q output of each stage is connected to the D input of the next stage and 4.9, the Q output of each stage is feed back to the input of first stage. The $\overline{CLR}$ followed by $\overline{PRE}$ makes the stage of fist stage to '1' and remaninig outputs are Zero, i.e. QA is one and Q_B $Q_{C'}$ $Q_{D'}$ $Q_{E'}$ $Q_{F'}$ $Q_{G'}$ $Q_{H'}$ $Q_{I'}$ Q_J are zero.

The first clock pulse produces $Q_B = 1$ and remaning outputs are Zero. According to the clock pulse s applied at the clock input C, a sequence of ten staes is produced. These states are summarized in Table 9.10.

Table 4.2: Ring-counter sequence (10-bits)

Clock Pulse	Q_A	Q_B	Q_C	Q_D	Q_E	Q_F	Q_G	Q_H	Q_I	Q_J
0	0	0	0	0	0	0	0	0	0	0
1	1	1	0	0	0	0	0	0	0	0
2	0	0	1	0	0	0	0	0	0	0
3	0	0	0	1	0	0	0	0	0	0
4	0	0	0	0	1	0	0	0	0	0
5	0	0	0	0	0	1	0	0	0	0
6	0	0	0	0	0	0	1	0	0	0
7	0	0	0	0	0	0	0	1	0	0
8	0	0	0	0	0	0	0	0	1	0
9	0	0	0	0	0	0	0	0	0	1

Fig. 4.10 gives the timming sequence for a ten-bit ring counter.

The ring counter can be used for counting the number of pluses. The number of pluses counted is read by noting which flip-flip is in state 1. No decoding circuirtry is requrired. since there is one pluse at the output for each of the N clock pluses, this. circuit is also referred to as a divied-by-N- counter or an N:L 1 scalar. Ring counters can be instructed for any desired *MOD* number, that is *MOD N* ring counter requires Nflip-flops.

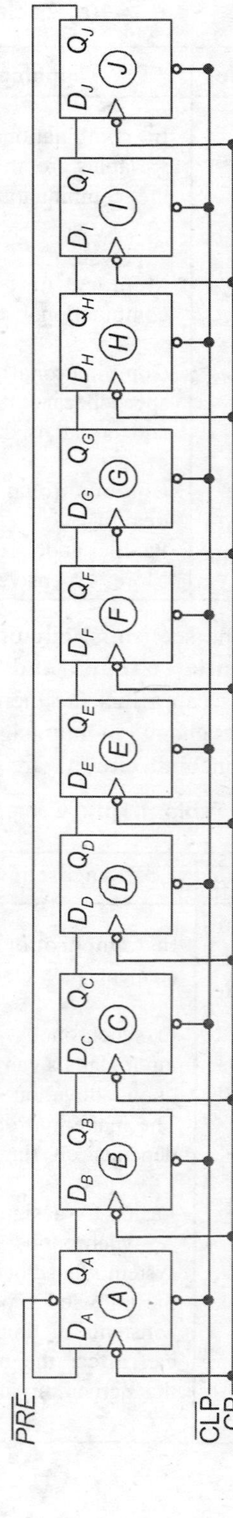

Fig.4.9: A ten bit ring counter

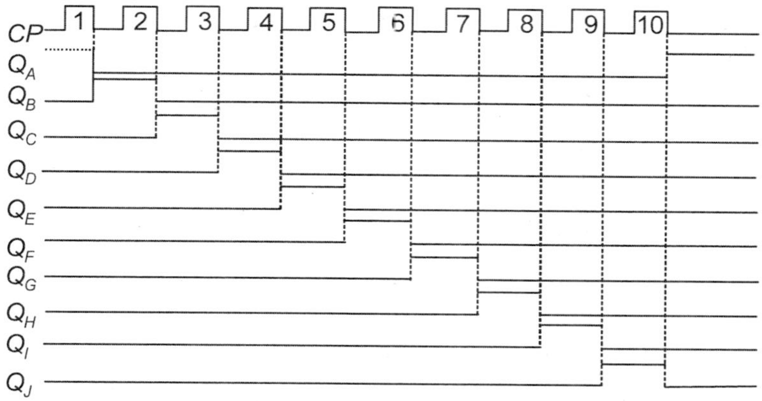

Fig. 4.10: Timing sequence for a ten-bit ring counter

4.12 THE JOHNSON COUNTER OR TWISTED RING COUNTER

In a Johnson counter, the Q output of each stage of flip-flop is connected to the D input of the next stage. The single exception is that the complement output of the last flip-flop is connected back to the D-input of the first flip-flop as shown in Fig. 4.12.

[Note: Johnson counter can be implemented with *SR* or *JK* flip-flops as well.]

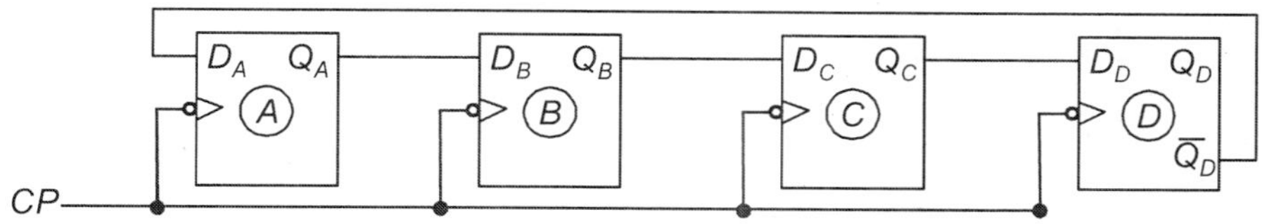

Fig. 4.12: Four-bit Johnson counter

As shown in Fig. 4.12 there is a feedback from the rightmost flip-flop complement output to the leftmost flip-flop input. This arrangement produces a unique sequence of states.

Initially, the register (all flip-flops) is cleared. So all the outputs, Q_A, Q_B, Q_c, Q_D are zero. The output of last stage, Q_D is zero. Therefore complement output of last stage, Q_D is one. This is connected back to the D input of first stage. So D_A is one. The first falling clock edge produces $Q_A = 1$ and $Q_B = 0$, $Q_c = 0$, $Q_D = 0$ since D_B, D_c, D_D are zero. The next clock pulse produces $Q_A = 1$, $Q_B = 1$, $Q_c = 0$, $Q_D = 0$. The sequence of states is summarized in Table 4.3. After 8 states the same sequence is repeated.

Table 4.3. Four-bit Johnson sequence

Clock Pulse	Q_A	Q_B	Q_c	Q_D
0	0	0	0	0
1	1	0	0	0
2	1	1	0	0
3	1	1	1	0
4	1	1	1	1
5	0	1	1	1
6	0	0	1	1
7	0	0	0	1

In this case, four bit register is used. So the four bit sequence has a total of eight states. Fig.4.13 gives the timing sequence for a four-bit Johnson counter.

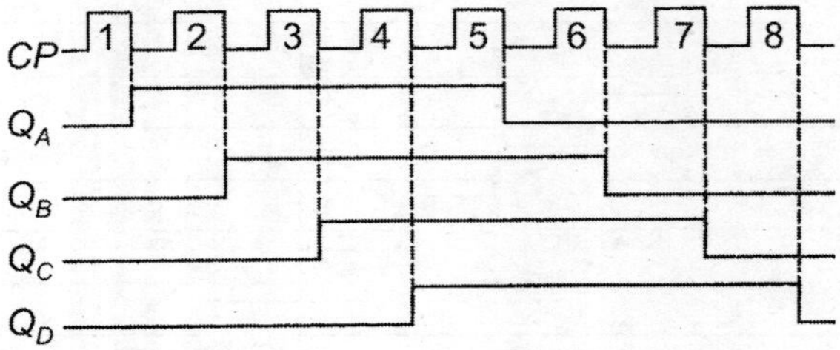

Fig. 4.13: Timing sequence for a four-bit Johnson counter.

If we design a counter of five bit sequence, it has a total of ten states, as shown in Table-4.4 below:

Table 4.4: Five-bit Johnson sequence

Clock Pulse	Q_A	Q_B	Q_C	Q_D	Q_E
0	0	0	0	0	0
1	1	0	0	0	0
2	1	1	0	0	0
3	1	1	1	0	0
4	1	1	1	1	0
5	1	1	1	1	1
6	0	1	1	1	1
7	0	0	1	1	1
8	0	0	0	1	1
9	0	0	0	0	1

So in general we can say that, an *n*-stage Johnson counter will produce a modulus of 2 x *n* where *n* is the number of stages (i.e. flip-flops) in the counter. As shown in tables, the counter will 'fill up' with 1s from left to right and then it will 'fill up' with 0s again. One advantage of this type of sequence is that it is readily decoded with two input AND gates. Table 4.5 Count sequence 4.5 gives the count sequence and required decoding.

Table 4.5: Count sequence and required decoding

Clock Pulse	Q_A	Q_B	Q_C	Q_D	AND Gate required for output
0	0	0	0	0	$\overline{Q}_A \overline{Q}_D$
1	1	0	0	0	$\overline{Q}_A \overline{Q}_B$
2	1	1	0	0	$\overline{Q}_B \overline{Q}_C$
3	1	1	1	0	$\overline{Q}_C \overline{Q}_D$
4	1	1	1	1	$\overline{Q}_A \overline{Q}_D$
5	0	1	1	1	$\overline{Q}_A \overline{Q}_B$
6	0	0	1	1	$\overline{Q}_B \overline{Q}_C$
7	0	0	0	1	$\overline{Q}_C \overline{Q}_D$

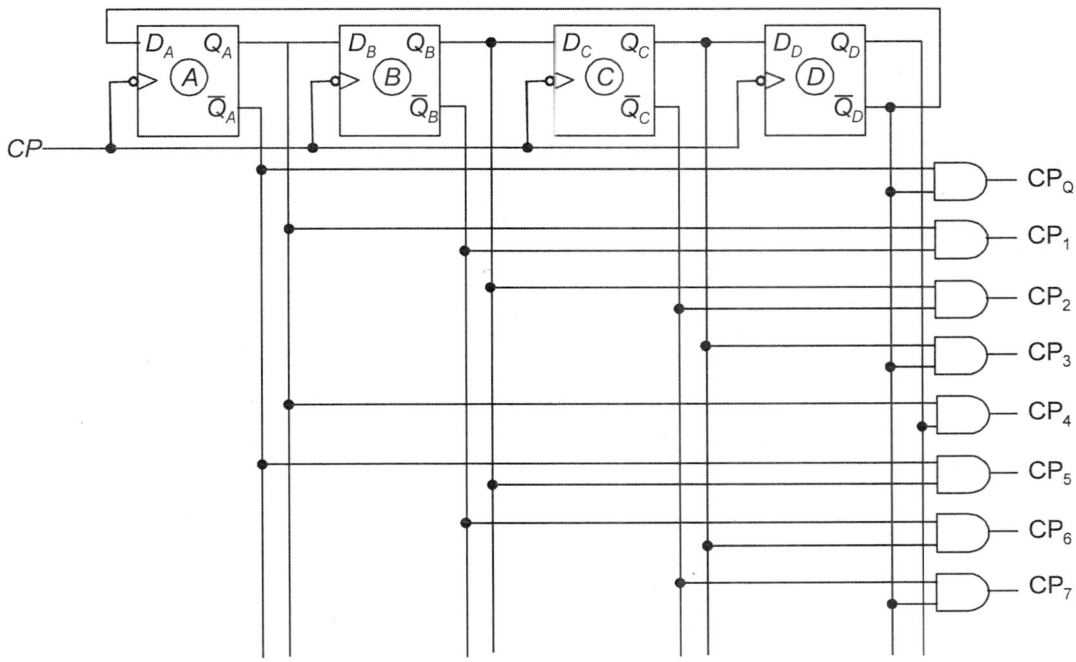

Fig. 4.14: Johnson counter with decoder

PROBLEMS

4.1. Explain the following in brief
(i) Register
(ii) Shift register
(iii) Counters
(iv) Ripple counters

4.2. Write VHDL code for modulo-3 counter in behavioural modelling.

4.3. Write VHDL code for modulo-14 counter in structural modelling.

4.4. Write VHDL code for modulo-5 counter in dataflow modelling.

4.5. Write VHDL code for 2 bit down counter.

4.6. Write down the truth table and VHDL code for the following. Draw the circuit and output waveform of
(a) Binary counter
(b) 8-bit register with asynchronous.

4.7. Write down truth-table, VHDL code for the following. Also draw the circuit and output waveform
(a) n-bit register with parallel load
(b) 4-bit up/down counter.

4.8. Design and implement a counter using VHDL, which counts upto 9 and down counts again form 9 to 0. 4.9. Model and simulate 4-bit bidirectional shift register. Assume the counter is :

(i) SIPO

(ii) PISO mode.

4.10. Draw and explain look ahead carry circuit.

4.11. Write a short note on simulation of Johnasan counter.

4.12. Differentiate between combinational and sequential circuit.

4.13. Differentiate between Latch and flip flop.

UNIT 5

SPECIFICATION AND IMPLEMENTATION OF A MICROCOMPUTER

5.1 MICROCOMPUTER—AN INTRODUTION

A microcomputer is a **general-purpose system;** that is, the system is organized in sucl a way that it can perform a large variety of computations. To achieve this objective, each computation is represented by a sequence of instructions (a **program)** that is stored in a read-write memory and is **executed** by the microcomputer. The execution of the instructions produces transformations on the data. In this way, changing the computation consists just of loading a different program in the memory.

The main advantage of a microcomputer with respect to the single-computation systems described in previous chapters is its generality. However, for a specific computation, such a generality results in more silicon area in an implementation (more modules) than the single-computation counterpart, and in a degradation of the speed of execution of the computation.

The implementation of a computation using a microcomputer is usually called a **software solution,** because it consists of writing a program and executing it in a computer. In contrast, the implementation of a computation using a dedicated digital system is called a **hardware solution.**

Modern microcomputers are very complex systems, so it is out of the scope of this text to present a comprehensive discussion of their specification, implementation, programming, and use. This chapter presents the basic concepts, and shows that the methods discussed in previous chapters are applicable to digital computers. Thus, this chapter should provide a suitable starting point for further detailed studies.

Instead of presenting the concepts in a general fashion, we develop them around the description of a simple microcomputer system. We begin with the specification of the microcomputer system, and the description of how computations are represented for execution by this system. This level of description is called the **architecture** of the system. Then, we present an implementation.

5.1.1 Basic Components of a Microcomputer

A microcomputer system consists of three subsystems, as illustrated in Figure 5.1:

- a **processor;**
- a **memory subsystem;** and
- an **input/output (I/O) subsystem.**

The **processor** is the computing engine of the microcomputer: it executes the sequence of instructions, that is, the program. Because it is the fundamental component of a computer system, a processor is also known as the **central processing unit** (CPU).

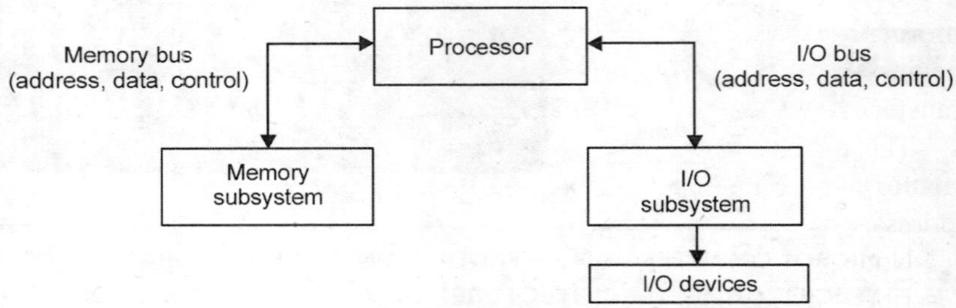

Figure 5.1: Microcomputer system.

The **memory subsystem** stores the program as well as the data used by the program. Some computers have separate memory subsystems for the program and for the data, although the prevalent approach used nowadays is one subsystem that is shared for both functions. The memory is divided into **locations,** such as bytes and words; these locations are identified by their **address.**

A memory operation is used to read or write one memory location. In a read operation, the processor provides the address and receives the data read from the location addressed. On the other hand, in a write operation the processor provides both the address and the data to be stored at the corresponding location.

Modern computer systems implement **virtual memory,** in which a very large memory is available to the programmer (2^{32} or 2^{64} locations, each of one byte); such a memory is implemented as a hierarchy of levels ranging from fast semiconductor devices up to disk storage (see Figure 5.2). Because large semiconductor memories have access time equivalent to several processor cycles, using a single level of semiconductor memory for instruction fetch or for load/store operations would degrade the performance of the system (the processor would have to spend cycles waiting for the memory accesses to complete); To reduce this degradation, the semiconductor memory is implemented as two levels of the hierarchy:

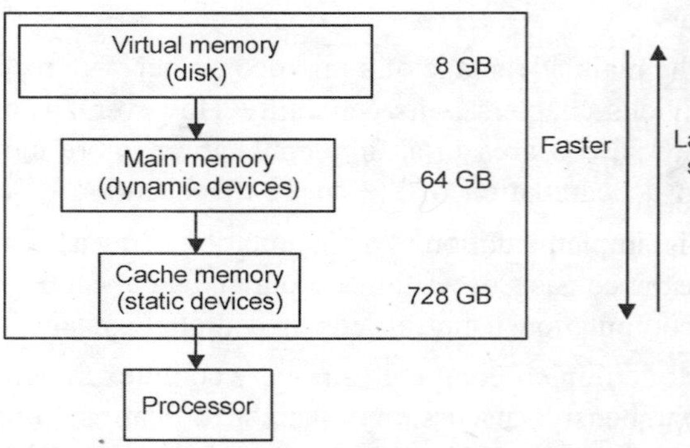

Figure 5.2: Memory hierarchy in a computer.

- a small (fast) memory (implemented by static RAM), called a **cache memory,** which contains the portions of a program (data and instructions) being used at a given point in time; and
- a large (slower) memory (implemented by dynamic RAM, which is slower but less expensive per bit), called the **main memory.**

With this structure, the processor finds instructions and data in the cache most of the time, resulting in an overall semiconductor memory subsystem whose size is determined by the dynamic devices but whose average access time approaches that of the static devices (the cache memory). In practice, contemporary systems may have several levels of cache memory.

The **I/O subsystem** contains the interfaces between the microcomputer system and external devices, such as keyboards, monitors, printers, and modems; these interfaces allow the exchange of information between the microcomputer and the external world. In a similar manner as done for memory locations, each specific interface—called an **I/O port**—is identified by an address and the transfer of data is done in units such as bytes or words.

The connections of the processor with the memory and with the I/O subsystem are called the **memory bus** and the **I/O bus,** respectively. These buses consist of sub-buses corresponding to the address, the data, and the control parts.

Note that each of the three subsystems is in itself a sequential system, so that each one can be described and implemented using the techniques for sequential systems studied in previous chapters. Because of the complexity of the systems, register-transfer level descriptions are most suitable.

In terms of these components, the basic operation of a microcomputer consists of the repetition of the following **instruction loop:**

1. The processor reads an instruction from the memory (instruction fetch), by providing an **instruction address.**
2. The processor
   a. performs the operations specified by the instruction, and
   b. determines the location in memory of the next instruction.

The operations might

- ❖ operate on data that the processor already has internally;
- ❖ bring data from the memory or the I/O subsystem; or
- ❖ Store data in the memory or transfer it to the I/O subsystem.

As stated earlier, the detailed description of the structure of a microcomputer system, including virtual memory, cache memory and I/O devices, is beyond the scope of this text. For simplicity, we do not consider virtual memory in the rest of this chapter; furthermore, we assume that the memory of the microcomputer system described next is composed entirely of static memory whose access time is shorter than one processor cycle. In addition, we do not described that I/O device attached to the microcomputer neither the signals connecting to/from those devices.

5.2 SPECIFICATION (ARCHITECTURE) OF A SIMPLE MICROCOMPUTER SYSTEM

We now illustrate the specification of a computer system by presenting a simple microcomputer which we call XMC(eXample Microcomputer). Figure 5.3 shows its components the processor, memory and input/output subsystems__and the connections bitween them. the memory bus and I/O bus are composed of the following signals:

- ❖ an **address bus,** Which identifies the memory location or I/O port benig accessed. The memory address bus 24 lines (to address up **224** locations) whereas the I/O address bus has 11 lines (to address **211** ports).
- ❖ **control signals,** Which indicate the type of access being perfomed (read or write), the length of the operand (byte or word), and an enabling signal;
- ❖ a **data bus,** which tranfers the data to/from the processor and the memorhy or I/O subsystem. These data busess have 32 lines for 32-bit data items.

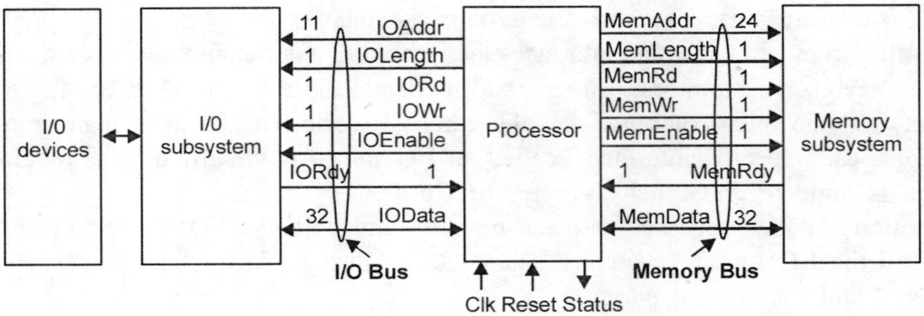

Figure 5.3: Structure of XMC.

The processor also has an output signal called status, which is used to indicate the activity within the instruction cycle (this signal is useful to follow the activity on the processor cycle by cycle).Because we do not describe the I/P devices, the only exteral inputs to XMC are the clock and reset signals.

The corresponding μ VHDL structure description is aas follows, in which we use the package ieee.std_logic_1164 as well as package comp_pkg defined below.

```
LIBRAY ieee;
USE iee.std_logic_1164all;

PACKAGE comp_pkg IS
  SUBTYPE WordT     IS STD_LOGIC_VECTOR (31 DOWNT0 0);
  SUBTYPE MAddrT    IS STD_LOGIC_VECTOR (23 DOWNT0 0);
  SUBTYPE I0 AddrT  IS STD_LOGIC_VECTOR (10 DOWNT0 0);
  SUBTYPE ByteT     IS STD_LOGIC_VECTOR ( 7 DOWNT0 0);
  TYPE     Statust  IS (undef, p_reset, fetch, execute, memop, ioop);
  FUNCTION get_carry(RA_Data,RB_Data,Imm,Opcode: STD_LOGIC_VECTOR)
           RETURN STD_LOGIC;
  FUNCTION get_ovf (RA_Data,RB_Data,Imm,Opcode: STD_LOGIC_VECTOR)
           RETURN STD_LOGIC;
  FUNCTION get_cc (RA_Data,RB_Data,Opcode: STD_LOGIC_VECTOR)
           RETURN STD_LOGIC_VECTOR;
END comp_pkg;
PACKAGE BODY comp_pkg IS
   FUNCTION get_carry(RA_Data,RB_Data,Imm,Opcode: STD_LOGIC_VECTOR)
           RETURN STD_LOGIC
   IS
     VARIABLE cy: STD_LOGIC:= '0';
   BEGIN
      -- description of carry generation included here
      RETURN(cy);
   END get_carry;

   FUNCTION get_ovf (RA_Data,RB_Data,Imm,Opcode: STD_LOGIC_VECTOR)
             RETURN STD_L0CIC
   IS
```

```
        VARIABLE ovf: STD_L0GIC:= '0';
    BEGIN
        --description of overflow generation included here
        RETURN(ovf);
    END get_ovf;

    FUNCTION get_cc (RA_Data,RB_Data,Opcode: STD_LOGIC_VECTOR)
                RETURN STD_LOGIC_VECTOR
    IS
          VARIABLE cc: STD_L0GIC_VECT0R(3 DOWNTO 0):= "0000";
    BEGIN
      --description of cc generation included here
      RETURN(cc);
   END get_cc;
END comp_pkg;


LIBRARY ieee;
USE ieee.std_logic_1164.ALL;
USE WORK.ALL, WORK.comp_pkg.ALL;

ENTITY Computer IS
    PORT (Reset, Clk : IN STD_LOGIC);
END Computer;

ARCHITECTURE structural OF Computer IS
    SIGNAL MemAddr             : MAddrT    ; -- memory address bus
    SIGNAL MemLength, MemRd    : STD_LOCIC; -- memory control signals
    SIGNAL MemWr, MemEnable    : STD_LOGIC;
    SIGNAL MemRdy              : STD_LOGIC; -- memory status signal
    SIGNAL MemData             : WordT     ; -- memory data bus
    SIGNAL IOAddr              : IOAddrT   ; -- I/O address bus
    SIGNAL IOLength, I0Rd      : STD_LOGIC; -- I/O control signals
    SIGNAL IOWr, IOEnable      : STD_LOGIC;
    SIGNAL lORdy               : STD_LOGIC; -- I/O status signal
    SIGNAL IOData              : WordT     ; -- I/O data bus
    SIGNAL Status              : StatusT;
BEGIN
    Ul: ENTITY  Memory
        PORT MAP (MemAddr, MemLength, MemRd, MemWr, MemEnabl
                     MemRdy, MemData);
    U2:ENTITY 10
        PORT MAP (IOAddr, IOLength, IORd, IOWr, IOEnable,
                     IORdy, IOData);
    U3:ENTITY Processor
        PORT MAP (MemAddr, MemData, MemLength, MemRd, MemWr,
                      MemEnable, MemRdy,
                      IOAddr, IOData, IOLength, IORd, IOWr,
                      IOEnable, IORdy,
                      Status, Reset, Clk);
END structural;
```

5.2.1 Memory Subsystem

The memory addresses produced by the processor have 24 bits. Consequently, the memory subsystem (see Figure 5.4*a*) is an array of 2^{24} locations, each of one byte, for a total of 16 MB; this is typical for a computer workstation (circa 1997). Memory addresses range from 0 to $2^{24}-1$.

Memory can be accessed as bytes or as 32-bit words, organized as indicated in Figure 5.4*b*. Successive byte addresses differ by 1, with successive addresses allocated from least-significant to most-significant positions within a word. The address of a word is the address of the least-significant byte within the word.

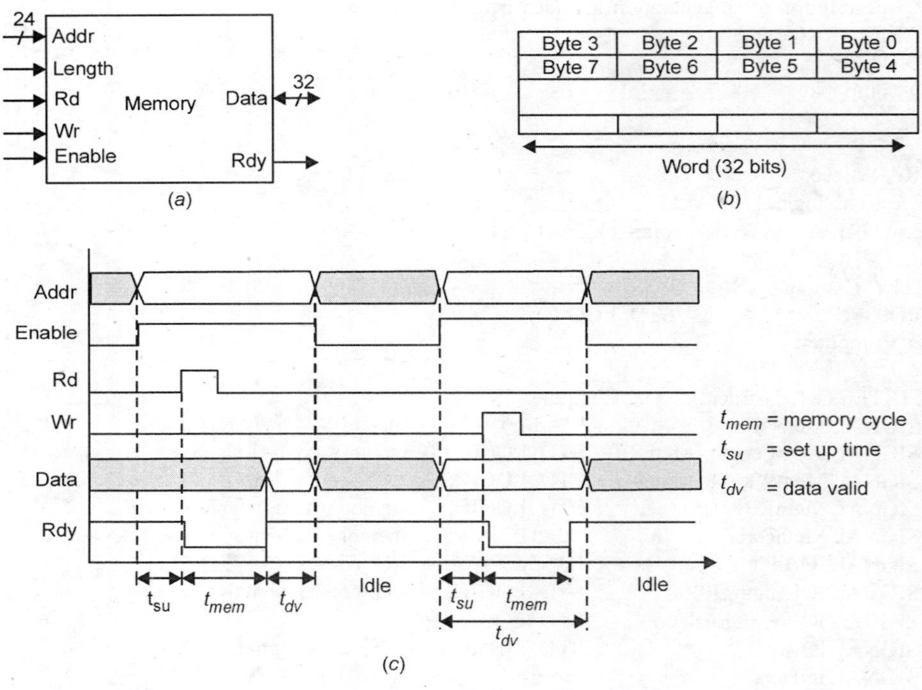

Figure 5.4: Memory subsystem. *(a)* External signals. *(b)* Internal organization. *(c)* Timing diagram.

The data bus to/from the memory is 32-bits wide. In the case of accessing a byte, as indicated by the input signal Length, the 8 least-significant bits of the bus are used to transfer the corresponding data.

The μVHDL entity declaration for this memory subsystem is as follows:

```
LIBRARY ieee;
USE ieee.std_logic_1.164.all;
USE WORK.comp_pkg.ALL;

ENTITY Memory IS
   PORT (Addr    : IN     MAddrT    ; --  memory address bus
         Length  : IN     STD_LOCIC; --  byte/word operand
```

```
        Rd, Wr  : IN     STD_LOGIC; -- access control signals
        Enable  : IN     STD_LOCIC; -- enable signal
        Rdy     : OUT    STD_LOCIC; -- access completion signal
        Data    : INOUT  WordT    ); -- memory data bus
    END Memory;
```

The operation of the memory is as follows (see Figure 5.4c):

- for the memory subsystem to be active, the Enabl e signal has to be set to 1;
- the subsystem responds to an event in either one of the control signals Rd or Wr (which are never active simultaneously); depending on which of these control signal is active, a memory read or a memory write operation is performed for an operand of the length specified in signal Length (0: byte, 1: word);
- the address of the memory location accessed is specified in signal Addr; in case of word accesses, the two lowermost bits of the address are ignored (they are assumed *vf* be 00);
- the data is transferred through signal Data;
- the completion of a memory operation is indicated by signal Rdy; this signal is set to 0 while the operation is in progress and to 1 when the operation finishes.

With respect to the timing of signals,

- signals Addr and Enable must be stable before signals Rd or Wr become active (setup time);
- for a read operation, signals Addr and-Enable must be stable until signal Rdy be-comes 1; the data accessed from memory remain stable in signal Data until signal Enable becomes 0 (data valid time);
- for a write operation, in addition to signals Addr and Enable, the signal Data must also be stable until Rdy becomes 1.

Finally, with respect to consecutive operations, the following situations have to be distin-guished:

- read followed by read, write followed by write, and write followed by read can be performed keeping signal Enable set to 1; for each operation, a pulse is required in the corresponding signal Rd or Wr;
- a read followed by a write requires that signal Enabl e is set to 0 and then set again to 1; this transition is required to reverse the direction of the data bus and to remove the data placed by the memory in the data bus during read operation.

Note that the operation of the memory subsystem is not synchronized by a clock signal; instead, memory operations are performed in response to the memory control signals. An active control signal (Rd or Wr) initiates the operation of the module; signal Rdy is set to 0 while the module is busy. Furthermore, the module does not accept further requests until the current memory access has completed.

The complete specification of a memory subsystem includes additional timing informa-tion such as hold times for the different signals; for simplicity, we do not include those specifications in this description.

A behavioral description of the memory subsystem is given next, which includes the verification of the timing constraints previously given.

```
LIBRARY ieee;
USE ieee.std_logic_unsigned.ALL;

ARCHITECTURE behavioral OF Memory IS
   CONSTANT Tmem: TIME :=   8 ns;  --  nanoseconds (static memory)
   CONSTANT Td    : TIME := 200 ps;  --  picoseconds
   CONSTANT Tsu   : TIME := 200 ps;  --  picoseconds
BEGIN
PROCESS (Rd, Wr, Enable)
   CONSTANT byte_l: STD_L0CIC:= '0';  --  constant declarations
   CONSTANT word_l: STD_L0GIC:= '1';
                                         -- memory declaration
   CONSTANT MaxMem : NATURAL := 16#FFFFFF#;  -- 2**24 bytes
   TYPE          MemArrayT IS ARRAY(0 TO MaxMem-l) OF ByteT;
   VARIABLE Mem   : MemArrayT;
                                          -- working variables
   VARIABLE tAddr : NATURAL;
   VARIABLE tData : WordT ;
   VARIABLE tCtrls: STD_L0GIC_VECT0R(2 D0WNT0 0);

BEGIN
   tCtrls:= Rd & Wr & Enable; -- group signals for simpler decoding
   CASE tCtrls IS
                                              -- output to tri-state
     WHEN "000" => Data <= (OTHERS => 'Z') AFTER Td;

     WHEN "011" =>                           -- write access;
                                             -- indicate module busy
           Rdy <= '0' AFTER Td, '1' AFTER Tmem;
           IF (Length = byte_l) THEN         -- compute address
             tAddr:= C0NV_INTEGER(Addr);   -- bit-vector to integer
                                             -- from pkg std_logic_unsigned
           ELSE
              tAddr:= CONV_INTEGER(Addr(23 D0WNT0 2) & "00");
           END IF;
           CASE Length IS
              WHEN byte_l    => Mem(tAddr)     :=  Data( 7 D0WNTO 0);
              WHEN word_l    => Mem(tAddr)     :=  Data( 7 DOWNTO 0);
                                Mem(tAddr+l)   :=  Data(15 DOWNTO 8);
                                Mem(tAddr+2)   :=  Data(23 DOWNTO 16);
                                Mem(tAddr+3)   :=  Data(31 DOWNT0 24);
              WHEN OTHERS => NULL;
           END CASE;

       WHEN "101" =>                              --read access
                                                  --indicate module busy
           Rdy <= '0' AFTER Td, '1' AFTER Tmem;
           IF (Length = byte_l) THEN              --compute address
              tAddr:= CONV_INTEGER(Addr);         --bit-vector to integer
           ELSE
              tAddr:= CONV_INTEGER(Addr(23 D0WNTO 2) & "00");
```

```
            END IF;
            CASE Length IS
                WHEN byte_l => tData( 7 DOWNTO 0) :=  Mem(tAddr);
                WHEN wordjl => tDataC 7 DOWNTO 0):=  Mem(tAddr);
                               tData(15 DOWNTO  8 := Mem(tAddr+l);
                               tData(23 DOWNTO 16):= Mem(tAddr+2);
                               tData(31 DOWNTO 24):= Mem(tAddr+3);
              WHEN OTHERS => NULL;
          END CASE;
          Data <= tData AFTER Tmem;       -- deliver data

      WHEN OTHERS => NULL;                -- memory not enabled
    END CASE;
  END PROCESS;
                                          -- timing verifications
  ASSERT NOT (Rd'EVENT AND Rd='l' AND NOT Addr'STABLE(Tsu))
      REPORT "Read address setup time violation";

  ASSERT NOT (Rd'EVENT AND Rd='l' AND NOT Enable'STABLE(Tsu))
      REPORT "Read enable setup time violation";

  ASSERT NOT (Wr'EVENT AND Wr='l' AND NOT Addr'STABLE(Tsu))
      REPORT "Write address setup time violation";

  ASSERT NOT (Wr'EVENT AND Wr='l' AND NOT Enable'STABLE(Tsu))
      REPORT "Write enable setup time violation";

  END behavioral ;
```

5.2.2 Input/output (I/O) Subsystem

The I/O subsystem contains the interfaces to devices that allow transferring data in/out to/from the computer. These interfaces are accessed as an array of 2048 ports (see Fig-ure 5.5), so that I/O addresses range from 0 to $2^{11} - 1$.

The μVHDL entity declaration for this I/O subsystem is as follows:

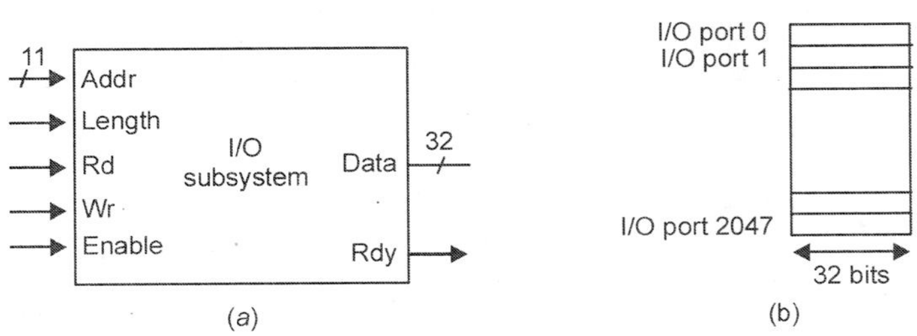

Figure 5.5: Input/output subsystem.

```
LIBRARY ieee;
USE ieee.std_logic_1164. all;
USE WORK. comp_pkg.ALL;
ENTITY I0 IS
```

```
      PORT (Addr     : IN       I0AddrT      ; -- I/0 address bus
            Length   : IN       STD_LOGIC ; -- byte/word control
            Rd, Wr   : IN       STD_LOGIC ; -- I/0 access control
            Enable   : IN       STD_LOGIC ; -- I/0 enable control
            Rdy      : OUT      STD_LOGIC ; -- I/0 completion signal
            Data     : INOUT    WordT     ) ; -- I/0 data bus
   END I0;
```

The general behavior of the I/O subsystem is the. same as that of the memory, except that the address spaces are different. The module responds to an event in either one of the control signals Rd,Wr (which we assume are never active simultaneously), after the I/O devices have been enabled by signal Enabl e; depending on which one of the control signals is active, either an I/O read or an I/O write operation is performed for an operand of the length specified in signal Length. The address of the external device accessed is specified in signal Addr, whereas the data are transferred through signal Data.

The behavioral specification of the I/O subsystem is similar to that of the memory, so we do not describe it here.

5.2.3 Processor

The processor architecture is described in terms of the processor state, which corresponds to all the registers that can be referenced (explicitly or implicitly) by the instructions. Moreover, the execution of an instruction is described by the changes it produces to this state.

PROCESSOR STATE

As shown in Figure 5.6, the processor state consists of the following registers:

- 32 general-purpose registers (32-bits wide), called RO, Rl, ..., R31;
- a 24-bit Program Counter register (PC);
- a 4-bit Condition Register (CR); and
- a 32-bit Instruction Register (IR)

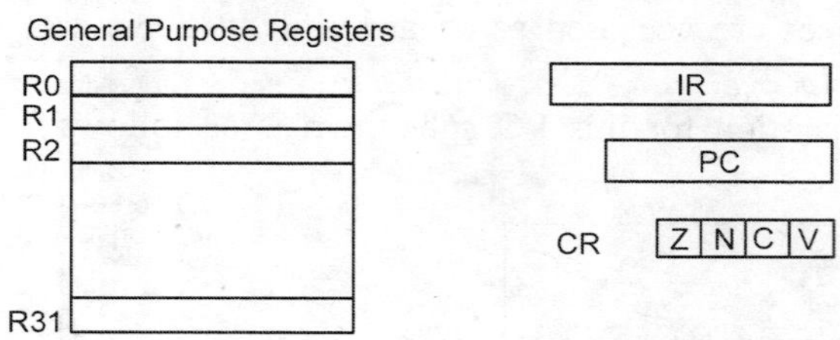

Figure 5.6: Processor state.

Basic Instruction Cycle

The behavior of the processor is as follows:

- Whenever input RESET is set to 1, the processor state is cleared (all registers are set to 0).
- Whenever input Clk is set to 1, an iteration of the **instruction loop** is performed (see Figure 5.7*a*):

  Fetch an instruction from the memory location specified by register PC. The memory access is completed when input MemRdy=l, at which point the instruction is placed in register IR.

Execute the instruction contained in IR, which includes

- performing the operations required by the instruction; and
- computing the address of the next instruction and placing that address in PC.

In terms of the two components of the instruction cycle, the timing behavior with respect to the memory bus is shown in Figure 5.7. Figure 5.7*b* depicts the timing for instructions that do not access data from memory, whereas Figure 5.7c illustrates the timing for a load operation that brings data from memory into a general-purpose register.

The execution of an instruction modifies the state of the processor; the specific modifica-tion depends on the instruction being executed. For example, as depicted in Figure 5.8a, executing the instruction ADD R7,R1,R5 puts in register R7 the sum of the contents of registers Rl and R5, and puts in CR the conditions corresponding to the result of the oper-ation; all other general-purpose registers remain with the same value as before executing the instruction.

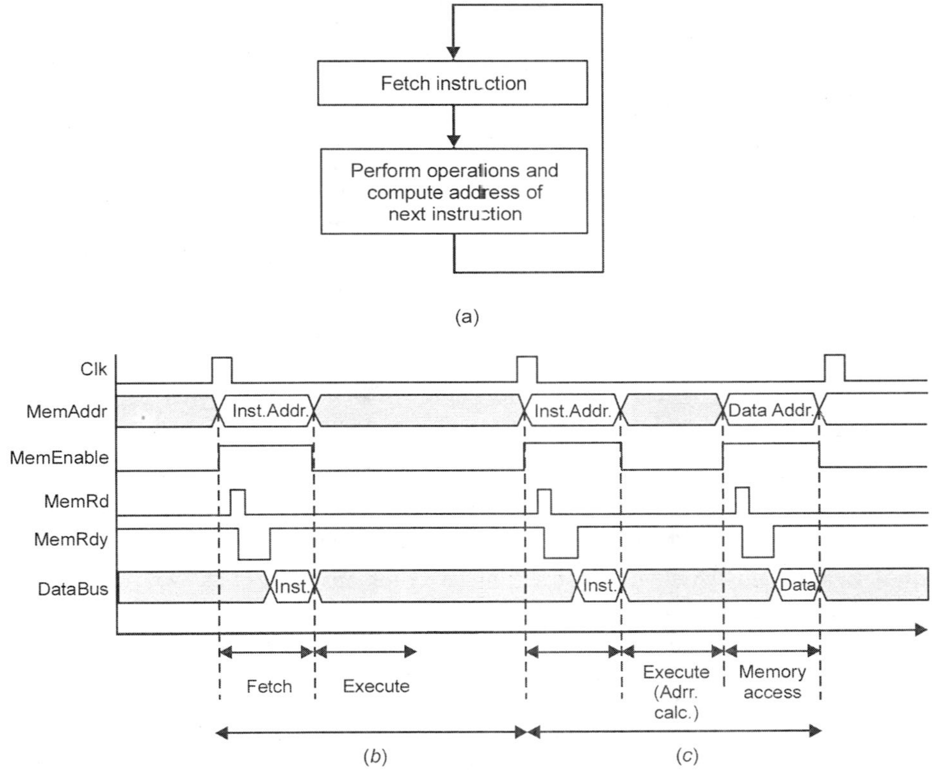

Figure 5.7: Behavior of the processor. (a) Instruction loop. *(b)* Memory bus behavior for register operation. (c) Memory bus behavior for load operation.

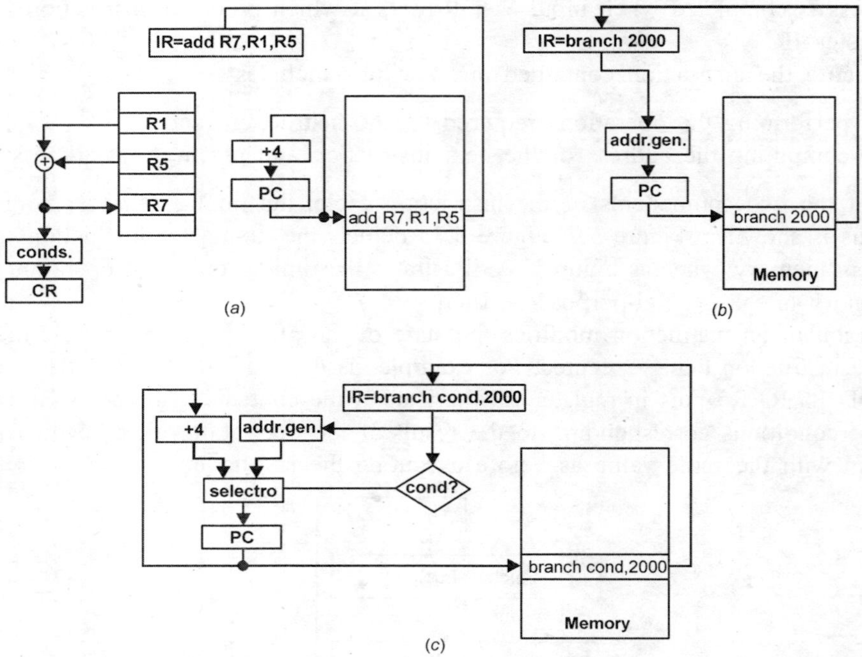

Figure 5.8: Behavior of instructions, (a) Add instruction. *(b)* Unconditional branch instruction. (c) Conditional branch instruction.

Instruction Sequencing

For all instructions except branches, the instructions are fetched sequentially from memory, in the order in which they are stored (increasing order); that is, if the instruction being executed has address *k,* the next instruction to be executed has address $k + 4$ (four bytes per instruction). This instruction sequencing is achieved by incrementing register PC by 4. The sequential execution order is modifiable by branch instructions. Such instructions specify the address of the next instruction, which is loaded in register PC (see Figure 5.8*b*). Two types of branches are typical:

Unconditional branch, in which the branch is always taken.

Conditional branch, in which the instruction specifies an address as well as a condition (the value of one specific bit of the Condition Register CR). If the value of the specified CR bit is equal to the value specified in the instruction, then the branch is **taken**, that is, the next instruction is fetched from the address specified in the branch instruction. Otherwise (i.e., if the value of the CR bit is not equal to the value specified in the instruction), the branch is not taken; in this case, the next instruction address is obtained by adding 4 to PC.

Conditional branches allow two types of important sequencing structures:

1. Data dependent sequencing, in which the instructions executed depend on the value of a data item.
2. Loops, in which a portion of the program is executed repeatedly, either a fixed number of times or until a data condition occurs.

Condition Setting

The execution of some instructions set CR (see Table 15.1). The bits of this register are set to reflect the result from the execution of arithmetic or logic operations, as follows:

Zero	Z	set to 1 if the result is 00... 00, set to 0 otherwise;
Negative	N	set to 1 if the leftmost bit of the result is 1 (negative in two's complement representation), set to 0 otherwise;
Carry	C	set to 1 if there is a carry-out from the most-significant bit of the adder, set to 0 otherwise; and
Overflow	V	set to the result of XOR between the carry-out from bit 30 and the carry-out from bit 31 (corresponds to overflow in two's complement addition).

Instruction Formats

An instruction is represented by one word (32 bits). These 32 bits have different meaning according to the formats depicted in Figure 5.9. In these formats, instructions are divided into the following **fields:**

- an **opcode field** (Opcode) which is six bits wide and specifies the operation performed by the instruction (up to 64 operations);
- two **register fields** (RT/RS and RA) which are five bits wide. Each specifies one of the 32 general-purpose registers. Registers RA and RS contain operands and register RT specifies the destination register; and
- a 16-bit field that may have different interpretations (depending on the opcode):
- a 5-bit **register field** (RB) which specifies an additional register used by the instruction (the rest of the bits are unused); or
- a 16-bit **value** (SI or UI), which specifies the value of an operand used by the instruction. This value can be interpreted as a signed integer, an unsigned integer, or a bit vector, depending on the opcode; or
- a 16-bit signed **displacement** (D) used to calculate an address in a memory instruction or in a branch; or
- an 11-bit **port number** (PN) used by I/O access instructions (the rest of the bits are unused).

Note that symbols RT, RS, RA, RB are used to denote the corresponding instruction fields as well as the contents of the registers whose address is specified in those instruction fields; the actual meaning of the symbol is inferred from the context. On the other hand, SI, UI, D, PN denote the contents of the corresponding fields within the instruction, whereas symbols PC, IR, CR denote the contents of those registers.

Instruction Set

The processor has six groups of instructions, which are listed in Table 5.1. For each instruction, the table lists the name, the opcode, the function, whether it sets the Condition

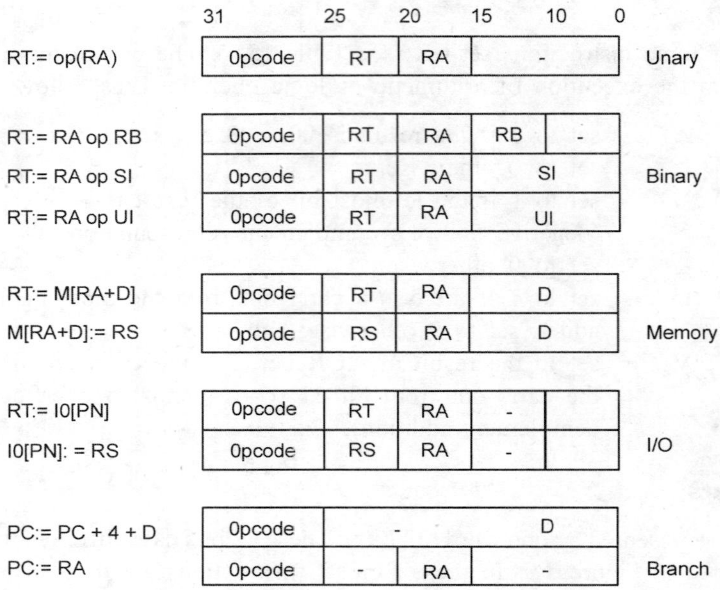

Figure 5.9: Instruction formats.

Register (CR), and the representation used in an assembly-language program, as described later. The corresponding instruction groups are:

Unary operations, which have a single operand and produce the corresponding result, including setting the condition register CR.

Binary operations, which have two operands and produce the corresponding result, also setting CR.

Memory operations, which access the memory location whose address (referred to as the **effective address** EA) corresponds to the lowermost 24 bits from the result of adding the sign-extended value specified in field D to the contents of the register specified in field RA. The length of the operand transferred can be one byte or one word (four bytes) (denoted as Mem(RA+D,l) and Mem(RA+D,4), respectively, in the table).

I/O operations, which access the I/O port whose address is specified in field PN.

Branch operations, which can be **conditional** or **unconditional.**

The (unconditional) B ranch instruction forces the next instruction to be fetched from address PC+4+D (24 bits), where D is the signed displacement field in the instruction. This mechanism for specifying the address of the next instruction is known as **PC relative** because the target instruction is located D+4 bytes from the current instruction (which is addressed by PC). Because field D is only 16-bits wide, this approach is limited to branching $\pm 2^{15} + 4$ bytes around the branch instruction.

The (unconditional) Branch indi rect forces the next instruction fetch to occur from the address contained in the lower 24 bits of the register specified by field RA. Thus, indirect branches are able to branch anywhere in the memory space but they require loading/using a register.

A conditional branch instruction (e.g., Branch i f N=0) tests a bit in the condition register according to the condition specified by the instruction; if the condition is fulfilled (in the example N=0), the next instruction is fetched from the address PC+4+D, otherwise the next instruction is tetched from address PC+4. Note that, as for unconditional direct branches, the branching range is limited to $\pm 2^{15}$ + 4 bytes from the branch itself.

The instruction-set listed in Tablet 5.1 provides a complete description of the capabilities of the processor; any computation sequence executed by this system must be represented as a program written using these instructions.

Table 5.1: Instruction set

Name	Opcode	Function	CR	Assembly	Language
No-op	000000	no operation		nop	
NOT	000010	RT:= not(RA)	Y	not	RT,RA
Left shift	000100	RT:= lshift(RA)	Y	lsh	RT,RA
Right shift	000110	RT: = rshift(RA)	Y	rsh	RT,RA
Left rotate	001000	RT:= lrot(RA)	Y	lrt	RT,RA
Right rotate	001010	RT:= rrot(RA)	Y	rrt	RT,RA
Add	010000	RT:= RA + RB	Y	add	RT,RA,RB
Add immed.	010001	RT:= RA + SI	Y	adi	RT,RA,SI
Subtract	010010	RT:= RA - RB	Y	sub	RT,RA,RB
Sub. immed.	010011	RT:= RA - SI	Y	sbi	RT,RA,SI
AND	010100	RT:= RA and RB	Y	and	RT,RA,RB
AND immed.	010101	RT:= RA and UI	Y	ani	RT,RA,UI
OR	010110	RT:= RA or RB	Y	or	RT,RA,RB
OR immed.	010111	RT:= RA or UI	Y	ori	RT,RA,UI
XOR	011000	RT:= RA xor RB	Y	xor	RT,RA,RB
XOR immed.	011001	RT:= RA xor UI	Y	xri	RT,RA,UI
Load byte	100000	RT:(7 to 0):= Mem(RA+D,l)		ldb	RT,D(RA)
Load word	100001	RT:(31 to 0):= Mem(RA+D,4)		ldw	RT,D(RA)
Store byte	100010	Mem(RA+D,l):=RS(7 to 0)		stb	RS,D(RA)
Store word	100011	Mem(RA+D,4):= RS(31 to 0)		stw	RS,D(RA)
I/O Rd byte	100100	RT(7 to 0):=IO(PN,l)		irb	RT,PN
I/O Rd word	100101	RT(31 to 0):= IO(PN,4)		irw	RT,PN
I/O Wr byte	100110	IO(PN,l):=RS(7 to 0)		iwb	RS,PN
I/O Wr word	100111	IO(PN,4):=RS(31 to 0)		iww	RS,PN
Branch	111poo	PC:= PC + 4 + D		br	D
Branch indirect	111001	PC:= RA		bri	RA
Branch if N=0	110000	If N=0 then PC:= PC+4+D		brp	D
Branch if N=l	110001	If N=l then PC = PC+4+D		brn	D
Branch if Z=0	110010	If Z=0 then PC:= PC+4+D		bnz	D
Branch if Z=l	110011	If Z=l then PC:= PC+4+D		brz	D
Branch if C=0	110100	If C=0 thenPC:= PC+4+D		bnc	D
Branch if C=l	110101	If C=l then PC:= PC+4+D		brc	D
Branch if V=0	110110	If V=0 thenPC:= PC+4+D		bnv	D
Branch if V=l	110111	If V=l then PC:= PC+4+D		brv	D

Data Representation

As it can be observed from the instruction types, a 32-bit data item can represent any of the following data types, the interpretation being given by the type of instruction that uses it:

- Logical data (bit-vectors) used for logical operations, such as AND and XOR.
- Integers (signed) in the two's complement notation; this is used for all arithmetic operations.
- Unsigned integers used for representation of addresses.

In addition, memory and I/O operations use 8-bit (byte) data as unsigned or logical values.

Binary Program and Assembly-language Program

Executing a computation sequence by XMC requires loading the corresponding program— in binary form—into the memory, and setting the program counter to the address of the first instruction. The program is usually extracted from some external device such as a disk, and transferred into the memory under control of another program known as a **loader** (in larger systems, there is an **operating** system that takes care of this as well as many other functions). The loader is always resident in the memory, either in a ROM or loaded when the computer is first powered up (booted).

Because the binary form of a program is difficult to handle by human beings, an **assembly language** is used to simplify writing, debugging, and modifying programs. A program called an **assembler** is used to translate the program into binary form. The assembly language form of the instructions in XMC is shown in the last column of Table 5.1.

Specification of the Processor in μ VHDL

We now give a uVHDL-based specification of the processor. The entity declaration is as follows (see Figure 5.3):

```
LIBRARY ieee;
USE ieee.std_logic_1164.all;
USE WORK.comp_pkg.ALL;

ENTITY processor IS
    PORT (MemAddr   : OUT    MAddrT     ; -- memory address bus
          MemData   : INOUT  WordT      ; -- data bus to/from memory
          MemLength : OUT    STD_LOCIC; -- memory operand length
          MemRd     : OUT    STD_LOCIC; -- memory read control signal
          MemWr     : OUT    STD_LOCIC; -- memory write control signal
          MemEnable : OUT    STD_LOCIC; -- memory enable signal
          MemRdy    : IN     STD_LOGIC; -- memory completion signal
          IOAddr    : OUT    IOAddrT    ; -- I/O address bus
          IOData    : INOUT  WordT      ; -- data bus to/from I/O
          IOLength  : OUT    STD_LOGIC; -- I/O operand length
          IORd      : OUT    STD_LOCIC; -- I/O read control signal
          IOWr      : OUT    STD_LOGIC; -- I/O write control signal
          IOEnable  : OUT    STD_LOGIC; -- memory enable signal
          IORdy     : IN     STD_LOGIC; -- I/O completion signal
          Status    : OUT    StatusT    ; -- processor status signal
          Reset     : IN     STD_LOGIC; -- reset signal
          Clk       : IN     STD_LOGIC);-- clock signal
END processor;
```

The behavior of the processor, summarized in the previous section'is described in u VHDL as follows. This description uses the package i eee. std_logi c_signed so that bit-vectors are regarded as signed values for arithmetic operations.

```
LIBRARY ieee;
USE ieee.std_logic_arith.all ;          -- use definitions and operations
USE ieee.std_logic_signed.all;          -- on signed values

ARCHITECTURE behavioral  OF processor IS
   -- registers (processor state)
   TYPE      RegFileT IS ARRAY(0 to 31) OF WordT;
   SIGNAL    CPR: RegFileT                  ; -- general registers
   SIGNAL    PC : MAddrT                    ; -- Program Counter register
   SIGNAL    CR : STD_LOGIC_VECTOR( 3 DOWNTO 0);  -- Condition Register
   SIGNA     IR  : STD_L0GIC_VECT0R(31 DOWNTO 0);   -- Instruction register

   -- signals used by output function
   SIGNAL    Phase: StatusT        ;   -- instr. cycle phase
   SIGNAL    tMemAddr: WordT  ;   -- memory address
   SIGNAL    tData       : WordT  ;   -- memory/io data

   ALIAS     Z  : STD_L0GIC IS CR(0)  ; -- Condition code Zero
   ALIAS     N  : STD_LOGIC IS CR(l)  ; -- Condition code Negative
   ALIAS     C  : STD_LOCIC IS CR(2) ; -- Condition code Carry
   ALIAS     V  : STD_L0GIC IS CR(3)  ; -- Condition code Overflow

   ALIAS     Opcode  : STD_L0GIC_VECT0R(5 DOWNTO 0)   IS  IR(31 DOWNTO 26);
   ALIAS     RT      : STD_LOGIC_VECT0R(4 DOWNTO 0)   IS  IR(25 DOWNTO 21);
   ALIAS     RA      : STD_LOGIC_VECT0R(4 DOWNTO 0)   IS  IR(20 DOWNTO 16);
   ALIAS     RB      : STD_LOGIC_VECT0R(4 DOWNTO 0)   IS  IR(15 DOWNTO 11);
   ALIAS     RS      : STD_LOGIC_VECTOR(4 DOWNTO 0)   IS  IR(15 DOWNTO 11);
   ALIAS     Imm     : STD_LOGIC_VECTOR(15 DOWNTO 0)  IS  IR(15 DOWNTO  0);
   ALIAS     D       : STD_LOGIC_VECTOR(15 DOWNTO 0)  IS  IR(15 DOWNTO  0);
   ALIAS     PN      : STD_LOGIC_VECTOR(10 DOWNTO 0)  IS  IR(10 DOWNTO  0);
   ALIAS     dlength : STD_LOGIC                      IS  IR(26)          ;

   -- other      declarations
  CONSTANT  delay         : TIME :=  200ps: -- register delay
  CONSTANT  Reset_delay : TIME :=    5 ns;
  CONSTANT  Exec_delay   : TIME :=   10 ns; -- Execute delay
  CONSTANT  Mdelay        : TIME :=  600ps: -- MemEnable signal delay
  CONSTANT  Pulse_Width : TIME :=  2.6 ns; -- memory signals width
  CONSTANT  Fetch_delay  : TIME :=    3 ns; -- disable memory after
                                            -- access completed
BEGIN
   PROCESS --transition function
   --working variables
   VARIABLE RS_data, RA_data,  RB_data              : WordT;
   VARIABLE RT_addr, RA_addr,  RB_addr , RS_addr   :  Natural;
   BEGIN
   WAIT ON Clk,Reset;
```

```
    IF (Reset'Event AND Reset = '1') THEN                    -- reset function
        PC <= (OTHERS => '0'); CR <= "0000"; IR <= (OTHERS => '0');

        FOR i IN 0 TO 31 LOOP
            GPR(i) <= (OTHERS => '0');
        END LOOP;
        Phase <= p_reset;
        Status <= p_reset;

        WAIT UNTIL (Reset = '0') AND (Clk = '1');
    END IF;

IF (Clk'Event AND Clk='l') THEN
    -- Instruction cycle
      Status   <= Fetch AFTER delay;
      Phase    <= Fetch AFTER delay;
                                                -- instruction fetch
      PC       <= PC + 4 AFTER Exec_delay;
      WAIT UNTIL MemRdy='l';      -- wait instr. fetch completed
      IR       <= MemData;
      WAIT FOR Fetch_delay;
                                                -- instruction execution
      Status   <= Execute;
      Phase    <= Execute;
      RA_addr := C0NV_INTECER('0' & RA); RB_addr := CONV_INTEGER('0' & RB);
                                  -- '0'to force bit-vector to positive value
      RA_data := CPR(RA_addr)   ; RB_data := CPR(RB_addr) ;
      RT_addr := C0NV_INTECER('0' & RT);
      RS_addr := CONV_INTECER('0' & RS); -- source reg. for store
      RS_data := CPR(RS_addr)  ; -- or I/0 write
WAIT FOR Exec_delay;

CASE Opcode IS
WHEN "000000" => null;                                   -- nop
WHEN "000010" => CPR(RT_Addr)<= not(RA_data);      -- not
WHEN "000100" => CPR(RT_Addr)<= RA_data(30 DOWNTO 0) & '0'; -- lshift
WHEN "000110" => GPR(RT_Addr)<= '0' & RA_data(31 DOWNTO 1); -- rshift
                                                                   -- lrotati
WHEN "001000" => GPR(RT_Addr)<= RA_data(30 DOWNTO 0) & RA_data(31);
                                                                   -- rrotatc
WHEN "001010" => GPR(RT_Addr)<= RA_DATA(0) & RA_data(31 DOWNTO 1);

WHEN "010000" => GPR(RT_Addr)<=RA_data + RB_data    ;   -- add
WHEN "010001" => GPR(RT_Addr)<=RA_data + Imm;
WHEN "010010" => GPR(RT_Addr)<=RA_data – RB_data    ;   -- sub
WHEN "010011" => GPR(RT_Addr)<=RA_data –Imm;
WHEN "010100" => GPR(RT_Addr)<=RA_data and RB_data  ;   -- and
WHEN "010101" => GPR(RT_Addr)<=RA_data and ext(Imm,RA_data'LENGTH) ;
                                      -- ext: zero extension from ieee pkg
WHEN "010110" => GPR(RT_Addr)<=RA_data or RB_data  ;     -- or
WHEN "010111" => CPR(RT_Addr)<= RA_data or extdmm, RA_data'LENGTH) ;
```

```
WHEN "011000" => GPR(RT_Addr)<=RA_data xor RB_data  ; — xor
WHEN "011001" => GPR(RT_Addr)<=RA_data xor ext(Imm,RA_data'LENGTH);
WHEN "100000" | "100001"  =>                    -- ldb, ldw
          Phase       <= MemOp;
          Status      <= MemOp;
          tMemAddr  <= RA_data + D;           -- mem.addr.
          WAIT until MemRdy = '1';

WHEN "100010" | "100011" =>                     -- stb, stw
          Phase       <= MemOp;
          Status      <= MemOp;
          tMemAddr  <= RA_data + D;           -- mem. addr.
          tData        <= RS_data;              -- mem. data
          WAIT until MemRdy = '1';

WHEN "100100" | "100101" =>                     -- irb, ire
          Phase  <= I0Op;
          Status <= I0Op;
          WAIT until I0Rdy = '1' ;

WHEN "100110" | "100111" =>                     -- iwb, iww
          Phase  <= I0Op;
          Status <= I0Op;
          tData  <= RS_date;                    -- io data
          WAIT until I0Rdy = '1' ;

WHEN "111000" => PC <= PC + D;                  -- branch
WHEN "111001" => PC <= RA_data(23 DOWNS 0);          -- br.ind.
WHEN "110000"  "110001"
                    => IF (N = 0pcodeC0)) THEN          -- br on N
                          PC <= PC + D;
                       END IF;
WHEN "110010" | "110011"
                  => IF (Z = 0pcode(0)) THEN            -- br on Z
                          PC <= PC + D;
                       END IF;
WHEN "110100"  "110101"
                  => IF (C = 0pcode(0)) THEN            -- br on C
                          PC <= PC + D;
                       END IF;
WHEN "110110" | "110111"
                  => IF (V = 0pcode(0)) THEN            -- br on V
                          PC <= PC + D;
                       END IF;
WHEN others  => null ;
END CASE;

IF (C0pcode(5 D0WNT0 4) = 0) or (0pcode(5 D0WNT0 4) = 1))
    and (0pcode /= 0) THEN
    -- set condition register
    IF (CPR(RT_Addr) = 0) THEN CR(0) <= '1';          -- zero result
```

```
    ELSE CR(0) <= '0';
    END IF;
    IF (GPR(RT_Addr)(31) = '1') THEN CR(1) <= '1';   -- negative result
    ELSE                             CR(1) <= '0';
    END IF;
    -- check if operation 0pcode generates carry out
    CR(2) <= get_carry(RA_Data,RB_Data,Imm,0pcode);
    -- check if operation 0pcode generates overflow
    CR(3) <= get_ovf(RA_Data,RB_Data,Imm,Opcode);
END IF;

WAIT FOR 0 ns;                    -- force signals to be updated
IF (Phase = MemOp) THEN
    IF (dlength = '1') THEN       -- ldw
        CPR(RT_addr) <= MemData;
    ELSE                          -- ldb
      CPR(RT_addr)( 7 D0WNT0 0)  <= MemData(7  D0WNT00);
      GPR(RT_addr)(31  D0WNT0 8)  <= (0THERS => '0');
    END IF;
    WAIT FOR Fetch_delay;
END IF;
IF (Phase = I00p) THEN
    IF (dlength = '1') THEN                   -- irw
      GPR(RT_addr) <=  I0Data;
    ELSE                                      -- irb
      GER.(Trader)( 7 D0WNT0 0) <= I0Data(7 D0WNT0 0);
      CPR(RT_addr)(31 D0WNT0 8) <= (0THERS => '0');
    END IF;
    WAIT FOR Fetch_delay;
 END IF;
END IF;
END PROCESS;

PROCESS                          -- output function
BEGIN
-- Instruction cycle
WAIT ON Phase;
IF (Phase = p_reset) THEN          -- reset
    MemRd <= '0'; MemWr <= '0'; MemEnable <= '0'; MemLength <= '0';
    MemData <= (0THERS => 'Z');
    I0Rd <= '0'; I0Wr <= '0'; I0Enable <= '0'; I0Length <= '0';
    I0Data <= (0THERS => 'Z');
ELSIF (Phase = Fetch) THEN       -- instruction fetch
    MemAddr   <= PC AFTER delay;
    MemEnable <= '1' AFTER delay;
    MemRd     <= '1' AFTER Mdelay, '0' AFTER Pulse_Width;
    MemLength <= '1' AFTER delay;
    WAIT UNTIL MemRdy='1';    -- wait instr. fetch completed
    MemEnable <= '0' AFTER Fetch_delay;
ELSIF (Phase = Execute) THEN NULL;       -- instruction execution
                                         -- no output signals
```

```
        ELSIF (Phase = Mem0p) THEN
            MemAddr   <= tMemAddr(23 D0WNT0 0) AFTER delay;
            MemEnable <= '1' AFTER delay;
            MemLength <= dlength AFTER delay;
            IF ((To_Bitvector(0pcode) = "100000") OR
                (To_Bitvector(0pcode) = "100001")) THEN  -- 1db, ldw
              MemRd     <= '1' AFTER Mdelay, '0' AFTER Pulse_Width;
              WAIT until MemRdy = '1';
              MemEnable <= '0' AFTER Fetch_de1ay;
              WAIT FOR Fetch_delay;
            END IF;
            IF ((To_Bitvector(0pcode) = "100010") OR
                (To_Bitvector(0pcode) = "100011")) THEN -- stb, stw
              MemWr   <= '1' AFTER Mdelay, '0' AFTER Pulse_Width;
              IF (dlength = '1') THEN                  -- stw
                MemData <= tData AFTER delay;
              ELSE                                     -- stb
                  MemData(7 DOWNTO 0) <= tData(7 D0WNT0 0) AFTER delay;
              END IF;
              WAIT until MemRdy = '1';
              MemEnable <= '0' AFTER delay;
              MemData   <= (OTHERS => 'Z') AFTER delay;
             WAIT FOR delay;
        END IF;
            ELSIF (Phase = I00p) THEN
                I0Addr    <= PN AFTER delay;
                I0Enable  <= '1' AFTER delay;
                I0Length  <= dlength AFTER delay;
                IF ((To_Bitvectbr(0pcode)  = "100100") 0R
                     (To_Bitvector(0pcode)  = "100101")) THEN      -- irb,   irw
                    I0Rd        <= '1' AFTER Mdelay,  '0'  AFTER Pulse_Width;
                    WAIT until  I0Rdy =  '1' ;
                    I0Enable  <= '0' AFTER Fetch_delay;
                    WAIT FOR Fetch_delay;
                END IF;
                IF ((To_Bitvector(0pcode)  = "100110") OR
                    (To_Bitvector(0pcode)  = "100111"))  THEN         -- iwb, iww
                    IF  (dlength =   '1') THEN                   -- iww
                      I0Data <= tData AFTER delay;
                    ELSE                                         --iwb
                      I0Data(7 D0WNT0 0)  <= tData(7 D0WNT0 0) AFTER delay;
                    END IF;
                    I0Wr        <=  '1' AFTER Mdelay, '0' AFTER Pulse_Width;
                    WAIT until  I0Rdy = '1';
                    I0Enable   <=   '0' AFTER delay.
                    I0Data     <=   (OTHERS =>   'Z') AFTER delay;
                    WAIT FOR delay;
                END IF;
            END IF;
            END PROCESS;
        END behavioral;
```

5.2.4 Specification of the Memory Contents

The μVHDL specification of XMC given in the previous section can be used to simulate the execution of programs. For these purposes, it is necessary to "load" a program in the memory. As described earlier, the actual operation of a computer uses a loader to achieve this function. For simulation purposes, a simpler alternative consists of specifying the program as the initial contents of the memory; moreover, it might be necessary to define a memory smaller than the maximum size indicated earlier. Taking these issues into account, the memory declaration becomes

```
-- memory declaration
CONSTANT  Maximum:  NATURAL:= 16#FFF#; --  4Kbytes
TYPE        MemArrayT IS ARRAY(0 to MaxMem-l) OF ByteT;
VARIABLE Mem : MemArrayT:=
(-- program
 3=>"01100000",  2=>"00000000",  1=>"00000000",  0=>"00000000",
 7=>"01000100",  6=>"00100000",  5=>"00000000",  4=>"00110010",
11=>"10000110", 10=>"10000001",  9=>"00000000",  8=>"00000000",
15=>"10000110", 14=>"10100001", 13=>"00000000", 12=>"00000100",
19=>"01000100", 18=>"01000000", 17=>"00000000", 16=>"00111111",
23=>"10001000", 22=>"01000001", 21=>"00000000", 20=>"00000000",
-- data
51=>"00110011", 50=>"00001111", 49=>"11110000", 48=>"11001100",
55=>"00110011", 54=>"00001111", 53=>"11110000", 52=>"11001100",
OTHERS  => "00000000");
```

where the memory contents corresponds to the following instructions (prefix Ox indicates hexadecimal value):

```
0×000000: xor R0,R0,R0    ; R0 = 0
0×000004: adi  R1,R0,50   ; R1 = 50
0×000008: Idw R20,0(R1)   ; R20= Mem(50,4)= Mem(48,4)
                          ; ignore 2
lowermost address bits
0×00000C: ldw R21,4(R1)   ; R21= Mem(54,4)= Mem(52,4)
0×000010: adi R2,R0,63    ; R2 = 63
0×000014: stb R2,0(R1)    ; Mem(50,l) = 63

0x000048: 0x330FF0CC
0x000052: 0x330FF0CC
```

In this case, the program starts at memory location 0, so that the program begins execution when input Reset changes from **1** to 0.

5.3 IMPLEMENTATION OF A SIMPLE MICROCOMPUTER SYSTEM

According to the specification given in Section 5.2, XMC consists of three subsystems. For an implementation, we need to focus on the structure of these subsystems. For simplicity, we do not discuss the implementation of the I/O subsystem.

5.3.1 Memory Subsystem

The memory subsystem is composed of storage modules (chips), a selector/distributor, and a controller. Let us consider storage chips having a capacity of 4 Mb (2^{22}), with a single-bit output (i.e., each memory chip is organized as an array of 2^{22} x 1 bits). Because the processor accesses both words and bytes, the memory must be able to deliver 32-bits-wide data. Consequently, the memory is organized as 32 parallel modules, for a total of 128 Mb (16 MB), as depicted in Figure **15.10.** A four-input selector/distributor, controlled by address bits 1,0 and input Length, is used to select the entire 32-bits word accessed by address bits 23 down to 2, or a byte within these 32-bits, which is transferred through the lowermost 8 bits in the data bus. The memory controller produces control signals for the read and write operations, as well as the MRdy signal.

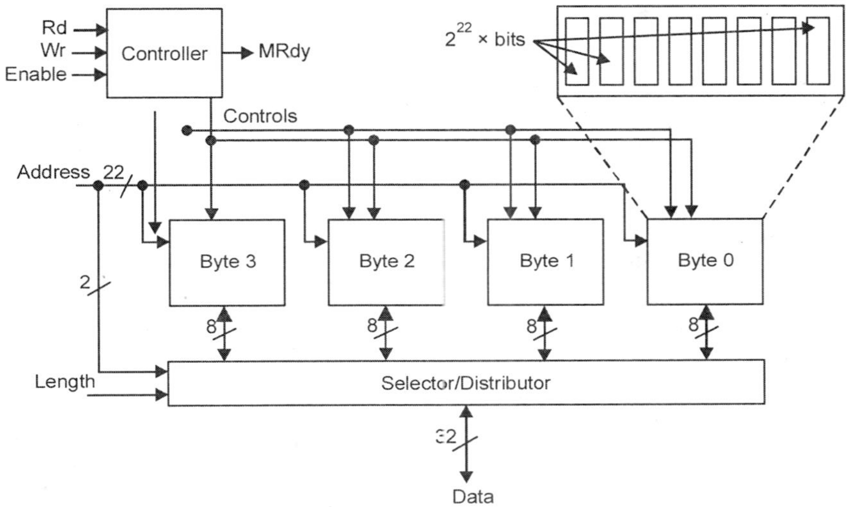

Figure 5.10: Implementation of the memory subsystem.

5.3.2 Processor

The processor is a rather complex sequential system, so it is better implemented as a RTL system consisting of two subsystems (see Figure **5.**11): the **data subsystem** and the **control subsystem.** The data subsystem contains the registers for the processor state and the operators for data modification, whereas the control subsystem controls the operation of the data subsystem.

The control subsystem generates the **control signals** for the data subsystem as well as for the memory and I/O subsystems; on the other hand, the data subsystem as well as the memory and I/ O subsystems generate the **condition signals** used by the control subsystem. In addition, the data subsystem connects with the memory and I/O subsystems via the corresponding buses. The data part of these buses are implemented as bidirectional channels.

We now give an informal description of the implementation of the processor in terms of these two subsystems. The uVHDL description of the implementation is given in Sec-tion 5.5. For simplicity, we do not include the I/O signals.

Data Subsystem

A network that implements the requirements of the data subsystem is depicted in Fig-ure **15.12.** This network consists of a register file that contains the general-purpose regis-ters, dedicated registers for PC, CR, and IR, an arithmetic-logic unit (ALU), as well as other supporting modules. Each of these modules has control signals that determine its operation. For example, signal ALUop determines the function performed by the ALU on the current input values, and signal WrCR produces the update of register CR with the value at its input. All register write operations are synchronized with signal Clk.

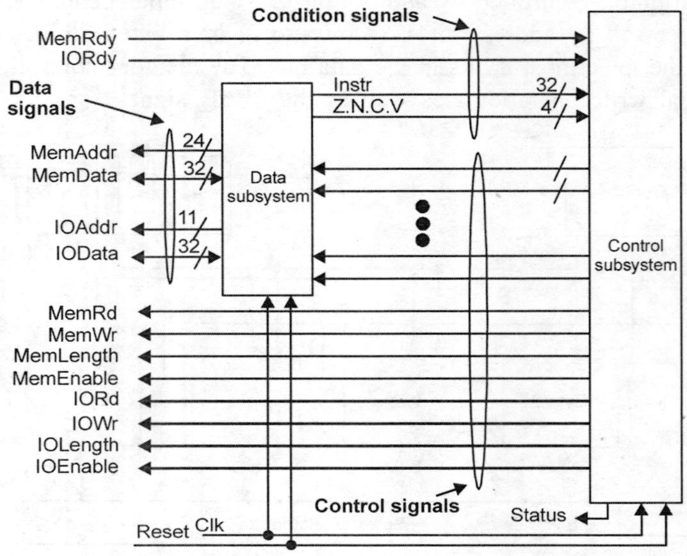

Figure 5.11: Implementation of the processor.

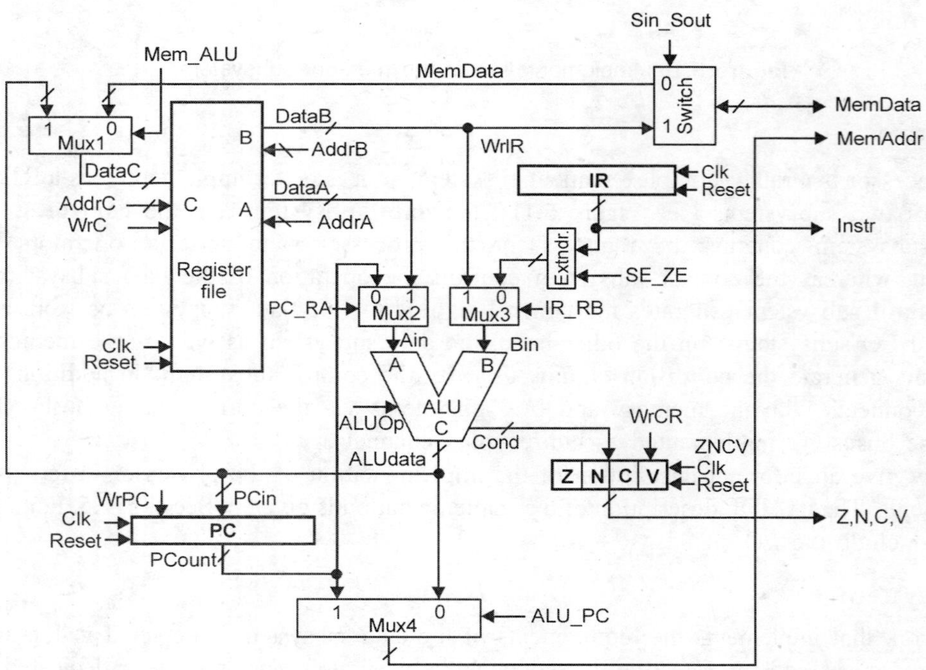

Figure 5.12: Implementation of data subsystem (I/O signals not shown).

Register File

The register file (CPR) contains 32 registers of 32 bits each. **It** has two **read ports** named A, B, which are used to read up to two operands required by an instruction, and one **write port** named C used to write the result. Each pot *i* has its own address (5 bits) and data signals (32 bits); the write port also has a control signal. As a result, two read operations and one write operation can be performed simultaneously. Each read port delivers the data stored in the register selected by the corresponding address signal. The write port stores the data present at the DataC input into the register selected by the Add rC signal, synchronized with signal Clk. Both operations have a delay of RF_delay. All registers are set to zero when the signal Reset is active.

Dedicated Registers

The data subsystem contains dedicated registers for IR, PC, and CR (32 bits, 24 bits, and 4 bits, respectively). To simplify the implementation, register PC is extended at the left with 8 bits that are set to zero; consequently, within the processor PC appears as a 32-bit register. The dedicated registers are loaded whenever the corresponding control input is set to 1 (WrIR, WrPC, and WrCR, respectively), synchronized with signal Clk. The new value is delivered at the output after a register read delay (Reg_de1ay). The registers are set to zero whenever signal Reset is active.

Arithmetic-logic Unit

The arithmetic-logic unit (ALU) is a combinational module that performs the arithmetic and logic operations; it has two data inputs, one data output, one control input, and one condition output. The data inputs are labeled A and B, whereas the data output is labeled C. The control input ALUop specifies which operation is performed according to the following encoding:

ALUop	Operation
0000	Zero_32
0001	A + B
0010	A – B
0011	– B
0100	A AND B
0101	A OR B
0110	A XOR B
0111	NOT(B)
1000	unused
1001	B
1010	SHIFTL(A)
1011	SHIFTR(A)
1100	ROTL(A)
1101	ROTR(A)
1110	A + 4
1111	unused

The arithmetic operations A + B, A – B, and –B are performed using signed integers in two's complement representation. This is consistent with the representation of the arithmetic instructions Add and Sub (for the operations with an immediate value, SI is sign-extended to 32 bits).

However, the ALU is also used to calculate the effective address in memory operations and the address of the next instruction in branches. In these cases, the content of the registers RA and PC are interpreted as positive integers, and the displacement D is sign-extended to 32 bits. The address is calculated with the ALU operation A + B, and the result is interpreted as a positive integer, which corresponds to (RA+D) mod 2**32 and (PC+4+D) mod 2**32, respectively (the operation PC+4 is done during the Fetch part of the instruction). The address sent to memory correspond to the lowermost 24 bits from the result of this operation.

The logical operations use 32-bit vectors. Therefore, for these operations the 16-bit UI is extended with zeros.

The 4-bit condition output Z, N, C, V reflects the value of the result in terms of the conditions zero, negative, carry, and overflow, respectively.

The data output and condition output are delivered after a delay (ALU-del ay) relative to a change in any of the input signals (either data or control).

Other Modules

The data subsystem also contains multiplexers, a zero/sign-extender and a switch; all these are combinational modules. The zero/sign-extender extends a 16-bit input to 32 bits, according to its control signal (0: zero, 1: sign extension); and the switch connects a bidirectional signal either to an input or an output. Each of these modules is characterized by its own delay (Mux_delay, Ext_delay, Switch _delay, respectively).

As shown in Figure 5.12, the interconnections among the modules, as well as with the memory subsystem, have the following characteristics:

- the connection to memory consists of the data bus (MemData) and the address bus (MemAddr); the address corresponds to the lower 24 bits either from the ALU output (for accessing data) or from register PC (for fetching instructions);
- the RF write port is shared by the paths from memory and from the ALU;
- the RF read port B is shared by the paths to memory and to ALU input B;
- ALU input A is shared by RF and PC;
- ALU input B is shared by RF and IR (the 16-bit fields D, SU, IU from IR are sign-extended or zero-extended)

Events and Flow of Data During Execution

Let us consider the events and flows of data that take place in the data subsystem during the Execute part of the instruction loop (e.i., the instruction fetch has already been performed and the instruction has been stored in IR).

Let us first consider an arithmetic-logic instruction; the sequence of events and their timing (as depicted in Figures 5.13 and 5.14) are as follows:

1. Signal Instr (the output from IR) is sent to the control subsystem for decoding.
2. After the instruction decode delay, the signals that control the modules are activated by the control subsystem; this includes the signals to access RF and the signals to control the ALU,

the multiplexers, and the storing of the results into the registers.

3. After the RF read delay, the contents of the registers used as operands by the instruction become available and are fed to the ALU through multiplexers Mux2 and Mux3.

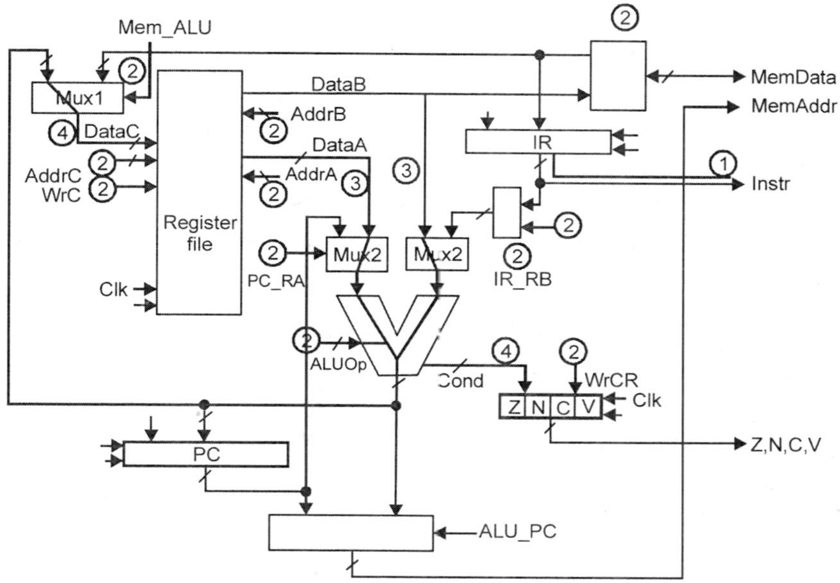

Figure 5.13: Sequence of events in data subsystem for arithmetic-logic instruction.

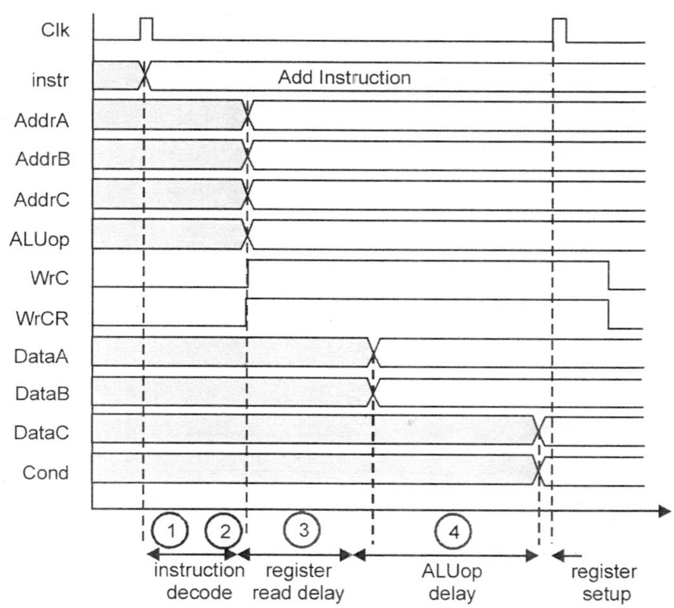

Figure 5.14: Timing diagram in data subsystem for arithmetic-logic instruction.

4. After the ALU delay, the ALU delivers the results from the operation (data and con-dition); the data result is stored in the RF (through Mux1), and the condition result is stored in CR. These storing operations are synchronized with the clock signal, so that the values must be stable earlier than the setup time of the registers.

The flow is similar for instructions that have an immediate operand SI or UI, except that the second operand is obtained from IR and extended accordingly.

In branch instructions, the branch address is calculated using the ALU and stored in PC instead of in the register file.

In memory operations, the address is calculated in the ALU and sent through the address bus. For a Load instruction the data come through the data bus and are written in RF through port C. In a Store operation, the data are obtained from RF through port B.

All these steps comprise the **critical path** in the execution step of an instruction: the instruction is decoded, the operands read, the operation performed, and the results trans-mitted to the input of the registers before the arrival of the next clock pulse (complying with the setup time of the registers). Moreover, in memory operations the access to mem-ory has to be included. Depending on the implementation of the control subsystem, this critical path might be required to fit in one or in several clock cycles. We consider one such implementation now.

Control Subsystem

The control subsystem produces the control signals so that the instruction loop is performed in the data subsystem. The control subsystem is a canonical sequential network, as depicted in Figure 5.15.

The first question to be answered in the design of the control subsystem is the number of clock cycles for the instruction loop. Activities are separated into different cycles (control states) for the following reasons:

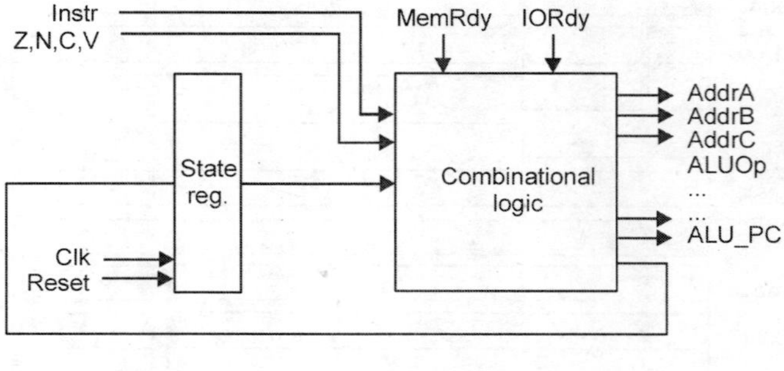

Figure 5.15: Control subsystem.

1. A resource is used more than once in the instruction. This is the case of the ALU, which is used to increment PC and also to perform the operation. It is also the case of the memory

in Load and Store operations as it is used twice, once for instruction fetch and once for data read/write.

2. Because different instructions take different times, dividing the execution in several cycles reduces the execution time of the faster instructions. For example, the main delays involved in the execution part of an Add instruction are the delay of reading the operands from RF, the delay of performing the operation in ALU, and the delay of storing the result in RF. On the other hand, the memory instruction Load has these same delays plus the delay of reading the operand from memory. Consequently, it is convenient to divide the execution of the Load instruction into two parts, so that the execution of the Add instruction is done in one cycle whereas the Load is done in two cycles.

The exact number of cycles used for an implementation and the activities that are in-cluded in each cycle are complex design decisions. The objectives are using the resources effectively and achieving the best execution time (for the instruction mix of typical pro-grams). In our illustration implementation we use three states, as shown by the state diagram depicted in Figure 5.16. The activities in each of the states are:

Fetch: The instruction is fetched from memory and stored in IR. The address of the next sequential instruction is calculated in the ALU and stored in PC.

Execute: The instruction is decoded, the operands are fetched from registers, the op-eration is performed in the ALU and the result is stored in the destination register. In memory instructions (Load and Store) as well as in branches, the effective address is calculated. In the case of branch instructions, the effective address is placed in PC.

Memop: In memory instructions, a memory operation (read, write) is performed. More-over, in Load the data are stored in RF.

As a consequence of this division, two clock cycles are required for instructions other than memory instructions, and three clock cycles for memory instructions.

The state register encodes the three states shown in Figure f5.16 using a l-out-of-3 code; that is, the state register consists of three flip-flops, with only one of these flip-flops being set to 1 at any given time.

The combinational logic receives as inputs the value in the state register as well as the condition signals from the data section, the memory, and the I/O subsystems (signals Instr,Z,N,C,V, MemRdy, and I0Rdy). Based on these inputs, this subsystem produces the signals that control the operation of the entire system, including the generation of the input to the state register (value of the next state). The values of these signals are shown in the state diagram depicted in Figure 5.16. Most of these signals depend only on the value of the state register, so they are available a short delay after the state register changes value; let us call this Ctrl _delay. On the other hand, decoding the instruction requires a few levels of logic, so that the corresponding delay is longer; we refer to this as Dec_delay.

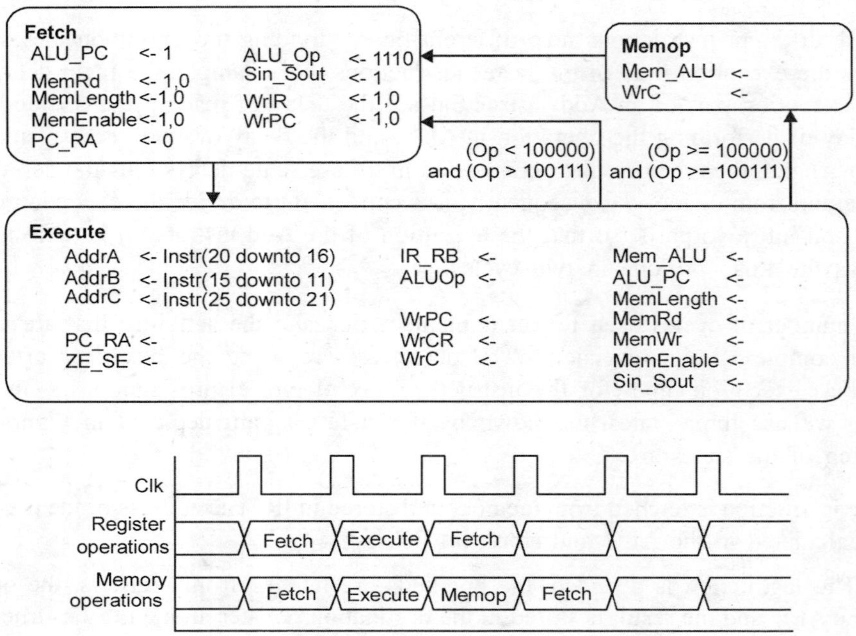

Figure 5.16: State diagram and timing for control subsystem of the processor (signal values not shown depend on instruction executed).

5.4 OPERATION OF THE COMPUTER AND CYCLE TIME

Let us consider now the detailed execution of the instruction iteration loop. For simplicity, we ignore the delays arising from transmitting the signals between the modules (we assume that such delays have been incorporated into the modules which generate/use the signals, see Figure 5.12). A dependence diagram is given for each state, indicating the signals that become active at the time shown.

1. Fetch (see Figure 5.17):

- **a.** The iteration begins at time $t_0 = t_{clk}$+Reg_delay, when the output of the state register has been set to the value Fetch, that is, SR_out = (001).
- **b.** At $t_1 = t_0$ + Ctrl_delay, the control subsystem activates the control signals to fetch the next instruction (ALU_PC=1, MemLength=l, MemEnable=l, Sin_Sout = 0, WrIR= l), and to increment PC (PC.RA=0, ALUop=1110, WrPC=l). In addition, the control logic produces the input to SR (next state = Execute = 010). Signal MemRd=l becomes stable at t_0+MCtrll_delay.
- **c.** At $t_2 = t_1$+ Mux_delay, Mux4 transmits the value in PC to MemAddrBus, and Mux2 transmits the value in PC to the ALU input A.

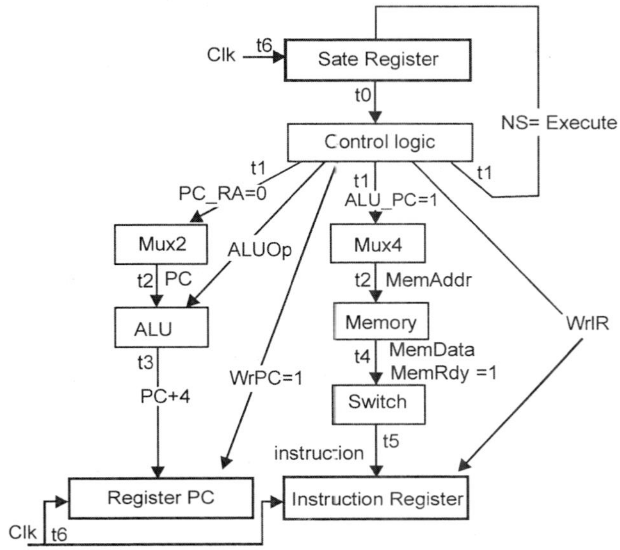

Figure 5.17: Dependencies for state Fetch.

d. At $t_3 = t_2$ + ALU-delay, the ALU delivers the result from the operation A+4. This value is now available at the input of PC.

c. At $t_4 = t_2$ + Mem_delay, signal MemRdy=l and MemData= Instruction.

f. At $t_5 = t_4$ + Switch_del ay, the instruction appears at the output of the switch module and is ready to be loaded in IR.

g. At $t_6 \geq t_5$ the clock pulse arrives, triggering the update of IR, PC, and SR. Note that the signal MemRdy =1 is not used, because we have assumed that the access to memory always fits in one cycle.

2. Execute (see Figure 5.18):

a. At $t_7 = t_{clk}$ + Reg_delay, signals IR_out = Instruction, PCout = updated PC, and SR_out=Execute.

b. At $t_8 - t_7$ + Ctrl,delay, control signals AddrA, AddrB, and AddrC are transmitted to the data section. Note that, because the register address fields in the instruction are always in the same position (if they exist), these signals can be extracted from IR_out without first decoding the opcode; this is possible even for those instructions that do not use all the register fields, in which case the value(s) read from RF that are not needed are just not used.

c. At $t_9 = t_7$ + Dec_delay, the decoding of the instruction is completed. Signals PC_RA, ZE_SE, IR_RB, ALUop, ALU_PC, WrCR, WrC, and WrPC are set according to the instruction being executed. Also the input value for SR (next state value = Memop = 100 or Fetch = 001) is produced, depending whether the instruction is a memory operation or not.

d. At $t_{10} = t_8$ + RF_delay, signals DataA and DataB receive the values stored in the registers addressed by signals AddrA and AddrB.

e. At $t_{11} = t_9$ + Ext_delay, the output of the Zero/Sign Extender becomes stable.

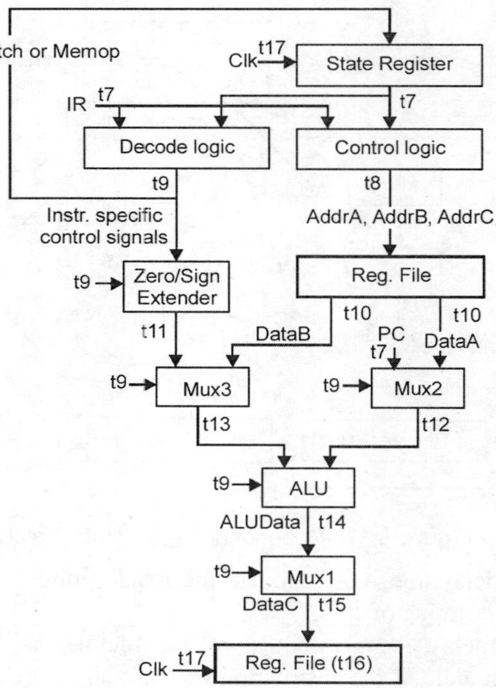

Figure 5.18: Dependencies for state Execute.

f. At $t_{12} = f_{10}$ + Mux_delay the output from Mux2 becomes stable; this assumes that $t_{10} > t_9$.

g. At $t_{13} = t_{11}$ + Mux_delay, the output from Mux3 becomes stable; as previously, this assumes that $t_{10} > t_9$.
This event completes the signals required by the ALU to perform its operation.

h. At $t_{14} = t_{13}$ + ALU_delay, the ALU delivers the results from its operation (data and condition). At this time the value is available also at the input of PC.

i. At $t_{15} = t_{14}$ + Mux_delay, Mux1 transmits signal ALUdata to the input of RF (signal DataC). At this time it is also available at MemAddr.

j. At $t_{16} - t_{15}$ + RF_del ay the data are ready to be written in RF.

k. At $t_{17} \geq t_{16}$, the next clock pulse arrives, triggering the update of CR, RF, or PC (for branches), and SR.

3. If the next state is **Memop** (see Figure 5.19):

a. At $t_{18} = t_{clk}$ + Reg_delay, SR_out= Memop.

b. At $t_{19} = t_{18}$ + Ctrl_delay, the control logic sets signals ALU_PC, Mem_ALU, MemLength, MemEnable, Si n_Sout, and WrC, depending whether the operation is a Load or a Store. MemRd or MemWr becomes stable at $t_{20} = t_{18}$ + MCtrl_delay.

c. At $t_{21} = t_{20}$ + Mem_delay, signal MemRdy=l and the memory access is completed.

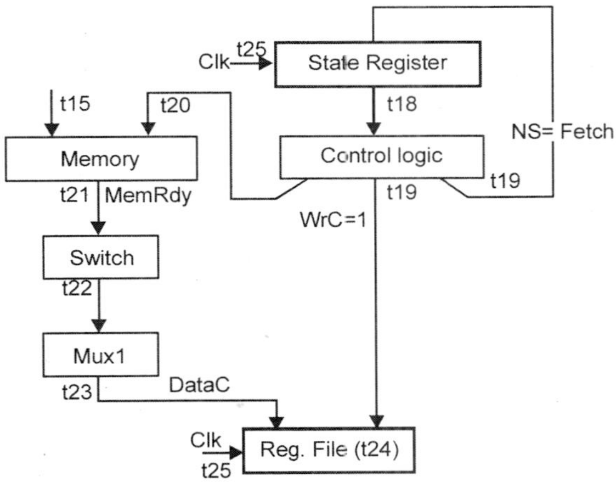

Figure 5.19: Dependencies for state Memop.

d. At $t_{22} = t_{21}$ + Switch_delay, the output from the switch module becomes stable.
e. At $t_{23} = t_{22}$ + Mux_del ay, the output from Muxl becomes stable.
f. At $t_{24} = t_{23}$ + RF_del ay, the data are ready to be written in RF.
g. At $t_{25} > t_{24}$, the next clock pulse arrives, triggering the update of RF, if the instruction is Load, and SR.

The critical path that determines the minimum clock period corresponds to the maximum of the critical paths among all the states. We now show an example of the calculation of the minimum clock period.

EXAMPLE 5.1: We calculate the minimum clock period using the following values for the delay of the different modules:

Register	Reg_delay	t_R	2	ns(setup and propagation delay)
Register file	RF_delay	t_{RF}	4	ns
ALU	ALU_delay	t_{ALU}	6	ns
Multiplexer	Mux_delay	t_{mux}	0.5	ns
Zero/sign ext.	Ext_delay	t_{ZSF}	0.5	ns
Switch	Switch_delay	t_{sw}	0.5	ns
Control delay	Ctrl_delay	t_{tl}	0.5	ns
Decode delay	Dec_delay	t_{dec}	3	ns
Memory	Mem_delay	t_{mem}	8	ns (static memory)

From the expressions given before, the following critical paths are obtained:

$$t_{\text{fetch}} = t_R + t_{ctl} + t_{mux} + t_{mem} + t_{sw}$$
$$= 2 + 0.5 + 1 + 8 + 0.5$$
$$= 12 \text{ ns}$$
$$t_{\text{exec}} = t_R + t_{ctl} + t_{\text{RF}} + t_{\text{mux}} + t_{ALU} + t_{mux} + t_{RF}$$

$$= 2 + 0.5 + 4 + 0.5 + 6 + 0.5+4$$
$$= 17.5 \ ns$$

$$t_{memop} = t_{P} + t_{ctl} + t_{mem} + t_{SW} + t_{mux} + t_{RF}$$
$$= 2 + 0.5 + 8 + 0.5 + 0.5 + 4$$
$$= 15 \ ns$$

Note that the delay in decoding the instruction is not present in these expressions, because the addresses for the RF access are obtained directly from IR and, therefore, the decoding is performed at the same time as the RF read.

From the expressions above, the critical path is in state Execute; as a result, the clock cycle is 17.5 ns, the clock frequency is 57 Mhz, and the instruction execution time is 2× 17.5 = 35 ns for register operations (approximately 28 million instructions per second (MIPS)) and 3 × 17.5 = 52.5 ns for memory operations (19 MIPS).

The actual instruction rate in the execution of a particular program depends on the instruction mix, that is, on the fraction of memory instructions. For example, for a program that has 20% memory instructions, the average execution instruction time is 35 × 0.8 + 52.5 × 0.2 = 38.5 ns, so the instruction rate is about 26 MIPS. %

To achieve a higher instruction rate, high-performance processors decode and execute more than one instruction at a time **(multiple-issue processors),** and/or use a technique called **pipelining,** which approaches a maximum rate of one instruction per clock cycle. A pipelined processor divides the instruction iteration loop into stages and overlaps the execution of several instructions, so that at a given time there can be one instruction executing per stage. These stages are similar to the states of the control subsystem discussed earlier. However, to achieve a larger instruction rate, the implementation might have more stages (five or more is typical) so that the critical path per stage is reduced.

5.5 DESCRIPTION OF THE PROCESSOR IMPLEMENTATION IN UVHDL

We now give the structural description of the processor and the data subsystem, the descrip tion of the modules that compose the data subsystem, and the behavioral description of the control subsystem. In these descriptions, the delays are those considered for Examplet 5.1.

5.5.1 STRUCTURE OF PROCESSOR

The structure of the processor is described in /XVHDL as follows:

```
LIBRARY ieee;
USE ieee.std_logic_1164.ALL;
USE WORK.comp_pkg.ALL, WORK.ALL;
ARCHITECTURE structural OF Processor IS
    SIGNAL Instr                      : WordT;
    SIGNAL ZE, NG, CY, 0V             : STD_L0GIC;
    SIGNAL AddrA, AddrB, AddrC        : STD_L0GIC_VECT0R(4 DOWNTO 0);
    SIGNAL ALU0p                      : STD_L0GIC_VECT0R(3 D0WNT0 0);
    SIGNAL WrC, WrPC, WrCR, WrIR      : STD_L0GIC;
    SIGNAL Mem_ALU, PC_RA, IR_RB      : STD_L0GIC;
    SIGNAL ALU_PC, ZE_SE, SinSout     : STD_L0GIC;
```

```
BEGIN
 P1:  ENTITY  Data_Subsystem
      PORT  MAP (MemAddr, MemData, I0Addr, I0Data,
                 Instr, ZE, NG, CY, 0V, Addax, Adder, Adder, ALU0p,
                 WrC, WrPC, WrCR, WrIR, Mem_ALU, PC_RA, IR_RB, ALU_PC,
                 ZE_SE, SinSout, Clk, Reset);

 P2:  ENTITY  Ctrl_Subsystem
      PORT  MAP (Instr, ZE, NG, CY, 0V, AddrA, AddrB, AddrC, ALU0p,
                 WrC, WrPC, WrCR, WrIR, Mem_ALU, PC_RA, IR_RB, ALU_PC,
                 ZE_SE, SinSout, MemRd, MemWr, MemLength, MemEnable,
                 MemRdy, I0Rd, I0Wr, I0Length, I0Enable, I0Rdy, Status,
                 Clk, Reset);
 END  structural;
```

5.5.2 *Data Subsystem*

The μ VHDL structural description of the data subsystem is as follows:

```
LIBRARY ieee;
USE ieee.std_logic_1164.ALL;
USE W0RK.comp_pkg.ALL, W0RK.ALL;
ENTITY Data_Subsystem IS
   P0RT (MemAddr                  : 0UT    MAddrT    ;
         MemData                  : IN0UT  WordT     ;
         I0Addr                   : 0UT    I0AddrT   ;
         I0Data                   : IN0UT  WordT     ;
         Instr                    : 0UT    WordT     ;
         ZE, NG, CY, 0V           : 0UT    STD_L0GIC ;
         Addax, Adder, Adder      : IN     STD_LOGIC_VECTOR(4 D0WNT00) ;
         ALU0p                    : IN     STD_L0GIC_VECT0R(3 D0WNT0 0) ;
         WrC, WrPC, WrCR, WrIR    : IN     STD_L0GIC ;
         Mem_ALU, PC_RA, IR_RB    : IN     STD_L0GIC ;
         ALU_PC, ZE_SE, Sin_Sout  : IN     STD_L0GIC ;
         Clk, Reset               : IN     STD_L0GIC);

END  Data_Subsystem;

ARCHITECTURE  structural  0F  Data_Subsystem  IS
   SIGNAL DataA, DataB, DataC   : WordT  ;
   SIGNAL Ain , Bin             : WordT  ;
   SIGNAL ALUdata, IRdata       : WordT  ;
   SIGNAL tMemdata              : WordT  ;
   SIGNAL Cond, CRout           : STD_L0GIC_VECT0R(3 D0WNT0 0);
   SIGNAL IRreg, IRext          : WordT  ;
   SIGNAL PCout                 : WordT:= (0THERS => '0');

BEGIN
   ALU1:  ENTITY ALU
          PORT  MAP(Ain,Bin,ALUop,ALUdata,Cond);
    GPR:  ENTITY Reg_File
          P0RT  MAP(AddrA,AddrB,AddrC,DataA,DataB,DataC,WrC,Reset,Clk);
```

```
     PC:   ENTITY Reg
           P0RT MAP(ALUdata(23 D0WNT0 0),PCout(23 D0WNT0 0),WrPC,Reset,Clk);
     CR:   ENTITY Reg
           P0RT MAPCCond,CRout,WrCR,Reset,Clk);

           ZE <= CRout(0);   CY <= CRout(l);
           NC <= CRout(2);   0V <= CRout(3);
     IR:   ENTITY Reg
           P0RT MAP(tMemData,IRreg,WrIR,Reset,Clk);

           Instr <= IRreg;

    MX1:   ENTITY Mux
           P0RT MAP(tMemData,ALUdata,Mem_ALU,DataC);
    MX2:   ENTITY Mux
           P0RT MAP(PCout,DataA,PC_RA,Ain);
    ZSE:   ENTITY Extender
           P0RT MAP(IRreg,ZE_SE,IRext);
    MX3:   ENTITY Mux
           P0RT MAP(IRext,DataB,IR_RB,Bin);
    MX4:   ENTITY Mux
           P0RT MAP(ALUdata(23 DOWNTO 0),PCout(23 DOWNTO 0),ALU_PC,MemAddr);
     SL:   ENTITY Switch
           P0RT MAP(MemData,tMemData,DataB,Si n_Sout);
    END structural;
```

5.5.3 Register File

The μVHDL description of the register file is as follows:

```
LIBRARY ieee;
USE ieee.std_logic_1164.ALL;
USE ieee.std_logic_unsigned.ALL;
USE W0RK.comp_pkg.ALL;

ENTITY Reg_File IS
   P0RT (AddrA, AddrB, AddrC    : IN STD_LOGIC_VECTOR(4 DOWNT0 0);
   DataA, DataB                 : OUT WordT;
   DataC                        : IN WordT;
   WrC                          : IN STD_LOGIC ;
   Reset, Clk                   : IN STD_LOGIC ;
END Reg_File;

ARCHITECTURE behavioral OF Reg_File IS
   TYPE RegFileT IS ARRAY(0 to 31) OF WordT;
   SIGNAL  GPR  :  RegFileT   ;
BEGIN
   PROCESS (AddrA,AddrB)              -- output function
      CONSTANT RF_delay     : TIME  := 4 ns;
BEGIN
   DataA <= GPR(CONV_INTEGER(AddrA)) AFTER RF_delay;
```

```
   DataB <= GPR(CONV_INTEGER(AddrB)) AFTER RF_delay;
END PROCESS;

 PROCESS(Reset,Clk)                 -- transition function
   BEGIN
     IF (Reset'EVENT and (Reset =  '1')) THEN
       FOR i  IN 0 TO 31 LOOP
         GPR(i) <= (OTHERS => '0');
   END LOOP;
 END IF;

 IF (Clk'EVENT AND Clk = '1' AND WrC = '1') THEN
       GPR.(C0NV_INTECER(AddrC)) <= DataC;
     END IF;
 END PR0CESS;
END behavioral;
```

5.5.4 Arithmetic-logic Unit

The μ VHDL description of the ALU is as follows:

```
LIBRARY ieee;
USE ieee.std_logic_1164.ALL;
USE ieee.std_logic_signed.ALL;
USE W0RK.comp_pkg.ALL;

ENTITY ALU IS
  P0RT(A, B: IN STD_L0CIC_VECT0R(31 D0WNT0  0);
       0p   : IN STD_L0GIC_VECT0R( 3 D0WNT0   0);
       C    : 0UT STD_L0GIC_VECT0R(31D0WNT0  0) ;
       Cond : 0UT STD_L0GIC_VECT0R( 3 D0WNT0 0));
END ALU;

ARCHITECTURE behavioral 0F ALU IS
BEGIN
  PR0CESS(A,B,0p)
    C0NSTANT ALU_delay : TIME := 6 ns;
  BEGIN
      CASE 0p IS
          WHEN "0000" => C <= (0THERS => '0')  AFTER ALU_delay;
          WHEN "0001" => C <= A + B            AFTER ALU_delay;
          WHEN "0010" => C <= A – B            AFTER ALU_delay;
          WHEN "0011" => C <= (OTHERS => '0')  AFTER ALU_delay;
          WHEN "0100" => C <= A and B          AFTER ALU_delay;
          WHEN "0101" => C <= A or B           AFTER ALU_delay;
          WHEN "0110" => C <= A xor B          AFTER ALU_delay;
          WHEN "0111" => C <= (0THERS => '0')  AFTER ALU_delay;
          WHEN "1000" => C <= A                AFTER ALU_delay;
          WHEN "1001" => C <= B                AFTER ALU_delay;
          WHEN "1010" => C <= A(30 D0WNT0 0) & '0'   AFTER ALU_delay;
          WHEN "1011" => C <= '0' & A(31 D0WNT0 1)   AFTER ALU_delay;
          WHEN "1100" => C <= A(30 D0WNT00) & A(31)  AFTER ALU_delay;
```

```
            WHEN "1101" => C <= A(0) & A(31 D0WNT0 1)   AFTER ALU_delay;
            WHEN "1110" => C <= A + 4                  AFTER ALU_delay;
            WHEN "1111" => C <= not(A)                 AFTER ALU_delay;
            WHEN OTHERS => NULL;
        END CASE;
        Cond <= get_cc(A,B,0p)  AFTER ALU_delay;
      END PR0CESS;
    END behavioral;
```

5.5.5 Registers

Registers are described in u VHDL as **follows:**

```
LIBRARY ieee;
USE ieee.std_logic_1164.ALL;

ENTITY Reg IS
P0RT(Data_in   : IN STD_L0GIC_VECT0R;
     Data_out  : 0UT STD_L0GIC_VECT0R;
     Wr        : IN STD_L0GIC   ;
     Reset     : IN STD_L0GIC   ;
     Clk       : IN STD_L0GIC) ;
END Reg;

ARCHITECTURE behavioral 0F Reg IS
BEGIN
   PR0CESS(Wr,Reset,Clk)
   C0NSTANT Reg_delay: TIME := 2 ns;
   VARIABLE BVZero: STD_L0GIC_VECT0R(Data_in'RANGE):= (0THERS => '0');
BEGIN
   IF (Reset = '1') THEN
      Data out <= BVZero AFTER Reg_delay;
   END IF;

   IF (Clk' EVENT AND Clk =  '1' AND Wr =  '1') THEN
      Data_out <= Data_in AFTER Reg_delay;
   END IF;
  END PR0CESS;
END behavioral ;
```

5.5.6 Other Modules in the Data Subsystem

The remaining modules in the data subsystem are described in uVHDL as follows:

```
LIBRARY ieee;
USE ieee.std_logic_1164.ALL;

ENTITY Mux IS
   P0RT (A_in,B_in : IN  STD_L0GIC_VECT0R;
         Sel       : IN  STD_L0GIC  ;
         Data_out  :  0UT STD_L0GIC_VECT0R);
END Mux;
```

```
ARCHITECTURE behavioral  0F Mux IS
BEGIN
   PR0CESS (A_in,  B_in,  Sel)
      C0NSTANT Mux_delay: TIME := 500 ps;
   BEGIN
      IF (Sel = '0') THEN
         Data_out <= A_in AFTER Mux_delay;
      ELSE
         Data out <= B_in AFTER Mux_delay;
      END IF;
   END PR0CESS;
END behavioral;
LIBRARY ieee;
USE ieee.std_logic_1164.ALL;

ENTITY Extender IS
   P0RT(X_in : IN STD_L0CIC_VECT0R(31 D0WNT0 0);
        ZE_SE : IN STD_L0CIC  ;
        X_out : 0UT STD_L0GIC_VECT0R(31 D0WNT0 0));
END Extender;

ARCHITECTURE behavioral OF Extender IS
BEGIN
   PROCESS(X_in, ZE_SE)
      C0NSTANT Ext_delay: TIME := 500 ps;
   BEGIN
      IF (ZE_SE = '0') THEN
         X_out(31 D0WNT0 16) <= (0THERS => '0') AFTER Ext_delay;
         X_out(15 D0WNT0 0) <= X_in(15 D0WNT0 0) AFTER Ext_delay;
      ELSE
         X_out(31 D0WNT0 16) <= (0THERS => X_in(15)) AFTER Ext_delay;
         X_out(15 D0WNT0 0) <= X_in(15 D0WNT00) AFTER Ext_delay;
      END IF;
   END PR0CESS;
END behavioral;

LIBRARY ieee;
USE ieee.std_logic_1164.ALL;

ENTITY Switch IS
   PORT(A       : IN0UT  STD_L0GIC_VECT0R;
        B_out : 0UT    STD_L0GIC_VECT0R;
        C_in   : IN     STD_L0GIC_VECT0R;
        Sel     : IN     STD_L0GIC         );
END Switch;

ARCHITECTURE behavioral OF Switch IS
BEGIN
   PROCESS(A, C_in, Sel)
      C0NSTANT Switch_delay: TIME := 500 ps;
      C0NSTANT dataZ: STD_L0GIC_VECT0R(A'RANGE):= (0THERS => 'Z');
```

```
    BEGIN
        IF (Sel = '0') THEN
            B_out <= A AFTER Switch_delay;
            A  <= dataZ;
        ELSE
            A  <= Cain AFTER Switch_delay;
        END IF;
    END PR0CESS;
END behavioral;
```

5.5.7 Control Subsystem

The behavioral description of the control subsystem is expressed in u VHDL as follows:

```
LIBRARY ieee;
USE ieee.std_logic_1164.ALL;
USE W0RK.comp_pkg.ALL;

ENTITY Ctrl_Subsystem IS
    P0RTCInstr                  : IN    WordT       ;
    ZE, NC, CY, 0V              : IN    STD_L0GIC   ;
    AddrA, AddrB, AddrC         : 0UT   STD_L0GIC_VECT0R(4 D0WNT0'0);
    ALU0p                       : 0UT   STD_L0GIC_VECT0R(3 D0WNT0 0);
    WrC, WrPC, WrCR, WrIR       : 0UT   STD_L0GIC   ;
    Mem_ALU, PC_RA, IR_RB       : 0UT   STD_L0GIC   ;
    ALU_PC, ZE_SE, Sin_Sout     : 0UT   STD_L0GIC   ;
    MemRd.MemWr                 : 0UT   STD_L0GIC   ;
    MemLength                   : 0UT   STD_L0GIC   ;
    MemEnable                   : 0UT   STD_L0GIC   ;
    MemRdy                      : IN    STD_L0CIC   ;
    I0Rd, I0Wr                  : 0UT   STD_L0GIC   ;
    I0Length                    : 0UT   STD_L0CIC   ;
    I0Enable                    : 0UT   STD_L0GIC   ;
    I0Rdy                       : IN    STD_L0GIC   ;
    Status                      : 0UT   StatusT     ;
    Clk, Reset                  : IN    STD_L0GIC );
END Ctrl_Subsystem;

LIBRARY ieee;
USE ieee.std_logic_signed.ALL;

ARCHITECTURE behavioral 0F Ctrl_Subsystem IS
    SIGNAL State: StatusT;
BEGIN
PR0CESS                                     -- transition function
     ALIAS       0pcode : STD_L0GIC_VECT0R(5 D0WNT0 0) IS Instr(31 D0WNT0 26);
     C0NSTANT    Reset_delay: TIME:= 500 ps ;
     C0NSTANT    Ctrl_delay : TIME:= 500 ps ;
BEGIN
    WAIT 0N Clk,Reset;
    IF (Reset'EVENT AND Reset = '1') THEN
        State  <= p_reset AFTER Reset_delay;
```

```
            Status <= p_reset AFTER Reset_delay;
            WAIT UNTIL Clk = '1';
     END IF;

     IF (Clk'EVENT) AND (Clk = '1') THEN
        CASE State IS
        WHEN p_reset      => Status <= fetch   AFTER Ctrl_delay;
                             State  <= fetch   AFTER Ctrl_delay;
        WHEN fetch        => Status <= execute AFTER Ctrl_delay;
                             State  <= execute AFTER Ctrl_delay;
        WHEN execute      => CASE 0pcode IS
            WHEN "100000" | "100001"  => State  <= memop AFTER Ctrl_delay;
                                         Status <= memop AFTER Ctrl_delay;
            WHEN "100010" | "100011"  => State  <= memop AFTER Ctrl_delay;
                                         Status <= memop AFTER Ctrl_delay;
            WHEN 0THERS               => State  <= fetch AFTER Ctrl_delay;
                                         Status <= fetch AFTER Ctrl_delay;
                             END CASE;
      WHEN memop | ioop =>  Status <= fetch AFTER Ctrl_delay;
                            State  <= fetch AFTER Ctrl_delay;
      WHEN undef           => NULL;
      END CASE;
     END IF;
   END PROCESS;
     PROCESS(State,Instr,MemRdy)  —'output function
          ALIAS  0pcode : STD_L0CIC_VECT0R( 5 D0WNT0 0) IS Instr(31 D0WNT0 26);
          ALIAS  Imm    : STD_L0GIC_VECT0R(15 D0WNT0 0) IS Instr(15 D0WNT0 0);
          ALIAS  D      : STD_L0GIC_VECT0R(15 D0WNT0 0) IS Instr(15 D0WNT0 0);
          ALIAS  PN     : STD_L0CIC_VECT0R(10 D0WNT0 0) IS Instr(10 D0WNT0 0);

          C0NSTANT Dec_de1ay     : TIME:= 3 ns;
          C0NSTANT Ctrl_delay    : TIME:= 500 ps;
          C0NSTANT MemRd_delay   : TIME:= 2500 ps;
          C0NSTANT MemRd_pulse   : TIME:= MemRd_de1ay + 3 ns ;
          C0NSTANT MemWr_delay   : TIME:= 2500 ps;
          C0NSTANT MemWr_pulse   : TIME:= MemWr_delay + 3 ns ;

          TYPE  Ctrl_LineT  IS
            REC0RD
               Mem0p, WrMem      : STD_L0CIC;
               RS_RB, IR_RB      : STD_L0CIC;
               WrC, WrPC, WrCR   : STD_L0CIC;
               ZE_SE             : STD_L0GIC;

               Aloud             : STD_L0CIC_VECT0R(3 D0WNT0 0);
          END REC0RD;

      VARIABLE  Ctrl_Line : Ctr"l_LineT;

      TYPE        Ctrl_TableT IS ARRAY(NATURAL RANGE 0 TO 63) OF Ctrl_LineT;

      CONSTANTCtrl_Tab1e: Ctr"l_Tab1eT: =
```

```
--      Mem     Wr     RS     IR     Wr     Wr     Wr     ZE     ALU
--      0p      Mem    RB     RB     C      PC     CR     SE     0p
(0  => ('0',    '0',   '1',   '1',   '0',   '0',   '0',   '0',   "0000"), -- nop
 2  => ('0',    '0',   '1',   '1',   '1',   '0',   '1',   '0',   "1111"), -- not
 4  => ('0',    '0',   '1',   '1',   '1',   '0',   '1',   '0',   "1010"),          — ____ lsh
 6  => ('0',    '0',   '1',   '1',   '1',   '0',   '1',   '0',   "1011"),          — ____ rsh
 8  => ('0',    '0',   '1',   '1',   '1',   '0',   '1',   '0',   "1100"),          — ____ lrt
10  => ('0',    '0',   '1',   '1',   '1',   '0',   '1',   '0',   "1101"),          — ____ rrt

16  => ('0',    '0',   '1',   '1',   '1',   '0',   '1',   '0',   "0001"), — add
17  => ('0',    '0',   '1',   '0',   '1',   '0',   '1',   '1',   "0001"), — adi
18  => ('0',    '0',   '1',   '1',   '1',   '0',   '1',   '0',   "0010"), — sub
19  => ('0',    '0',   '1',   '0',   '1',   '0',   '1',   '1',   "0010"), — sbi
20  => ('0',    '0',   '1',   '1',   '1',   '0',   '1',   '0',   "0100"), — and
21  => ('0',    '0',   '1',   '0',   '1',   '0',   '1',   '0',   "0100"), — ani
22  => ('0',    '0',   '1',   '1',   '1',   '0',   '1',   '0',   "0101"), — or
23  => ('0',    '0',   '1',   '0',   '1',   '0',   '1',   '0',   "0101"), — ori
24  => ('0',    '0',   '1',   '1',   '1',   '0',   '1',   '0',   "0110"), — xor
25  => ('0',    '0',   '1',   '0',   '1',   '0',   '1',   '0',   "0110"), — xri

32  => ('1',    '0',   '0',   '0',   '1',   '0',   '0',   '1',   "0001"),          — ____ ldb
33  => ('1',    '0',   '0',   '0',   '1',   '0',   '0',   '1',   "0001"),          — ____ ldw
34  => ('1',    '1',   '0',   '0',   '0',   '0',   '0',   '1',   "0001"),          — ____ stb
35  => ('1',    '1',   '0',   '0',   '0',   '0',   '0',   '1',   "0001"),          — ____ stw
36  => ('1',    '0',   '0',   '1',   '1',   '0',   '0',   '0',   "1001"),          — ____ irb
37  => ('1',    '0',   '0',   '1','1',      '0',   '0',   '0',   "1001"),          — ____ irw
38  => ('1',    '1',   '0',   '1",   '0',   '0',   '0',   '0',   "1001"),          — ____ iwb
39  => ('1',    '1',   '0',   '1',   '0',   '0',   '0',   '0',   "1001"),          — ____ iww
56  => ('0',    '0',   '1',   '0',   '0',   '1',   '0',   '1',   "0001"), — br
57  => ('0',    '0',   '1',   '0',   '0',   '1',   '0',   '1',   "1000"), — bri
48  => ('0',    '0',   '1',   '0',   '0',   '1',   '0',   '1',   "0001"), — brp
49=>   ('0',    '0',   '1',   '0',   '0',   '1'.   '0',   '1',   "0001"),          — ____ brn
50=>   ('0',    '0',   '1',   '0',   '0',   '1',   '0',   '1',   "0001"),          — ____ bnz
51=>   ('0',    '0',   '1',   '0',   '0',   '1',   '0',   '1',   "0001"),          — ____ brz
52=>   ('0',    '0',   '1',   '0',   '0',   '1',   '0',   '1',   "0001"),          — ____ bnc
53=>   ('0',    '0',   '1',   '0',   '0',   '1',   '0',   '1',   "0001"),          — ____ brc
54=>   ('0',    '0',   '1',   '0',   '0',   '1',   '0',   '1',   "0001"),          — ____ bnv
55=>   ('0',    '0',   '1',   '0',   '0',   '1',   '0',   '1',   "0001"),          — ____ brv
OTHERS => ('0', '0',   '1',   '1',   '0',   '0',   '0',   '1',   "0000")
);
BEGIN
IF (State'EVENT) THEN
   CASE State IS
   WHEN undef   => NULL;
   WHEN p_reset => ALU0p      <= "0000";
                   MemRd      <= '0';    MemWr      <= '0';
                   MemEnable  <= '0';    MemLength  <= '0';
                   I0Rd       <= '0*;    I0Wr       <= '0';
                   I0Enable   <= '0';    I0Length   <= '0';
   WHEN fetch  =>
          -- disable write signals from previous cycle
```

```
            WrCR    <= '0' AFTER Ctrl_delay;
            WrC     <= '0' AFTER Ctrl.delay;

             -- fetch instruction
            ALU_PC     <= '1' AFTER Ctrl_delay;
            MemLength <= '1' AFTER Ctrl_delay;
            MemEnable <= '1' AFTER Ctrl_delay;
            MemRd      <= '1' AFTER MemRd_delay,  '0' AFTER MemRd_pulse;
            Sin_Sout   <= '0' AFTER Ctrl_delay;                -- switch in

          -- increment PC
          PC_RA        <= '0' AFTER Ctrl_delay;
          ALU0p        <= "1110" AFTER Ctrl_delay;            -- PC + 4
          WrIR         <= '1' AFTER Ctr"l_delay;
          WrPC         <= '1' AFTER Ctrl_delay;
WHEN execute  =>
         -- disable signals from fetch cycle
         WrIR          <= '0' AFTER Ctrl_delay;
         WrPC          <= '0' AFTER Ctrl_delay;
         MemEnable     <= '0' AFTER Ctrl_delay;
         -- other actions done by Instr'EVENT

WHEN memop    ioop    =>
         -- initiate memory access
      ALU_PC         <=    '0' AFTER Ctrl_delay;                -- address to memory
      MemEnable      <=    '1' AFTER Ctrl_delay;
      MemLength      <=    0pcode (0) AFTER Ctrl_delay;         -- operand length
      WrC            <=    Ctrl_Line.WrC AFTER Ctrl_delay;
      IF (Ctrl_Line.WrMem = '0') THEN
          MemRd      <= '1' AFTER MemRd_delay,  '0' AFTER MemRd_pulse;
          MenuALU    <= '0' AFTER Ctrl_delay;
      ELSE
          MemWr      <= '1' AFTER MemWr_delay,  '0' AFTER MemWr_pulse;
          Sin_Sout   <= '1' AFTER Ctrl_delay;
      END IF;
END CASE;

END IF;
IF (Instr'EVENT) THEN
    -- decode opcode
    Ctrl_Line: = Ctrl_Tab1e(C0NV_INTEGER('0' & 0pcode));

    -- decode registers
    AddrA   <=  Instr (20  D0WNT0  16) AFTER Dec_delay;
    IF (Ctrl_Line.RS_RB  =   '0') THEN
        AddrB <= Instr(25  D0WNT0  21) AFTER Dec_delay;
    ELSE
        AddrB <= Instr(15D0WNT0  11) AFTER Dec_delay;
    END IF;
        AddrC <= Instr(25  D0WNT0  21) AFTER Dec_de1ay;
        -- decode control  signals
        PC_RA  <= not(Ctrl_Line.WrPC) AFTER Ctrl_delay;
```

```
        ZE_SE   <= Ctrl_Line.ZE_SE AFTER Ctrl_delay;
        IR_RB   <= Ctrl_Line.IR_RB AFTER Ctrl_delay;
        ALU0p   <= Ctrl_Line.ALUop AFTER Ctrl_delay;
        WrPC    <= Ctrl_Line.WrPC  AFTER Ctrl_delay;
        WrCR    <= Ctrl_Line,WrCR  AFTER Ctrl_delay;

        IF (Ctrl_Line.Mem0p =  '0') THEN
           WrC <= Ctrl_Line.WrC AFTER Ctrl_delay;
           Mem_ALU <= '1' AFTER Ctrl_delay;
        END IF;
    END IF;

    IF (MemRdy'EVENT AND MemRdy='l') THEN
       CASE State IS
          WHEN memop => IF (Ctrl_Line.WrMem = '1') THEN
                                          -- deactivate data bus
                                  Sin_Sout <= '0' AFTER Ctrl_delay;
                               END IF;
          WHEN OTHERS => NULL;
       END CASE;
     END IF;
     END PROCESS;
    END behavioral;
```

PROBLEMS

5.1. A memory subsystem has the following timing characteristics:

- the enable signal becomes active at time 0;
- the address must be valid 5 ns before the read or write signal is activated;
- the read signal is active for 30 ns;
- the data become available after 50 ns;
- the data remain available for 10 ns.

Determine the time when Rdy becomes high and when Enable becomes low. Give an upper bound on the number of memory accesses per second.

5.2. Give the binary representation of the following XMC instructions:

add R5,R7,R9
ldw R7,1200(R8)
brn 1000

5.3. Determine the effective address for the following XMC instructions

stw R5,1300(R7)
brn 1250
bri R7

with the following contents of registers:

PC = 5321
R1 = 10
R5 = –1250
R7 = –110

5.4. Write a sequence of branches to branch-on-negative a distance of 2^{20} locations from the location of the current instruction.

5.5. Write a program in assembly language and translate it to binary, for adding an array of 105 integers stored consecutively beginning in location 1000 and storing the result in location 2000. Try reducing the number of memory accesses and the number of instructions.

5.6. Use the u *VHDL* specification to simulate the behavior of the instructions given in Exercise 5.2.

5.7. Give the u VHDL specification of an instruction that is similar to 1 dw but, in addition, it increments by four the register used to calculate the effective address.

5.8. Show modifications to Figure 5.10 for implementing a 64 MB memory.

5.9. Show the sequence of events and values of the signals in Figure 5.12 during execution of the following instructions. The value of register PC is 1500 in all cases.

```
add  R7,R5,R8
sw   R3,11300(R2)
bn   10000
```

5.10. Show timing diagrams similar to that of Figure35.14 for a load word instruction and for a branch instruction.

5.11. Show an implementation of the control unit using the one flip-flop per state approach. Give all control signals required to execute the instructions of Exercise 5.2.

5.12. Calculate the cycle time using the same values as in Example (5.1, except for the delay of the arithmetic-logic unit and the memory, which now are 4 ns and 10 ns, respectively.

5.13. Explain the following in brief

(a) CPU (b) ALU

(c) Control unit (d) Memory unit.

5.14. What is the difference between micro computer and mainframe computer?

5.15. What do you understand by semi-conductor memory?

5.16. How will you differentiate primary memory and secondary memory?

5.17. Explain architecture of computer?

5.18. Describe the specification (architecture) of simple micro computer system. Describe with VHDL the input/output sub-system.

5.19. What are the basic component of computer. Write down VHDL code for memory subsystem.

5.20. Explain how a simple micro computer system works. Explain its implementation using VHDL.

5.21. Write short note on : Description of processor implementation in VHDL.

5.22. Implement a simple micro computer system.

5.23. Implement a simple micro computer using VHDL.

UNIT 6

INTRODUCTION TO PROGRAMMABLE LOGIC DEVICES

6.1 BRIEF OVERVIEW OF PROGRAMMABLE LOGIC DEVICES

Designers have always liked programmable logic devices such as PALs and FPGAs for implementation of digital circuits. First, there is reasonable integration ability, allowing implementation of a significant amount of functionality into one physical chip. Programmable logic devices remove the use of multiple off-the-shelf devices and the inconvenience and unreliability associated with external wires. Second, there is the increased ability to change designs. Many of the programmable devices allow easy reprogramming. In general, it is easier to change the design in case of errors or changes in design specifications. Nowadays, programmable logic comes in different types: devices that can be programmed only once and those that can be reprogrammed many times.

Figure 6.1 illustrates a classification of popular programmable logic devices. Programmable logic can be considered to fall into field programmable logic and factory programmable logic. The term *field* indicates that this type of device is programmed in the user's "field" rather than in a semiconductor fab. Often, many may refer to programmable logic to mean devices that are field programmable. However, there are factory programmable devices, too. These are generic devices which can be programmed at the factory to meet customers' requirements. The programming technology uses an irreversible process; hence, programming can be done only once. Examples of factory programmable logic are mask programmable gate arrays (MPGAs) and read-only memories (ROMs). The earliest generations of many programmable devices were programmable only at the factory.

Read-only memories can be considered as an early form of programmable logic. While primarily meant for use as memory, ROMs can be used to implement any combinational circuitry. MPGAs are traditional gate arrays, which require a mask to be designed. MPGAs are often simply called **gate arrays** and have been a popular technology for cre-ating application-specific integrated circuits (ASICs).

User programmable logic in the form of AND-OR circuits was developed at the beginning of the 1970s. By 1972-1973, one-time field programmable logic arrays that permitted instant cus-tomizations by designers were available. Some referred to these devices as field programmable logic arrays or FPLAs. Monolithic Memories Inc. (MMI), a company that was bought by Advanced Micro Devices (AMD), created integrated circuits called programmable logic arrays (PLAs) in 20- and 24-pin packages that could yield the same functionality as 5 to 20 off-the-shelf chips. A similar device is the programmable array logic or PAL.

PALs and PLAs contain arrays of gates. In the PLA, there is a programmable AND array and a pro-grammable OR array, allowing users to implement combinational functions in two levels of gates. The PAL is a special case of a PLA, in that the OR array is fixed and only the AND array is programmable. Many PALs also contain flip-flops.

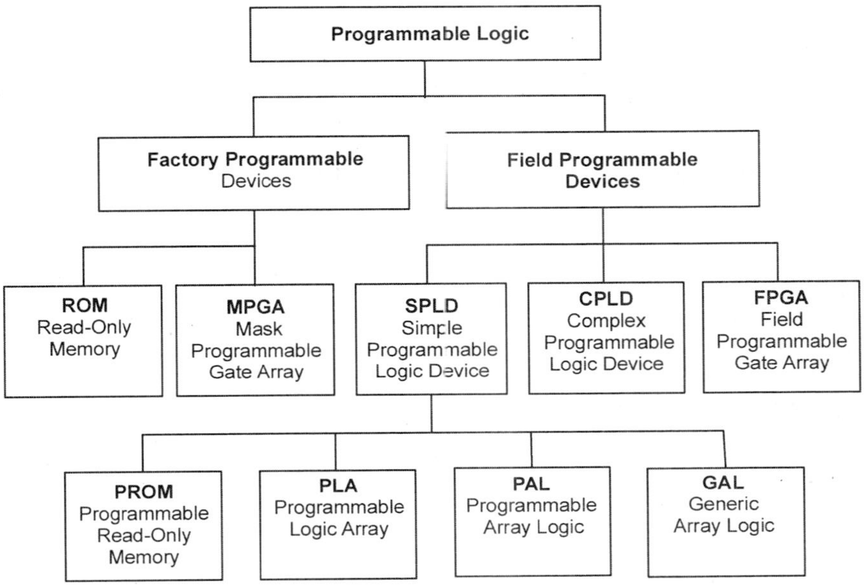

Figure 6.1: Major Programmable Logic Devices

In the 1970s and 1980s, PALs and PLAs were very popular. Part of the popularity was due to the ease of design. MMI and Advanced Micro Devices created a simple programming language, called PALASM, to easily convert Boolean equations into PLA configurations. PALASM made program-ming PALs and PLAs relatively simple.

The early programmable devices allowed only one-time programming. The next technological innovation that helped programmable logic was advancement in erasure of programmable devices. In early days, erasure of programmable logic used ultraviolet light. With ultraviolet light, erasing the configuration of a device meant removing the device from the circuit and placing it in an ultraviolet environment. Hence, in-circuit erasure was not possible. Ultraviolet erasers were slow; typically 10 or 15 minutes were required to perform erasures. Then electrically erasable technology came along.

This led to the creation of field programmable logic arrays that can be easily and quickly erased and reprogrammed without removing the chip from the board.

The early PALs and PLAs were soon followed by CMOS electrically erasable programmable logic devices (PLDs). While the term *PLDs* can be used to refer to any programmable logic devices, there are a set of devices, including the popular PALCE22V10, that are often referred to as PLDs. PLDs contain macroblocks with arrays of gates, multiplexers, flip-flops, or other standard building blocks. Several of these macroblocks appear in a PLD. Lattice Semiconductor created similar devices with easy reprogrammability and called its line of devices **GALs** or **generic array logic.**

Now, many refer to PLAs, PALs, GALs, PLDs, and PROMs collectively as simple PLDs (SPLDs) in contrast to another type of product that has come on the market, complex PLDs (CPLDs). As the name suggests, CPLDs have more integration capability than SPLDs. They come in sizes ranging from 500 to 16,000 gates. CPLDs essentially put multiple PLDs into the same chip with some kind of an interconnection circuit, typically a crossbar switch.

In the late 1980s, Xilinx started using static random-access memory (RAM) storage elements to hold configuration information for programmable devices and created devices called FPGAs that can integrate a fairly large amount of logic. Contrary to their names, the basic building blocks in these devices were not arrays of gates but were bigger and complex blocks containing static RAMs and multiplexers. Several PLD vendors and gate array companies soon jumped into the market, creating a variety of FPGA architectures, some of which used reprogrammable technologies and others of which used one-time programmable fuse technologies. The FPGA technology has con-tinually improved in the last 15 years. Now, there are FPGAs that can contain more than 5 million gates.

Programmable logic devices basically contain an array of basic building blocks which can be used to implement whatever functionality one desires. Different programmable devices differ in the building blocks or the amount of programmability they provide. Table 6-1 illustrates a com-parison of various programmable logic devices. FPGAs are bigger and more complex than CPLDs. The routing resources in FPGAs are more complex than those in simple programmable devices. The variety of alternate routes that can be taken causes the paths taken by signals to be unpre-dictable. FPGAs are more expensive than CPLDs and SPLDs. They contain more overhead for programming. In this chapter, we describe various programmable devices, including SPLDs, CPLDs, and FPGAs.

Table 6.1: A Comparison of Programmable Devices

	SPLD	**CPLD**	**FPGA**
Density	Low Few hundred gates	Low to Medium 500 to 12,000 gates	Medium to High 3,000 to 5,000,000 gates
Timing **Cost**	Predictable Low	Predictable Low to Medium	Unpredictable Medium to High
Major Vendors	Lattice Semiconductor Cypress AMD	Xilinx Altera	Xilinx Altera Lattice Semiconductor Actel
Example Device Families	**Lattice Semiconductor** GAL16LV8 GAL22V10 **Cypress** PALCE16V8 **AMD** 22V10	**Xilinx** CoolRunner XC9500 **Altera** MAX	**Xilinx** Virtex Spartan **Altera** Stratix **Lattice** Mach ECP **Actel** Accelerator

> Many names and abbreviations in this field have historically been used to refer to specific types of programmable devices; however, one may not find the name to be meaningful. Consider PALs and PLAs. Both are arrays of logic. The fact that PLAs contain programmable AND and OR arrays and PALs contain only programmable AND arrays is due to nothing but historical reasons. PALs and PLAs could very well be named the other way around. But it is important for students to understand what these names popularly refer to because they will need to communicate with fellow designers and other design teams. Conventions are important in facilitating communication.

6.2 SIMPLE PROGRAMMABLE LOGIC DEVICES

With the advent of CPLDs and FPGAs, the early generation programmable logic devices, such as ROMs, PALs, PLAs, and PLDs, can be collectively called simple programmable logic devices (SPLDs). In this section, we describe the implementation of digital circuits in simple PLDs.

6.2.1 READ-ONLY MEMORIES

A read-only memory (ROM) consists of an array of semiconductor devices that are interconnected to store an array of binary data. Once binary data is stored in the ROM, it can be read out whenever desired, but the data that is stored cannot be changed under normal operating conditions. Figure 6.2(a)shows a ROM that has three input lines and four output lines. Figure 6.2(b) shows a typical Trutn table, which relates the ROM inputs and outputs. For each combination of input values on the three input lines, the corresponding pattern of 0's and l's appears on the ROM output lines. For example, if the combination ABC = 010 is applied to the input lines, the pattern $F_0 F_1 F_2 F_3$ = 0111 appears on the output lines. Each of the output patterns that is stored in the ROM is called a *word.* Since the ROM has three input lines, we have 2^3 = 8 different combinations of input values. Each input combination serves as an *address,* which can select one of the eight words stored in the mem-ory. Since there are four output lines, each word is four bits long, and the size of this ROM is 8 words X 4 bits.

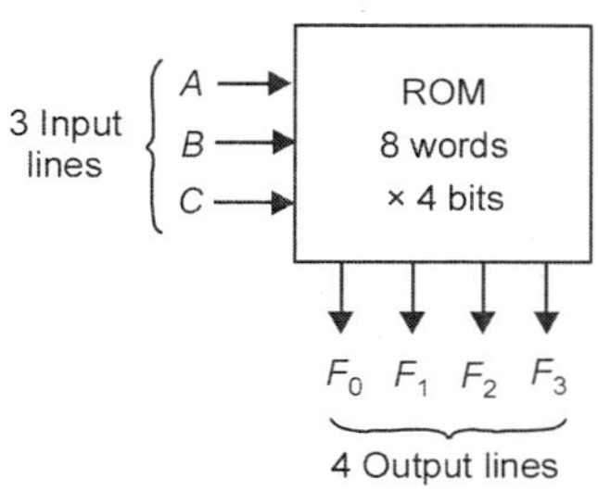

(a) Block diagram

A	B	C	F_0	F_1	F_2	F_3
0	0	0	1	0	1	0
0	0	1	1	0	1	0
0	1	0	0	1	1	1
0	1	1	0	1	0	1
1	0	0	1	1	0	0
1	0	1	0	0	0	1
1	1	0	1	1	1	1
1	1	1	0	1	0	1

Typical data stored in ROM (2^3 words of 4 bits each)

(b) Truth table for ROM

Figure 6.2: An 8-Word x 4-Bit ROM

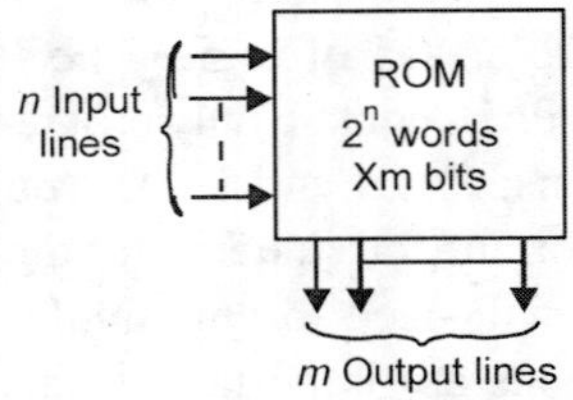

n Input Variables	*m* Input Variables
00 . . . 00	100 . . . 110
00 . . . 01	100 . . . 110
00 . . . 10	101 . . . 101
00 . . . 11	100 . . . 110
⋮	
11 . . . 00	001 . . . 011
11 . . . 01	110 . . . 110
11 . . . 10	011 . . . 000
11 . . . 11	111 . . . 101

Typical data array stored in ROM (2^n words) or m bits each

Figure 6.3: Read-Only Memory with *n* Inputs and *m* Outputs

A ROM which has *n* input lines and *m* output lines (Figure6.3) contains an array of 2" words, and each word is *m* bits long. The input lines serve as an address to select one of the 2" words. When an input combination is applied to the ROM, the pattern of 0's and l's stored in the corresponding word in the memory appears at the output lines. For the example in Figure 6.3' if 00 ... 11 is applied to the input (address lines) of the ROM, the word 110 .. . 010 will be selected and transferred to the out-put lines. A 2^n X *m* ROM can realize *m* functions of *n* variables since it can store a truth table with 2^n rows and *m* columns. Typical sizes for commercially available ROMs range from 32 words X 4 bits to 512K words X 8 bits, or larger.

A ROM basically consists of a decoder and a memory array. When a pattern of *n 0* 's and l's is applied to the decoder inputs, exactly one of the 2" decoder outputs is 1. This decoder output line selects one of the words in the memory array, and the bit pattern stored in this word is transferred to the memory output lines.

Basic types of ROMs include mask programmable ROMs, user programmable ROMs (PROMs), erasable programmable ROMs (usually called EPROMs), electrically erasable and programmable ROMs (EEPROMs), and flash memories. In the mask programmable ROM, the data array is per-manently stored at the time of manufacture. This is accomplished by selectively including or omitting the switching elements at the row-column intersections of the memory array. This requires prepara-tion of a special "mask," which is used during fabrication of the integrated circuit. Preparation of this mask is expensive, so use of mask programmable ROMs is economically feasible only if a large quantity (typically several thousand or more) is required with the same data array. There are also one-time user programmable ROMs or PROMs.

Modification of the data stored in a ROM is often necessary during the developmental phases of a digital system, so EPROMs are used instead of mask programmable ROMs. EPROMs use a special charge-storage mechanism to enable or disable the switching elements in the memory array. An EPROM programmer is used to provide appropriate voltage pulses to store electronic charges in the memory array locations. The data stored in this manner is generally permanent until erased using ultraviolet light. After erasure, a new set of data can be stored in the EPROM.

The EEPROM is similar to an EPROM, except that erasure is accomplished using electrical pulses instead of ultraviolet light. A traditional EEPROM can be erased and reprogrammed only a limited number of times, typically 100 to 1000 times. **Flash memories** are similar to EEPROMs, except that they use a different charge-storage mechanism. They usually have built-in programming and erasure capability so that data can be written to the flash memory while it is in a circuit without the need for a separate programmer.

A ROM can implement any combinational circuit. Essentially, if the outputs for all combinations of inputs are stored in the ROM, the outputs can be "looked up" in the table stored in the ROM. The ROM method is also called the **look-up table (LUT)** method for this reason.

Consider the implementation of a 2-bit adder in a ROM. This adder must add two 2-bit numbers. Since the maximum value of a 2-bit number is 3, the maximum sum is 6, necessitating 3 bits for the sum. The truth table for such an adder is illustrated in Figure 6.4 We could also design a 2-bit full adder assuming a carry input in addition to the two 2-bit numbers.

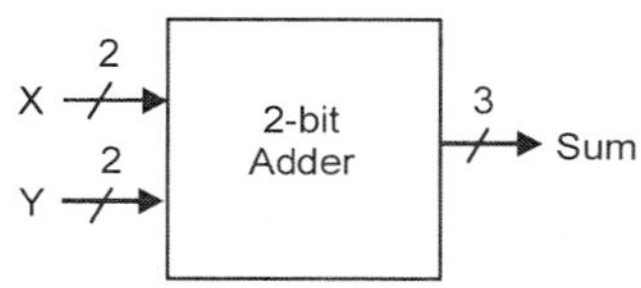

X_1	X_0	Y_1	Y_0	S_2	S_1	S_0
0	0	0	0	0	0	0
0	0	0	1	0	0	1
0	0	1	0	0	0	0
0	0	1	1	0	0	1
0	1	0	0	0	0	1
0	1	0	1	0	1	0
0	1	1	0	0	1	1
0	1	1	1	1	0	0
1	0	0	0	0	1	0
1	0	0	1	0	1	1
1	0	1	0	1	0	1
1	0	1	1	1	0	1
1	1	0	0	1	1	1
1	1	0	1	1	0	0
1	1	1	0	1	0	1
1	1	1	1	1	1	0

Figure 3.4: Block Diagram and Truth Table of a 2-Bit Adder

This 2-bit adder can be implemented with a 16 X 3 ROM. The input numbers (*X* and *Y*) must be connected to the four address lines, and the three data lines will produce the sum bits.

Figure 6.5 illustrates the ROM implementation of this 2-bit full adder. Assuming the connections that are snbwn, the contents of the ROM in its 16 locations should be 0,1,2, 3,1,2, 3,4,2,3,4,5, 3, 4,5, and 6, respectively (representing the digits in decimal). The LSB of the sum will come from the LSB of the data bus.

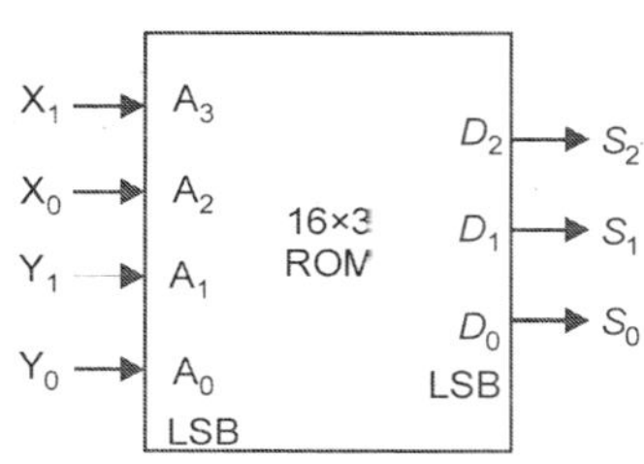

Figure 6.5: ROM Implementation of a 2-Bit Full Adder

6.2.2 Programmable Logic Arrays

A programmable logic array (PLA) performs the same basic function as a ROM. A PLA with *n* inputs and *m* outputs (Figure 6.7 can realize *m* functions of *n* variables. The internal organization of the PLA is different from that of the ROM. The decoder is replaced with an AND array that realizesselected product terms of the input variables. The OR array OR's together the product terms needed to form the output functions.

Table 6.2: ROM Truth Table

Q_3	Q_2	Q_1	X	Q_3^+	Q_2^+	Q_1^+	Z
0	0	0	0	0	0	1	1
0	0	0	1	0	1	0	0
0	0	1	0	0	1	1	1
0	0	1	1	1	0	0	0
0	1	0	0	1	0	0	0
0	1	0	1	1	0	0	1
0	1	1	0	1	0	1	0
0	1	1	1	1	0	1	1
1	0	0	0	1	0	1	1
1	0	0	1	1	1	0	0
1	0	1	0	0	0	0	0
1	0	1	1	0	0	0	1
1	1	0	0	0	0	0	1
1	1	0	1	0	0	0	0
1	1	1	0	0	0	0	0
1	1	1	1	0	0	0	0

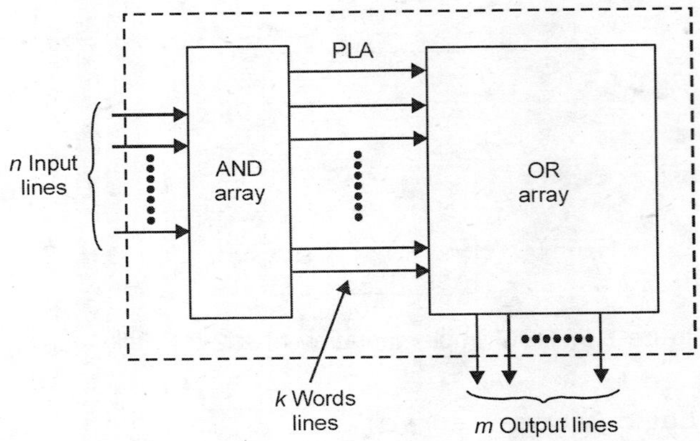

Figure 6.7: Programmable Logic Array Structure

Figure 6.8 shows a PLA that realizes the following functions:

$$F_0 = \Sigma m(0,1,4,6) = A'B' + AC$$
$$F_1 = \Sigma m(2,3,4,6,7) = B + AC$$
$$F_2 = \Sigma m(0,1,2,6) = A'B' + BC$$
$$F_3 = \Sigma m(2,3,5,6,7) = AC + B$$

The above logic functions contain three variables. In a PLA implementation, each product term in the equation is created first, and then required product terms are OR'ed using the OR gate. Hence, product terms can be shared while using the PLA. Instead of minimizing each function sep-arately, we want to minimize the total number of product terms. There are five distinct product terms in the above four equations. Figure 6.8 illustrates a PLA with three inputs, five product terms, and four outputs, implementing the above four equations. It should be noted that the number of terms in each equation is not important, as long as there are AND gates to generate all product terms required for all outputs together.

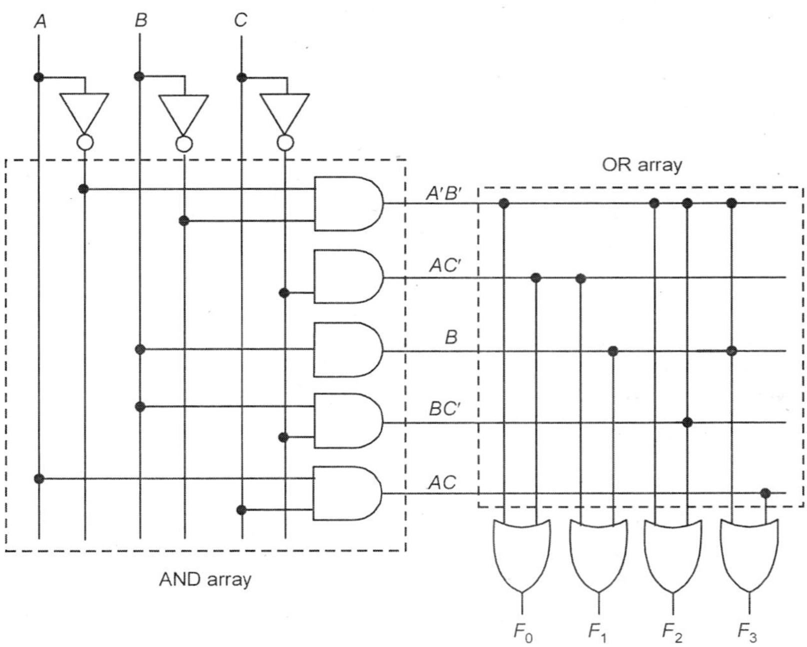

Figure 6.8: PLA with Three Inputs, Five Product Terms, and Four Outputs (Logic Level)

Internally, the PLA may use NOR-NOR logic instead of AND-OR logic. The array shown in Figure 6.8 is thus equivalent to the nMOS PLA structure of Figure 6.9 Logic gates are formed in the array by connecting nMOS switching transistors between the column lines and the row lines.

Source, drain, and gate are the names of the three terminals of the metal oxide semiconductor (MOS) transistor. The gate is the one that is used to control the ON/OFF action. There are two types of MOS transistors, *n-channel MOS (nMOS) and p-channel MOS (pMOS)*. The illustrations in this section use nMOS transistors. A popular technology since the 1990s is complementary MOS (CMOS), where nMOS and pMOS transistors are used together in a complementary fashion.

Figure 6.10 shows the implementation of a two-input NOR gate using nMOS transistors. The tran-sistors act as switches, so if the gate input is a logic 0, the transistor is off. If the gate input is a logic 1, the transistor provides a conducting path to ground. If $X_1 = X_2 = 0$, both transistors are off, and the pull-up resistor brings the Z output to a logic 1 level (+*V*). If either X_1 or X_2 *is* 1, the corresponding transistor is turned on, and Z = 0. Thus, $Z = (X + X_2)' = X'_1 X_2'$, which corresponds to a NOR gate.

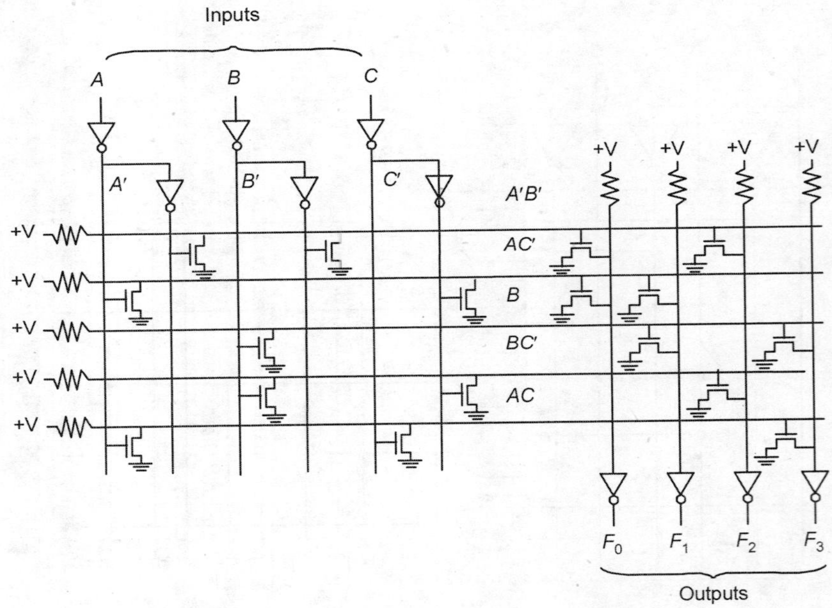

Figure 6.9: PLA with Three Inputs, Five Product Terms, and Four Outputs (Transistor Level)

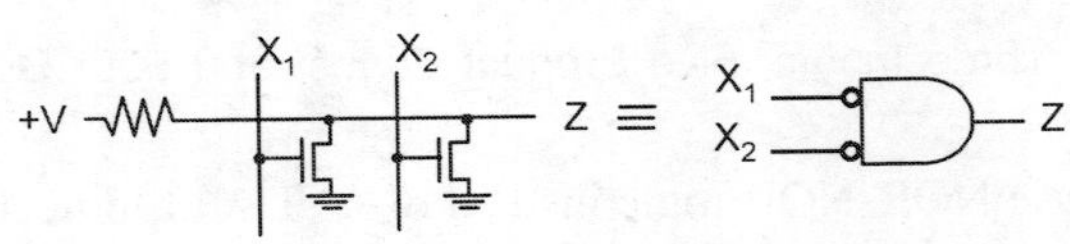

Figure 6.10: nMOS NOR Gate

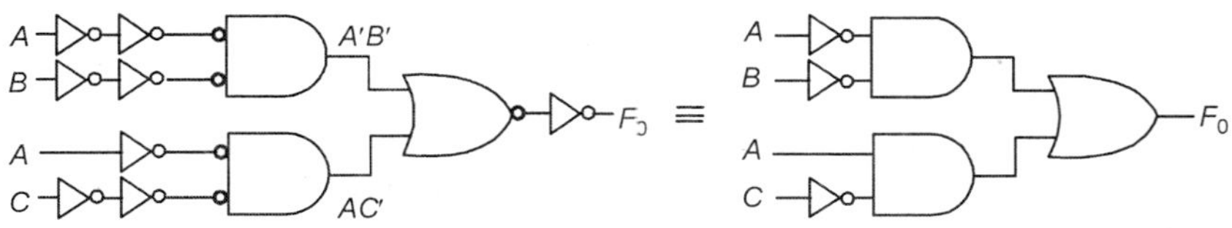

Figure 6.11: Conversion of NOR-NOR toAND-OR

The part of the PLA array that realizes F_0 is equivalent to the NOR-NOR gate structure shown in Figure 6.11 After canceling the extra inversions, this reduces to an AND-OR structure.

The contents of a PLA can be specified by a modified truth table. Table 6.3 specifies the PLA in Figure 6.8 The input side of the table specifies the product terms. The symbols 0,1, and – indicate whether variable is complemented, not complemented, or not present in the corresponding product term. The output side of the table specifies which product terms appear in each output function. A 1 or 0 indicates whether a given product term is present or not present in the corresponding output function. Thus, the first row of Table 6.3 indicates that the term *A'B'* is present in output functions F_0 and F_2, and the second row indicates that *AC* is present in F_0 and F_1'

Table 6.3: PLA Table for Equations

Product Term	**Inputs** *A*	*B*	C	**Outputs** F_0	F_1	F_2	F_3
A'B'	0	0	—	1	0	1	0
AC	1	—	0	1	1	0	0
B	—	1	0	1	0	1	1
BC	—	1	0	0	0	1	0
AC	1	—	1	0	0	0	1

Next, we will realize the following functions using a PLA:

$$\begin{aligned} F_I &= \Sigma m(2,3,5,7,8,9,10,11,13,15) \\ F_2 &= \Sigma m(2,3,5,6,7,10,11,14,15) \\ F_3 &= \Sigma m\ (6,7,8,9,13,14,15) \end{aligned} \quad \text{... (6.2)}$$

If we minimize each function separately, the result is

$$\begin{aligned} F_1 &= bd + b'c + ab' \\ F_2 &= c + a'bd \\ F_3 &= bc + ab'c' + abd \end{aligned} \quad \text{... (6.3)}$$

If we implement these reduced equations in a PLA, a total of eight different product terms (including *c*) are required.

Instead of minimizing each function separately, we want to minimize the total number of rows in the PLA table. In this case, the number of terms in each equation is not important, since the size of the PLA does not depend on the number of terms within an equation. Equations 6-3 are plotted on the Karnaugh maps shown in Figure 6.12 Since the term *ab'c'* is already needed for F_3, we can use it in F_1 instead of *ab'*. The other two l's in *ab'* are covered by the *b'c* term. This eliminates the need to use a row of the PLA table for *ab'*. Since the terms *a'bd* and *abd* are needed in F_2 and F_3, respectively, we can replace *bd* in F_1 with *a'bd* + *abd.This* eliminates the need for a row to implement *bd.* Since *b'c* and *bc* are used in F_1 and F_3, respectively, we can replace *c* in F_2 with *b'c* + *bc.* The resulting Equations (3=4) correspond to the reduced PLA table (Table 6-4). Instead of using Karnaugh maps to reduce the number of rows in the PLA, the Espresso algorithm can be used.

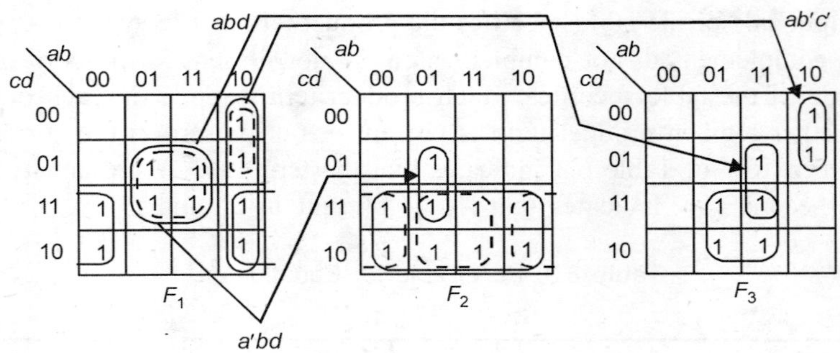

Figure 6.12: Multiple-Output Karnaugh Maps

Table 6-4: Reduced PLA Table

a	b	c	d	F_1	F_2	F_3
0	1	–	1	1	1	0
1	1	–	1	1	0	1
1	0	0	–	1	0	1
–	0	1	–	1	0	1
–	1	1	–	0	1	1

$$F_1 = a'bd + abd + ab'c' + b'c$$
$$F_2 = a'bd + b'c + bc \quad ...(6.4)$$

Equations 6-4 have only five different product terms, so the PLA table has only five rows. This is a significant improvement over Equations 6-3 which require eight product terms. Figure 6-13 shows the corresponding PLA structure, which has Tour inputs, five product terms, and three outputs. A dot at the intersection of a word line and an input or output line indicates the presence of a switching element in the array.

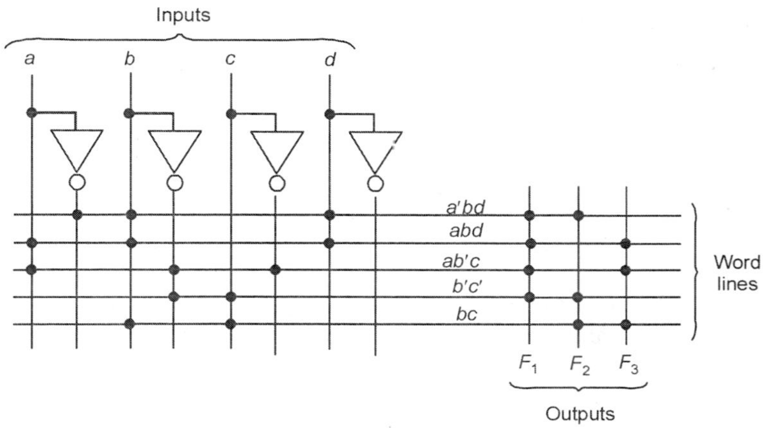

Figure 6-15: *PLA Realization of Equations (3-4)*

A PLA table is significantly different than a truth table for a ROM. In a truth table, each row represents a minterm; therefore, exactly one row will be selected by each combination of input values. The 0's and 1's of the output portion of the selected row determine the corresponding output values. On the other hand, each row in a PLA table represents a general product term. Therefore, zero, one. or more rows may be selected by each combination of input values. To determine the value of F for a given input combination, the values of F in the selected rows of the PLA table must be OR'ed together. The following examples refer to the PLA table of Table 6-4 If $abcd$ = 0001, no rows are selected, and all F_i's are 0. If $abcd$ = 1001, only the third row is selected, and $F_1 F_2 F_3 = 101$. If $abed$ = 0111, the first and fifth rows are selected. Therefore, $F_1 = 1 + 0 = 1$, $F_2 = 1 + 1 = 1$, and $F_3 = 0 + 1 = 1$.

6.2.3 Programmable Array Logic

The PAL (programmable array logic) is a special case of the programmable logic array in which the AND array is programmable and the OR array is fixed. The basic structure of the PAL is the same as the PLA shown in Figure 6-7 Because only the AND array is programmable, the PAL is less expensive than the more general PLA, and the PAL is easier to program. For this reason, logic designers frequently use PALs to replace individual logic gates when several logic functions must be realized.

Figure 6-14 represents a segment of an unprogrammed PAL. The symbol

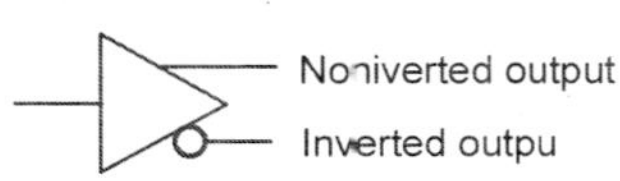

represents an input buffer, which is logically equivalent to

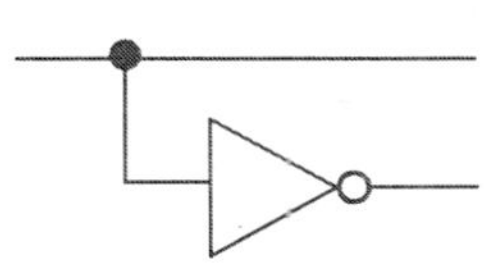

A buffer is used since each PAL input must drive many AND gate inputs. When the PAL is programmed, some of the interconnection points are programmed to make the desired connections to the AND gate inputs. Connections to the AND gate inputs in a PAL are represented by X's as shown in the following diagram:

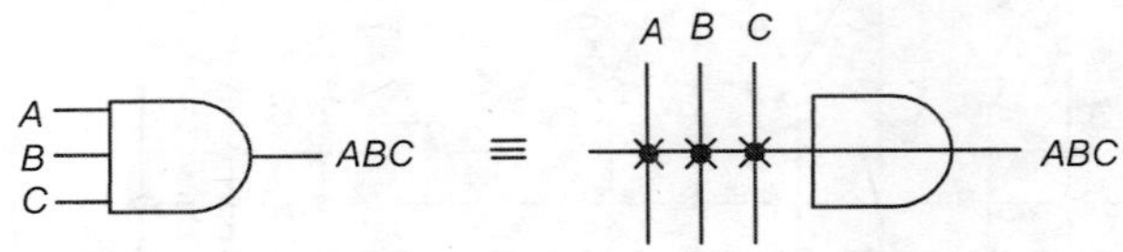

As an example, we will use the PAL segment of Figure 6–14 (a) to realize the function $I_1 I'_2 + I_1' I_2$. The X's in Figure 6-14(b) indicate that I_1 and I'_2 lines are connected to the first AND gate, and the I'_1 and I_2 lines are connected to the other gate.

When designing with PALs, we must simplify our logic equations and try to fit them into one (or more) of the available PALs. Unlike the more general PLA, the AND terms cannot be shared among two or more OR gates; therefore, each function to be realized can be simplified by itself without regard to common terms. For a given type of PAL, the number of AND terms that feed each output OR gate is fixed and limited. If the number of AND terms in a simplified function is too large, we may be forced to choose a PAL with more gate inputs and fewer outputs.

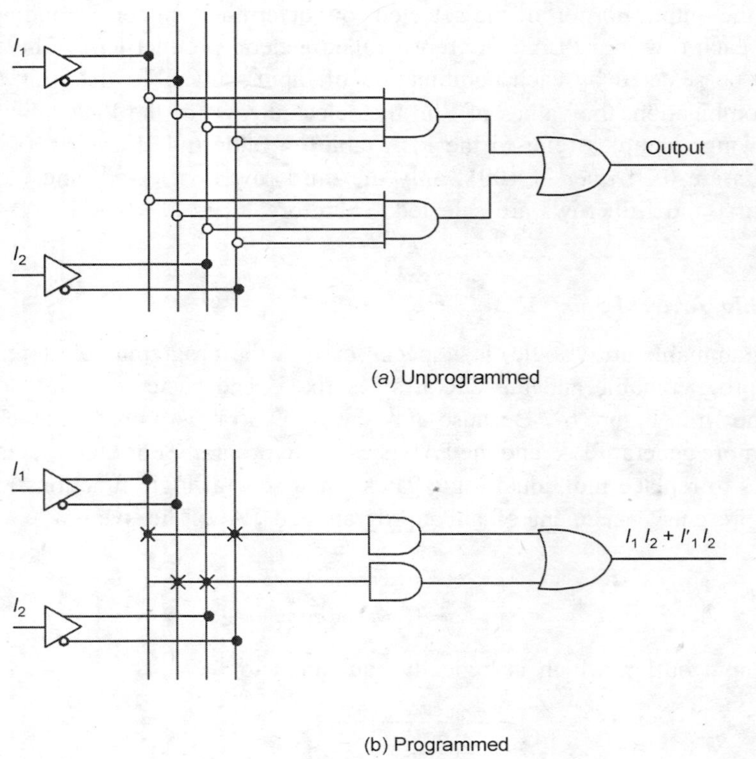

Figure 6.14: PAL Segment

As an example of programming a PAL, we will implement a full adder. The logic equations for the full adder are

$$Sum = X'Y'C_{in} + X'YC'_{in} + XY'C'_{in} + XYC_{in}$$
$$C_{out} = XC_{in} + YC_{in} + XY$$

Figure 6-15 shows a section of a PAL where each OR gate is driven by four AND gates. The X's on the diagram show the connections that are programmed into the PAL to implement the full adder equations. For example, the first row of X's implements the product term $X'Y'C_{in}$.

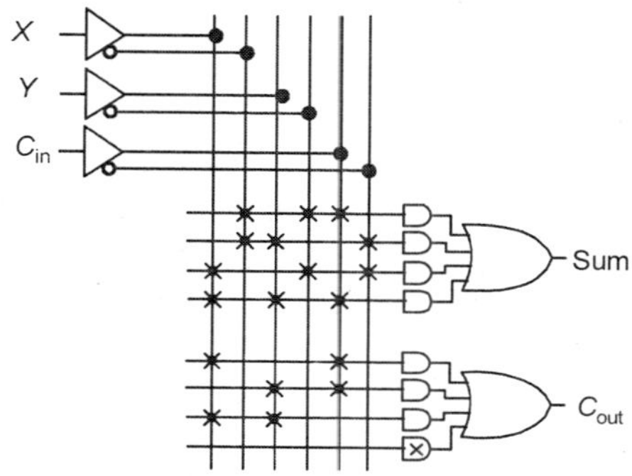

Figure 6.15: Implementation of a Full Adder Using a PAL

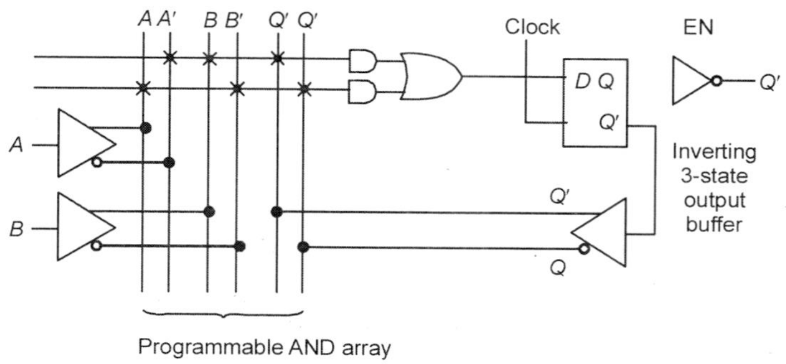

Figure 6.16: Segment of a Sequential PAL

Typical combinational PALs have from 10 to 20 inputs and from 2 to 10 outputs, with 2 to 8 AND gates driving each OR gate. PALs are also available that contain D flip-flops with inputs driven from the programmable array logic. Such PALs are called sequential PALs. They provide a convenient way of realizing sequential circuits. Figure 6-16 shows a segment of a sequential PAL. The D flip-flop is driven from an OR gate, which is fed by two AND gates. The flip-flop output is fed back to the pro-grammable AND array through a buffer. Thus, the AND gate inputs can be connected to *A*, *A'*, *B*, *B'*, *Q*, or *Q'*. The diagram shows the realization of the next state equation:

$$Q^+ = D = A'BQ' + AB'Q$$

The flip-flop output is connected to an inverting tristate buffer, which is enabled when EN = 1. A few decades ago, PALs were very popular among digital system designers. A very popular PAL was the 16R4. This PAL has an AND gate array with 16 input variables, and it has four D flip-flops. Nowadays, several other programmable devices, such as GALs CPLDs, and FPGAs, have arrived. PALs have practically disappeared; hence, we do not describe fur-ther any of the traditional PAL devices.

6.2.4 Programmable Logic Devices/Generic Array Logic

PALs and PLAs have been very popular for implementing small circuitry and interface logic often need-ed by designers. As integrated circuit technology has improved, a wide variety of other programmable logic devices have become available. Traditional PALs are not reprogrammable. However, there are flash erasable/reprogrammable PALs now. Often, these are referred to as PLDs.

The 22CEV10 (Figure 6-17 is a CMOS electrically erasable PLD that can be used to realize both combinational and sequential circuits. The abbreviation PLD has been used as a generic term for all programmable logic devices and also refers to specific devices such as the 22CEV10. In addition to the AND-OR arrays that the PALs have, most PLDs have some type of a mac-roblock that contains some multiplexers and some additional programmability. These PLDs are named with reference to their input and output capability. For instance, the 22CEV10 has 12 ded-icated input pins and 10 pins that can be programmed as either inputs or outputs. It contains 10 D flip-flops and 10 OR gates. The number of AND gates that feeds each OR gate ranges from 8 through 16. Each OR gate drives an *output logic macrocell.* Each macrocell contains one of the 10 D flip-flops. The flip-flops have a common clock, a common asynchronous reset (AR) input, and a common synchronous preset (SP) input. The name 22V10 indicates a versatile PAL with a total of 22 input and output pins, 10 of which are bidirectional I/O (input/output) pins.

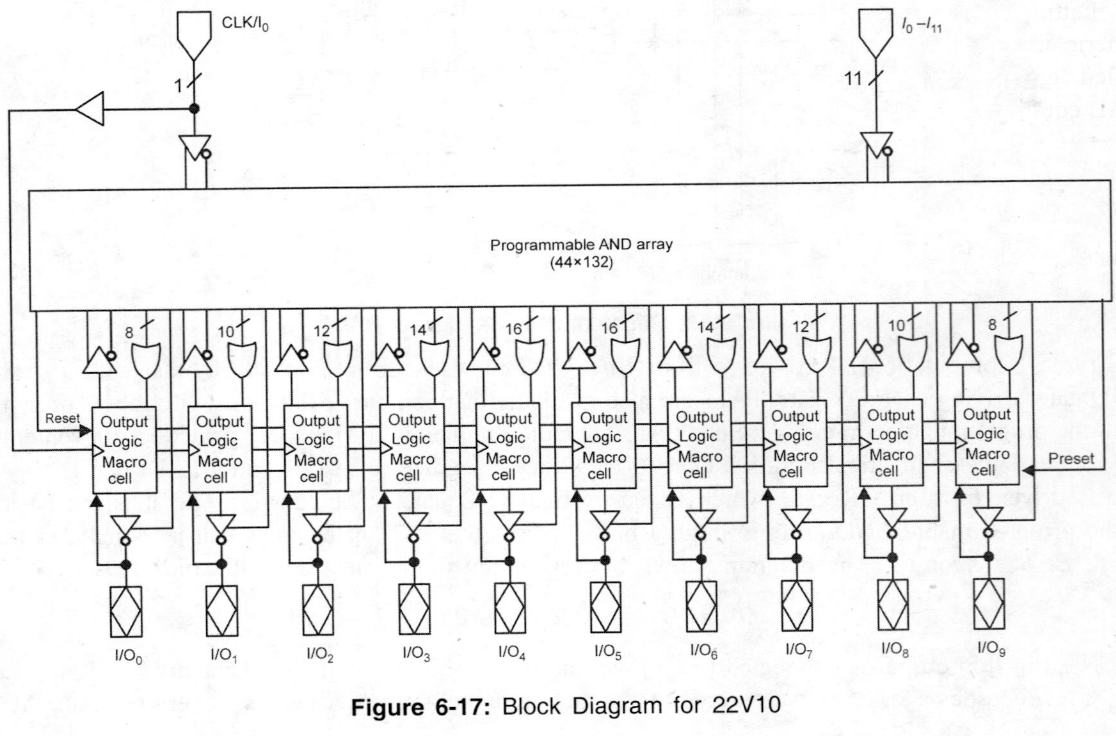

Figure 6-17: Block Diagram for 22V10

Figure 6-18 shows the details of a 22CE V10 output macroceli. The connections to the output pins are controlled by programming this macrocell. The output MUX control inputs S_1 and S_0 select one of the data inputs. For example, $S_1 S_0 = 10$ selects data input 2. Each macrocell has two programmable inter-connect bits. S_1 or S_0 is connected to ground (logic 0) when the corresponding bit is programmed. Erasing a bit disconnects the control line (S_1 or S_0) from ground and allows it to float to logic 1. When $S_1 = 1$, the flip-flop is bypassed, and the output is from the OR gate. The OR gate output is connected to the I/O pin through the multiplexer and the output buffer.The OR gate is also fed back so that it can be used as an input to the AND gate array. If $S_1 = 0$, then the flip-flop output is connected to the output pin, and it is also fed back so that it can be used for AND gate inputs. When $S_0 = 1$, the output is not invertec, so it is an active high. When $S_0 = 0$, the output is inverted, so it is an active low. The output pin is driven by a tristate inverting buffer. When the buffer output is in a high-impedance state, the OR gate and flip-flop are disconnected from the output pin, and the pin can be used as an input. The dashed lines in Figure6-18(a) show the path when both S_1 and S_0 are 0, and the dashed lines in Figure 6-18(*b*).show the path when both S_1 and S_0 are 1. Note that in the first case, the flip-flop output Q is inveriea oy the output buffer, and in the second case the OR gate output is inverted twice, so there is no net inversion.

Several PLDs similar to the 22V10 have been popular. Typically these PLDs had 8 to 12 I/O pins. Each output pin is typically connected to an output macrocell, and each macrocell has a D flip-flop. The I/O pins can be programmed so that they act as inputs or as combinational or flip-flop outputs. Some of the PLDs have a dedicated clock input, and the others have a dual-purpose pin that can be used either as a clock or as an input. All the PLDs typically have tristate buffers at the outputs, and some of them have a dedicated output enable ($\overline{OE}$).

Lattice Semiconductor created similar devices which are in-circuit programmable and called them generic array logic (GAL). GALs are perfect for implementing small amounts of interface logic, often called "glue" logic. Most of the common PLDs, like the PALCE22V10, PALCE20V8, and so on, have GAL equivalents, called GAL22V10, GAL20V8, and so on.

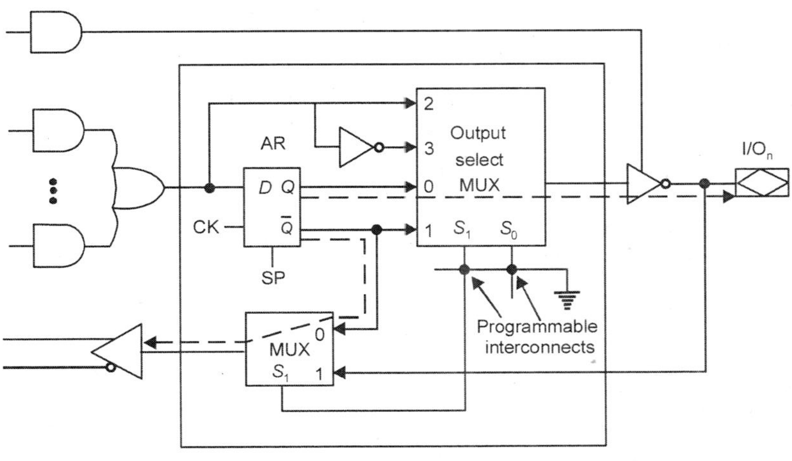

(a) Paths with $S_1 = S_0 = 0$

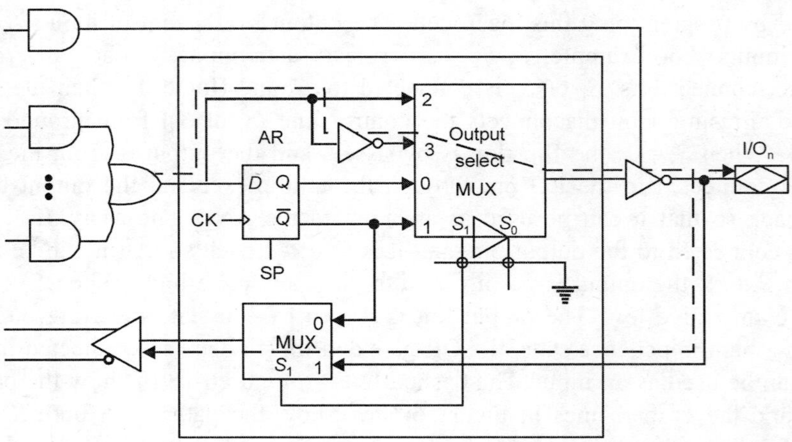

(b) Paths with $S_1 = S_0 = 1$

Figure 6.18: PLD Output Macrocell.

Design Flow for PLDs

Computer-aided design programs for PALs and PLDs are widely available. Such programs accept logic equations, truth tables, state graphs, or state tables as inputs and automatically generate the required bit patterns. These patterns can then be downloaded into a PLD programmer, which will create the necessary connections and verify the operation of the PAL. Many of the newer types of PLDs are erasable and reprogrammable in a manner similar to EPROMs and EEPROMs. Hence, in these newer devices, bit patterns corresponding to the required EEPROM content will be generated by the software.

PALASM and **ABEL** are examples of two languages that were popularly used with PALs and PLDs. PALASM is a PLD design language from MMI and AMD. ABEL is a PLD design language from DATA I/O. Intel used to manufacture PLDs and had a PLD language called PLDShell. While PALASM and ABEL can still be used, nowadays designs for GALs can be done using hardware description languages such as VHDL or Verilog.

6.3 COMPLEX PROGRAMMABLE LOGIC DEVICES

Improvements in integrated circuit technology have made it possible to create programmable ICs equivalent to several PLDs in the same chip. These chips are called complex programmable logic devices (CPLDs). When storage elements such as flip-flops are also included on the same IC, a small digital system can be implemented with a single CPLD.

CPLDs are an extension of the PAL concept. In general, a CPLD is an IC that consists of a number of PAL-like logic blocks together with a programmable interconnect matrix. CPLDs typically contain 500 to 10,000 logic gates. Essentially, several PLDs are interconnected using a crossbar-like switch and fabricated inside the same IC. An *N X M* crossbar switch is one in which each of the *N* input lines can be connected to any of the *M* output lines simultaneously. It is expensive to build these switches; however, use of such a switch results in predictable timing. Many CPLDs are electronically erasable and reprogrammable and are sometimes referred to as EPLDs (erasable PLDs).

A typical CPLD contains a number of macrocells that are grouped into function blocks. Connections between the function blocks are made through an interconnection array. Each macrocell contains a flip-flop and an OR gate, which has its inputs connected to an AND gate array. Some CPLDs are based on PALs, in which case each OR gate has a fixed set of AND gates associated with it. Other CPLDs are based on PLAs, in which case any AND gate output within a function block can be connected to any OR gate input in that block.

Xilinx, Altera, Lattice Semiconductor, Cypress, and Atmel are the major CPLD manufacturers in the market today. The major products available on the market are listed in Table 6-5 Some vendors specify their gate capacities in usable gates, and some specify it in terms of logic elements.

Table 6.5: Majo CPLDs and their Approximate Capacity

Vendor	CPLD family	Gate Count
Xilinx	CoolRunner-ll	750 to 12K
	CoolRunner XPLA3	750 to 12K
	XC9500XV	800 to 6400
	XC9500	800 to 6400
	XC9500XL	800 to 6400
Atmel	CPLD ATF15	750 to 3000 usable gates
	CPLD-2 22V10	500 usable gates
Cypress	Delta39K	30K to 200K
	Flash370i	800 to 3200
	Quantum38K	30K to 100K
	Ultra37000	960 to 7700
	MAX340 high-density EPLDs	600 to 3750
Lattice	ispXPLD 5000MX	75K to 300K
Semiconductor	ispMACH 4000B/C/V/Z	640 to 10,240
Altera	MAX II	240 to 2210 logic elements
	MAX3000	600 to 10K usable gates
	MAX7000	600 to 10K usable gates

6.3.1 An Example CPLD: The Xilinx CoolRunner

Xilinx has two major series of CPLDs, the CoolRunner and the XC9500. Figure 6.19 shows the basic architecture of a CoolRunner family CPLD, the Xilinx XCR3064XL. This CPLD has four function blocks, and each block has 16 associated **macrocells** (MC1, MC2, .. .). Each function block is a pro-grammable AND-OR array that is configured as a PLA. Each macrocell contains a flip-flop and multi-plexers that route signals from the function block to the input/output (I/O) block or to the **interconnect array** (IA).The interconnect array selects signals from the macrocell outputs or I/O blocks and connects them back to function block inputs. Thus, a signal generated in one function block can be used as an input to any other function block. The I/O blocks provide an interface between the bidirectional I/O pins on the IC and the interior of the CPLD.

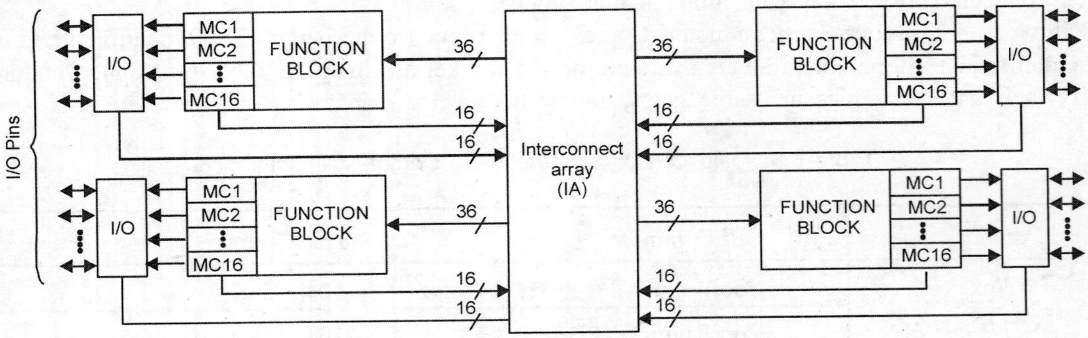

Figure 6-19: Architecture of Xilinx CoolRunner XCR$_3$06$_4$XL CPLD.

Figure 6-20 shows how a signal generated in the PLA (function block) is routed to an I/O pin through a macrocell. Any of the 36 inputs from the IA (or their complements) can be connected to any inputs of the 48 AND gates. Each OR gate can accept up to 48 product term inputs from the AND array. The macrocell logic in this diagram is a simplified version of the actual logic. The first mux (1) can be programmed to select the OR gate output or its complement. The mux (2) at the output of the macrocell can be programmed to select either the combinational output (G) or the flip-flop output (Q).This output goes to the interconnect array and to the output cell. The output cell includes a three-state buffer (3) to drive the I/O pin. The buffer enable input can be programmed from several sources. When the I/O pin is used as an input, the buffer must be disabled.

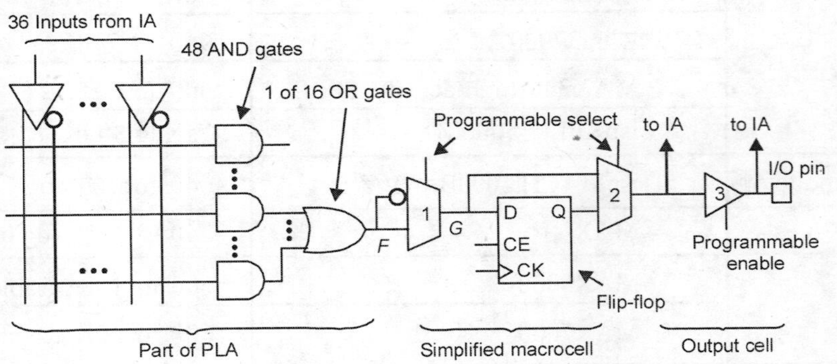

Figure 6-20 CPLD Function Block and Macrocell (Simplified Version of XCR$_{30}$6$_4$XL*)*

Figure 6-21shows how a Mealy sequential machine with two inputs, two outputs, and two flip-flops can be implemented by a CPLD. Four macrocells are required, two to generate the D inputs to the flip-flops and two to generate the Z outputs. The flip-flop outputs are fed back to the AND array inputs via the interconnection matrix (not shown). The number of product terms required depends on the complexity of the equations for the *D's* and the Z's

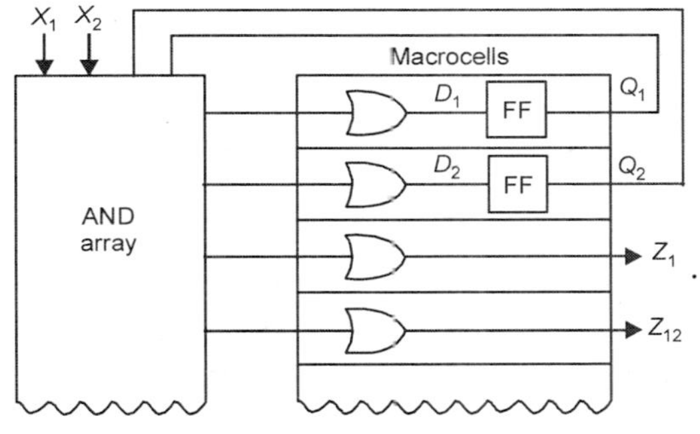

Figure 6-21: CPLD Implementation of a Mealy Machine

CPLD Implementation of a Parallel Adder with Accumulator

Assume that we need to implement an adder with an accumulator, as in Figure 3-34, in a CPLD. The accumulator register needs one flip-flop for each bit. Each bit also needs to generate the sun and carry bits corresponding to that bit.

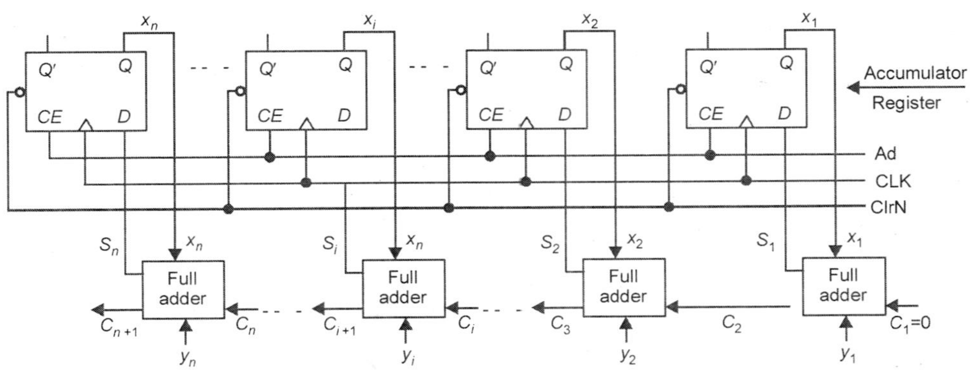

Figure 6-22: N-Bit Parallel Adder with Accumulator

Figure 6-22 shows how three bits of such a parallel adder with an accumulator can be implemented using a CPLD. Each bit of the adder requires two macrocells. One of the macrocells implements the sum function and an accumulator flip-flop. The other macrocell implements the carry, which is fed back into the AND array. The *Ad* signal can be connected to the enable input *(CE)* of each flip-flop via an AND gate (not shown). Each bit of the adder requires eight product terms (four for the sum, three for the carry, and one for *CE)*. For each accumulator flip-flop,

$$D_i = X_i^+ = S_i = X_i \oplus Y_i \oplus C_i$$

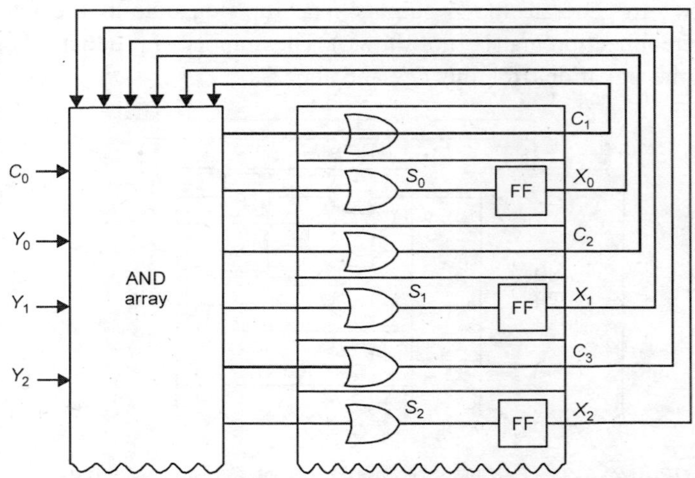

Figure 6-23 CPLD Implementation of a Parallel Adder with Accumulator

If the flip-flops are programmed as T flip-flops, then the logic for the sum can be simplified. For each accumulator flip-flop

$$X^{+}_{i} = X_i \oplus Y_i \oplus C_i$$

Therefore, the T input is

$$T_i = X^{+}_{i} \oplus X_i = Y_i \oplus C_i$$

The add signal can be AND'ed with the T_i input so that the flip-flop state only can change when $Ad = 1$:

$$T_i = Ad(Y_i \oplus C_i) = Ad\ Y_i C_i' + Ad\ Y'_i\ C_i$$

The equation for carry is

$$C_{i+1} = X_i Y_i + X_i C_i + Y_i C_i$$

6.4 FIELD PROGRAMMABLE GATE ARRAYS

In this section, we introduce field programmable gate arrays (FPGAs). FPGAs are ICs that contain an array of identical logic blocks with programmable interconnections. The user can program the functions realized by each logic block and the connections between the blocks. FPGAs have revolutionized the way prototyping and designing are done. The flexibility offered by reprogrammable FPGAs has enhanced the design process. While different kinds of programmable devices had been around, when Xilinx used static RAM (SRAM) storage elements to create programmable logic blocks and introduced

its family of XC2000 devices in 1985, the world received a totally new and powerful technology. There are a variety of FPGA products available in the market now. Xilinx, Altera, Lattice Semiconductor, Actel, Cypress, QuickLogic, and Atmel are examples of companies that design and sell FPGAs.

FPGAs provide several advantages over traditional gate arrays or mask programmable gate arrays (MPGAs). A traditional gate array can be used to implement any circuit but is programmable only in the factory. A specific mask to match the particular circuit is created in order to fabricate the gate array. The design time of a gate-array-based IC is a few months. FPGAs are standard off-the-shelf products. Manufacturing time reduces from months to hours as one adopts FPGAs instead of MPGAs. Design iterations become easier with FPGAs. This is a tremendous advantage when it comes to time-to-market. It becomes easy to correct mistakes that creep into designs. Mistakes and design specification changes become less costly. Prototyping cost is reduced. At low volumes, FPGAs are cheaper than MPGAs.

FPGAs have disadvantages, too. FPGAs are less dense than traditional gate arrays (MPGAs). In FPGAs, a lot of resources are spent to merely achieve the programmability. MPGAs have better performance than FPGAs. Programmable points have resistance and capacitance. They slow down signals, so FPGAs are slower than traditional gate arrays. Also, interconnection delays are unpredictable in FPGAs. PLDs, like PALs and GALs, are simple and inexpensive. CPLDs are faster than FPGAs and are cheaper. The overhead for programmability is fairly low in PALs and CPLDs. The main advantage of CPLDs over FPGAs is the lower cost and predictability in timing.

Several commercial FPGAs are listed in Table 6-6 As we notice, some of these chips contain logic equivalent to 5 million gates. The capacity of some FPGAs is specified in number of look-up tables

Table 6-6: Examples of Commercial FPGAs

Xilinx	Spartan-ll	15K to 200K
Vendor	FPGA Product	Capacity (Approx) in Gates/LUTs
Xilinx	Spartan-ll	15K to 200K
	Spartan-IIE	50K to 600K
	Spartan-3	50K to 5M
	Virtex-5	19,200 to 207,360 LUTs
	Virtex	57,906 to 1,124,022
Virtex-E	71,693 to 4,074,387	
	Virtex-ll	40K to 8M
Altera	ACEX 1K	56K to 257K
	APEX II	1.9M to 5.25M
	FLEX 10K	10K to 50K
	Stratix/Stratix II	10,570 to 132,540 logic elements
Lattice Semiconductor	LatticeECP2	6K to 68K LUTs
	Lattice SC	15.2K to 115.2KLUTS
	ispXPGA	139K TO 1.25M
	MachXO	256 to 2280 LUTs
	LatticeECP	6.1K to 32.8K LUTs
Actel	Axceierator	125K To 2M
	eX	3K to 12K
	ProASO	30K to 3M
	MX	3K to 54K
Quick Logic	Eclipse/EclipsePlus	248K to 662K
	Quick RAM	45K to 176K
	pASIC 3	5K to 75K
Atmel	AT40K	5K to 40K
	AT40KAL	5K to 50K

(LUTs). Due to the large capacity, it is possible to prototype or even manufacture large systems in a single FPGA. In this chapter, we describe the basic organization of FPGAs.

Organization of FPGAs

Figure 6-24 shows the layout of a typical FPGA. The interior of FPGAs typically contains three ele-ments that are programmable:

Programmable logic blocks
Programmable input/output blocks
Programmable routing resources

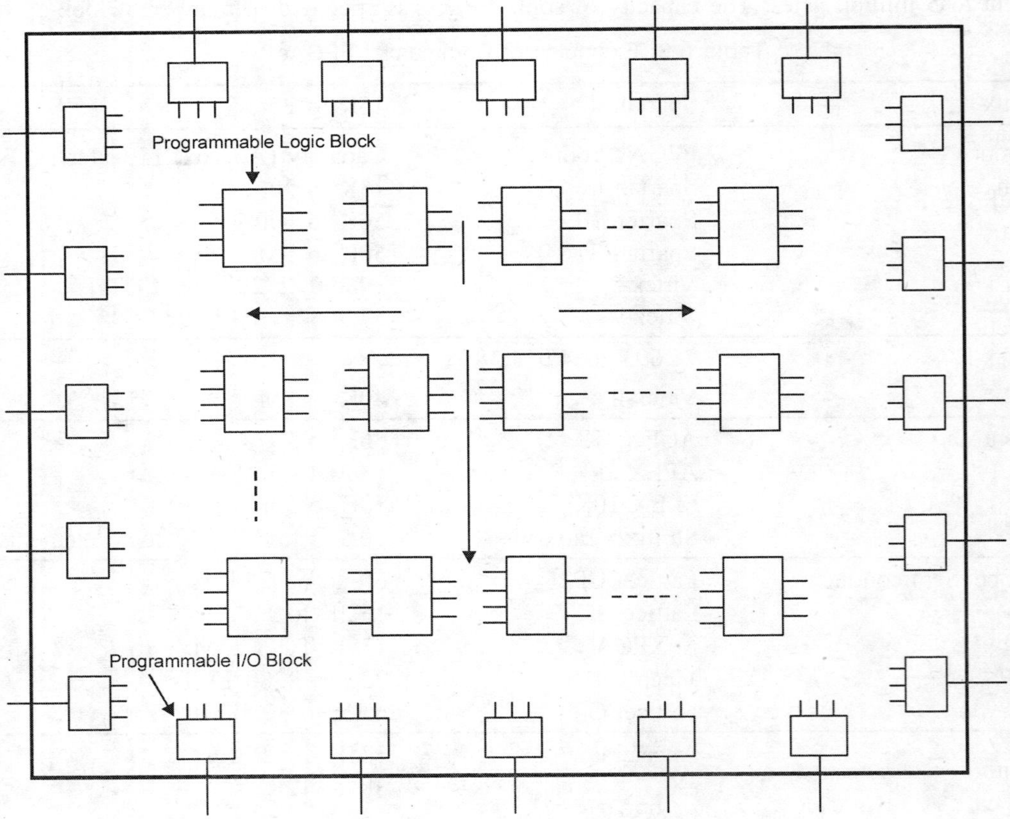

Figure 6.24: Layout of a Typical FPGA

Arrays of programmable logic blocks are distributed within the FPGA. These logic blocks are sur-rounded by input/output (I/O) interface blocks. These I/O blocks can be considered to be on the periphery of the chip. They connect the logic signals to FPGA pins. The space between the logic blocks is used to route connections between the logic blocks.

The "field" programmability in FPGAs is achieved by reconfigurable elements, which can be programmed or reconfigured by the user. As mentioned, there are three major programmable ele-ments in FPGAs: the logic block, the interconnect, and the input/output block. Programmable logic blocks are created by using multiplexers, look-up tables, and AND-OR or NAND-NAND arrays. "Programming" them means changing the input or control signals to the multiplexers, changing the look-up table contents, or selecting/not selecting particular gates in AND-OR gate blocks. For a programmable interconnect, "programming" means making or breaking specific con-nections. This is required to interconnect various blocks in the chip and to connect specific I/O pins to specific logic blocks. Programmable I/O blocks denote blocks which can be programmed to be input, output, or bidirectional lines. Typically, they can also be "programmed" to adjust the prop-erties of their buffers such as inverting/noninverting, tristate, passive pull-up, or even to adjust the **slew rate,** which is the rate of change of signals on that pin.

What makes an FPGA distinct from a CPLD is the flexible general-purpose interconnect. In a CPLD, the interconnect is fairly restricted. The general-purpose interconnect in an FPGA gives it a lot of flexibility, but it also has the disadvantage of being slow. A connection from one part of the chip to another part might have to travel through several programmable interconnect points, resulting in large and unpredictable signal delays.

While Figure 6-24 was used to illustrate the general structure of an FPGA, not all FPGAs look like that. Commercial FPGAs use a variety of architectures. The FPGA architecture or organization refers to the manner or topology in which the logic blocks and interconnect resources are distributed inside the FPGA. The organization that is presented in Figure 6-24 is often referred to as symmetrical array architecture. If we examine the various FPGAs that have been on the market since their inception in the late 1980s, we could classify them into four different basic architectures or topologies:

Matrix-based (symmetrical array) architectures
Row-based architectures
Hierarchical PLD architectures Sea-of-gates architecture

These architectures are illustrated in Figure 6-25.

Matrix-Based (Symmetrical Array) Architectures

The logic blocks in this type of FPGA are organized in a matrix-like fashion as illustrated in Figure 6-25(a).Most Xilinx FPGAs belong to this category. The logic blocks in these architectures are typically of a large granularity (capable of implementing four-variable functions or more). These architectures typically contain 8 x 8 arrays in the smaller chips and 100 X 100 or larger arrays in the bigger chips. The routing resources are interspersed between the logic blocks. The routing in these architectures is often called two-dimensional channeled routing since routing resources are general-ly available in horizontal and vertical directions.

Row-Based Architectures

These architectures were inspired by traditional gate arrays. The logic blocks in this architecture are organized in rows, as illustrated in Figure 6-25(b).Thus, there are rows of logic blocks and routing resources. The routing resources interspersed between the rows can be used to interconnect the var-ious logic blocks. Traditional mask programmable gate arrays use very similar architectures. The routing in these architectures is often called one-dimensional channeled routing because the routing resources are located as a channel in between rows of logic resources. Some Actel FPGAs employ this architecture.

Hierarchical Architectures

In some FPGAs, blocks of logic cells are grouped together by a local interconnect and several such groups are interconnected by another level of interconnect. For instance, in Altera APEX20 and APEX II FPGAs, 10 or so logic elements are connected to form what Altera calls a logic array block (LAB), and then several LABs are connected to form a MEGALAB. Thus, there is a hierarchy in the organization of these FPGAs. These FPGAs contain clusters of logic blocks with localized resources for interconnection. The global interconnect network is used for the interconnections between the clusters of logic blocks in these FPGAs.

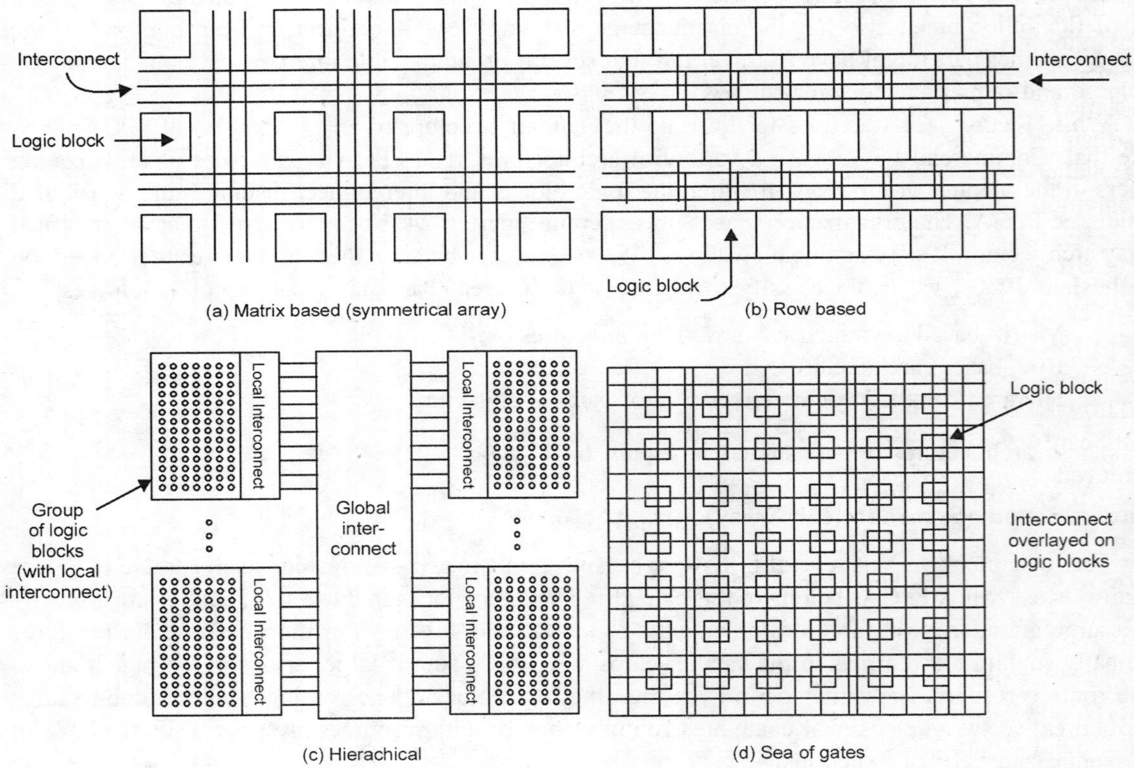

Figure 6.25: Typical Architectures for FPGAs

Sea-of-Gates Architecture

The sea-of-gates architecture is yet another manner to organize the logic blocks and **interconnect** in an FPGA. The general FPGA fabric consists of a large number of gates, and then there is an interconnect superimposed on the sea of gates as illustrated in Figure 6-25(d). Plessey, a manufacturer who was in the FPGA market in the mid-1990s, made FPGAs ofthis architecture. The basic cell they used was a NAND gate, in contrast to the larger basic cells used by manufacturers like Xilinx. While the terminology *sea of gates* is the most popular, there are also terminologies like **sea of cells** and **sea of tiles** to indicate the topology of FPGAs with a large number of fine-grain logic cells. The Actel Fusion FPGAs contain a sea of tiles, where each tile can be configured as a three-input logic function or a flip-flop/latch.

6.4.2 FPGA Programming Technologies

FPGAs consist of a large number of logic blocks interspersed with a programmable interconnect. The logic block is programmable in the sense that the same building block can be "programmed" or "configured" to create any desired circuitry. There is also programmability in the* interconnections between the logic blocks.

Several techniques have been used to achieve the programmable interconnections between FPGAs. The term **programming technology** is used here to denote the technology by which the programmability in an FPGA is achieved. In some devices, the reconfigurability is achieved by changing the contents of static RAM cells. In some devices, it is achieved by using flash memory cells. In others, it is achieved by fusing metal links. In general, FPGAs use one of the following programming methods:

StaticRAM programming technology EPROM/EEPROM/flash programming technology Antifuse programming technology

The SRAM Programming Technology

The SRAM programming technology involves creating reconfigurability by bits stored in static RAM (SRAM) cells. The logic blocks, I/O blocks, and interconnect can be made programmable by using con-figuration bits stored in SRAM. Reconfigurable logic blocks can easily be implemented as LUTs, which is the same approach as the ROM method described in Section 6.2.1 Sixteen SRAM cells can imple-ment any function of four variables. The programmable interconnect can also be achieved by SRAM. The key idea is to use pass transistors to create switches and then control them using the SRAM con-tent. Consider the arrangement in Figure 6-26(a).The SRAM cell is connected

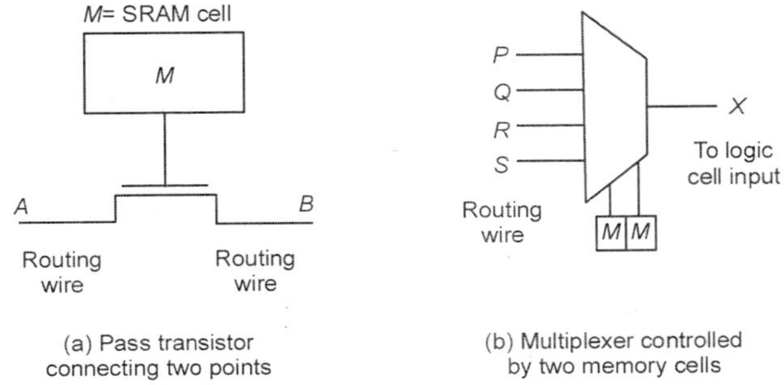

Figure 6-26: Routing with Static RAM Programming Technology.

to the gate of the pass transistor. When the SRAM cell content is 0, the pass transistor is OFF, and hence no connection exists between points *A* and *B*. *A* closed path can be achieved by turning the pass transistor ON by making the SRAM cell content 1. SRAM bits can be used to construct routing matrices by using multiplexers as in Figure 6-26(b).Changing the contents of the SRAM in the arrangement in Figure6-26(b) will allow the designer to change what is connected to point *X.* The bits that are stored in the SRAM for deciding the LUT functionality or interconnection are called **configuration bits.**

A SRAM cell usually takes six transistors, as illustrated in Figure 6-27 Four cross-coupled transistors are required to create a latch, and two additional transistors are used to control passing data bits into the latch. When the *Word Line* is set to high, the values on the *Bit Line* will be latched into the cell. This is the write operation. The read operation is performed by precharging the *Bit Line* and *Bit Line* to a logic 1 and then setting *Word Line* to high. The contents stored in the cell will then appear on the *Bit Line.* Some SRAM cell implementations only use five transistors. One advantage of using static RAM is that it is volatile and you can write new contents again and again. This provides flexibility during prototyping and development. Another advantage is that the fabrication steps for making SRAM cells are not different from the steps for making logic. The major disadvantage of the SRAM programming technology is that five or six transistors are used for every SRAM cell. This adds a tremendous cost to the chip. For example, if an FPGA has 1 million programmable points, it means that approximately 5 or 6 million transistors are spent in achieving this programmability.

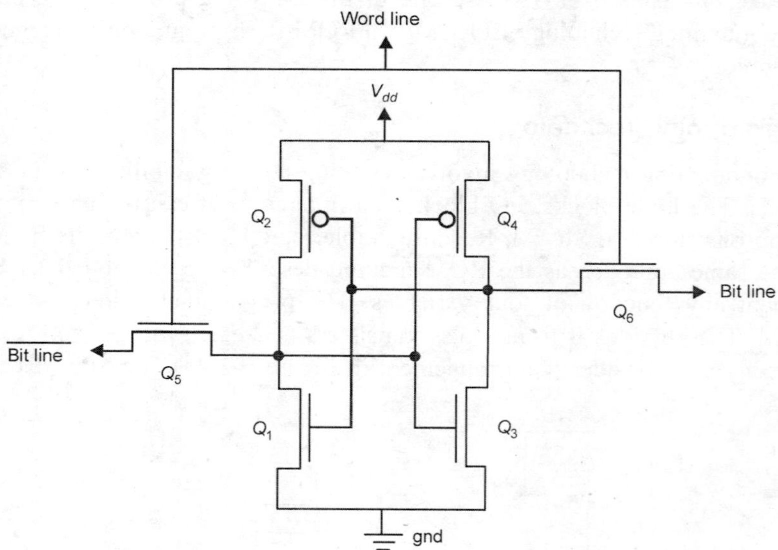

Figure 6.27: Typical Six-Transistor SRAM Cell

Being volatile can become a disadvantage when an FPGA is used in the final product. Hence, when SRAM FPGAs are used, a nonvolatile device such as an EPROM should be used to permanently store the configuration bits. Typically, what is done is to use the EPROM as a "boot ROM." The EPROM contents are transferred to the SRAM when power comes up.

Xilinx FPGAs were the first FPGAs to use SRAM as the programming technology. In fact, it is the flexibility and reprogrammability of SRAM FPGAs that caused FPGAs to become widely popular. Now, many companies use the SRAM programming technology for their FPGAs.

EPROM/EEPROM Programming Technology

In the EPROM/EEPROM programming technology, EPROM cells are used to control programmable connections. Assume that EPROM/EEPROM cells are used instead of the SRAM cells in Figure 6-26 transistor with two gates, a floating gate and a control gate, is used to create an EPROM cell. Figure6-28 illustrates an EPROM cell. The pull-up resistor connects the drain of the transistor to the power supply (labeled V_{dd} in the figure).To turn the transistor off, charge can be injected on the floating gate using a high voltage between the control gate and the drain of the transistor. This charge increases the threshold voltage of the transistor and turns it off. The charge can be removed by exposing the floating gate to ultraviolet light. This lowers the threshold voltage of the transistor and makes it function normally.

EPROMs are slower than SRAM; hence, SRAM-based FPGAs can be programmed faster. EPROMs also require more processing steps than SRAM. EPROM-based switches have high ON resistance and high static power consumption. The EEPROM is similar to EPROM, but removal of the gate charge can be done electrically.

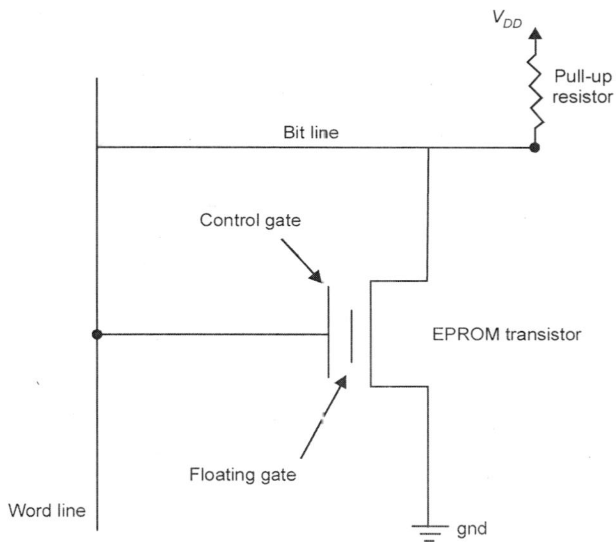

Figure 6.28: The EPROM Programming Technology

Flash memory is a form of EEPROM that allows multiple locations to be erased in one operation. Flash memory stores information in floating-gate transistors as in traditional EPROM. The floating gate is isolated by an insulating oxide layer, and hence any electrons placed there are trapped. The cell is read by placing a specific voltage on the control gate. When the voltage to read is placed, electrical current will or will not flow depending on the threshold voltage of the cell, which is controlled by the number of electrons trapped in the floating gate. In some devices, the information is stored as absence or presence of current. In some advanced devices, the amount of current flow is sensed, and hence mul-tiple bits of information can be stored in a cell. To erase, a large voltage differential is placed between the control gate and source, which pulls electrons off. Flash memory is erased in segments/sectors; all cells in a block are erased at the same time.

The Antifuse Programming Technology

In some FPGAs, the programmable connections between different points are achieved by what is called an "antifuse." Contrary to fuse wires that blow open when high current passes through them, the "anti-fuse" programming element changes from high resistance (open) to low resistance (closed) when a high voltage is applied to it. Antifuses are often built using dielectric layers between N+ diffusion and poly-silicon layers or by amorphous silicon between metal layers. Antifuses are normally OFF; permanently connected links are created when they are programmed. The process is irreversible, and hence antifuse FPGAs are only one-time programmable. Programming an antifuse requires applying a high voltage and currents in excess of normal currents. Special programming transistors larger than normal transistors are incorporated into the device in order to accomplish the programming. There are different antifuse tech-nologies; a popular one is the Via antifuse technology.

Antifuse technology has the advantage that the area consumed by the programmable switch is small. Another advantage is that antifuse-based connections are faster than SRAM- and EEPROM-based switches. The disadvantage of the antifuse technology is that it is not reprogrammable. It is a permanent connection; if an error or design change necessitates reprogramming, a new device is required.

Comparison of Programming Technologies

Table 6-7 compares the characteristics of the major programming technologies used by FPGAs. Only the SRAM and EEPROM programming technologies allow in-circuit programmability. In-circuit programmability means that an FPGA can be reprogrammed without removing it from the board in which it is used. In-circuit programmability is not possible in traditional EPROM-based devices, but EEPROM/flash technologies allow in-circuit reprogrammability.

Table 6.7: Characteristics of the Major FPGA Programming Technologies

Programming Technology	Volatility	Programmability	Area Overhead	Resistance	Capacitance
SRAM	Volatile	In-circuit reprogrammable	Large to high	Medium	High
EPROM	Nonvolatile	Out-of-circuit reprogrammable	Small	High	High
EEPROM	Nonvolatile	In-circuit reprogrammable	Medium to high	High	High
Antifuse	Nonvolatile	Not reprogrammable	Small	Small	Small

SRAM FPGAs have several disadvantages: high area overhead, large delays, volatility, and so on. However, the in-circuit programmability and fast programmability have made them very popular. SRAM FPGAs are more expensive than other types of FPGAs because each programmable point uses six transistors. This extra hardware contributes only to the reprogrammability but not to the actual circuitry realized with the FPGA. EEPROM/flash-based FPGAs are comparable to SRAM FPGAs in many aspects; however, they are not as fast as SRAM FPGAs.

6.4.3 Programmable Logic Block Architectures

FPGAs in the past have employed different kinds of programmable logic blocks as the basic building block. In this section, we present some generalized versions of typical building blocks in commercial FPGAs.

The logic blocks vary in the basic components they use. For instance, some FPGAs use LUT-based logic blocks, while others use multiplexers and logic gates to build their logic blocks. There also have been FPGAs where logic blocks simply consisted of transistor pairs (e.g., crosspoint FPGAs). Logic building blocks in early Altera FPGAs were PLD blocks. There were also FPGAs that used NAND gates as the building block (e.g., Plessey).

The logic blocks also vary in their architecture and size. Some FPGAs use large basic blocks, which can implement large functions (several five-variable or four-variable functions) and have several flip-flops in each basic block. In contrast, there are FPGA building blocks which only allow a three-variable function or a flip-flop in one block. Some FPGAs allow choices as to whether latched/unlatched or both kinds of outputs can be brought out. Some FPGAs allow one to control the type of flip-flop that is realized. Some allow positive edge/negative edge clock, direct set/reset inputs to the flip-flop, and so on. Different FPGA manufacturers use different names (often trade-marked) to denote their logic blocks. In the Xilinx literature, a programmable logic block is called a **Configurable Logic Block (CLB).** Altera calls their basic blocks **Logic Elements (LE)** and a collec-tion of 8 or 10 of them **Logic Array Blocks (LABs).** The basic cells in Actel Fusion FPGAs are referred to as **Versailles.**

Look-Up Table-Based Programmable Logic Blocks

Many LUT-based FPGAs use a four-variable look-up table plus a flip-flop as the basic element and then combine several of them in various topologies. Consider the structure in Figure 6.29 There are two four-variable look-up tables (often denoted by the short form **LUT4)** and two flip-flops in this programmable logic block. The LUT4 can also be called a **four-variable function generator** since it can generate any function of four variables. The two LUT4s can generate any two functions

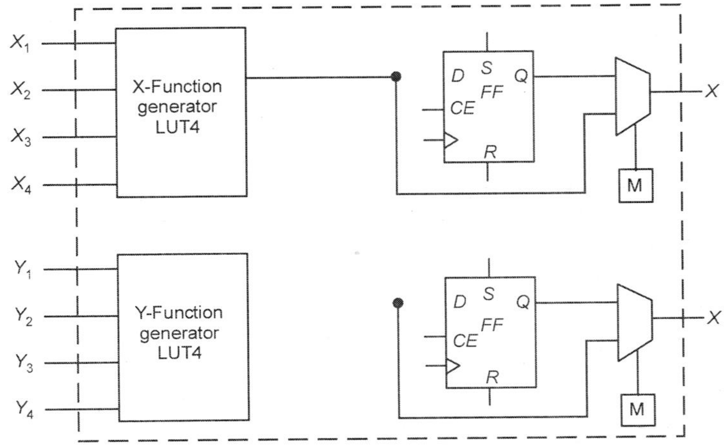

Figure 6.29: *A* Look-Up Table-Based Programmable Logic Block.

of four variables. The inputs to the *X-function* generator are called X_1, X_2, X_3, and X_4, and the inputs to the *Y-function* generator are called Y_1, Y_2, Y_3, and Y_4. The functions can be steered to the output of the block *(X* and *Y)* in combinational or latched form. There are two D flip-flops in the logic block. The D flip-flops are versatile in the sense that they have clock enable, direct set, and direct reset inputs. A multiplexer selects between the combinatorial output and the latched version of the output. The little box with "M" in it (beneath the multiplexer) indicates a memory cell that is required to provide appropriate select signals to select between the latched and unlatched form of the function. An early Xilinx FPGA, the XC3000, used building blocks very similar to this structure.

Let us assume that we want to implement the function $F_1 = A'B'C + A'BC + AB$ using an FPGA with programmable logic blocks as in Figure 6-29. Since this is a three-variable function, a four-input LUT is more than sufficient to implement the function. The path highlighted in Figure 6-30assumes that the X-function generator (top LUT) is used. Let us assume that X_1 is the LSB and X_4 is the MSB to the LUT. Since function F_1 only uses three variables, the X_4 input is not used. A truth table can be constructed to represent the function, and the LUT contents can be derived.

The LUT contents to implement function F_1 will be 0,1,1,0,0,0,1,1,0,1,1, 0,0,0,1,1. The first 8 bits in the LUT reflect the truth table outputs when the function is represented in a truth table form. Since input X_4 is not grounded, the first 8 bits are repeated to take care of the possibility that the X_4 input might stay at a logic 1 when it is unused. Since the functions are stored in LUT form, the number of terms in the function is not important. Common minimizations to reduce the number of terms are not relevant. The number of variables is what is important.

Many commercial FPGAs use LUTs. Examples are the Xilinx Spartan/Virtex, Altera Cyclone II/ APEX II, QuickLogic Eclipse/PolarPro, and Lattice Semiconductor ECP Many of these FPGAs put two or more four-input LUTs into a block in various topologies. Some FPGAs also provide multiplexers in addition to look-up tables.

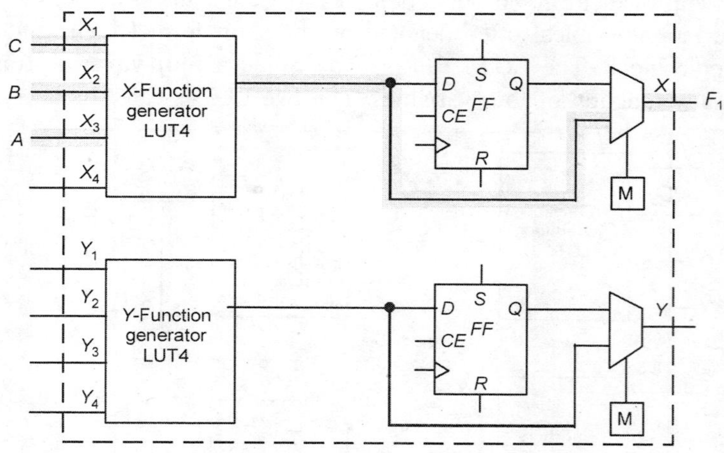

Figure 6.30: Highlighting Paths for Function F_1

Logic Blocks Based on Multiplexers and Gates

Some FPGAs use multiplexers as the basic building block. As you know, any combinational function can be implemented using multiplexers alone. In the most naive method, a 4-to-l multiplexer can gen-erate any two-input function. If inverted inputs can be provided, a 4-to-l multiplexer can generate any three-input function. Examples of multiplexer-based basic blocks are given in Figure 6-34 Logic blocks similar to these were used in early Actel FPGAs such as the ACT I and ACT II.

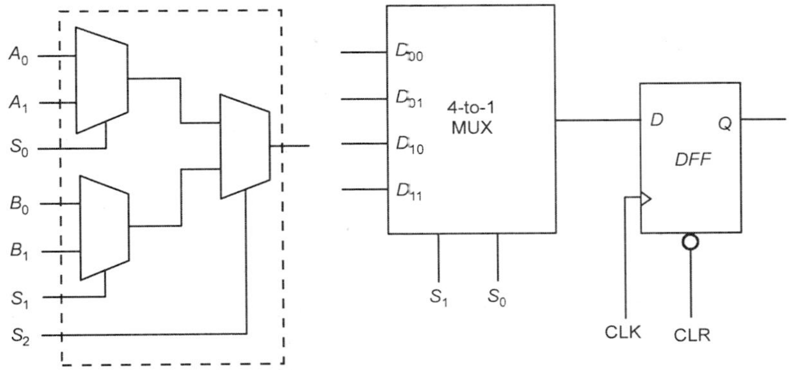

(a) (b)

Figure 3.31: *Multiplexer-Based Logic Blocks in FPGAs*

Let us assume that we want to implement the function $F_l = A'B'C + A'BC + AB$ using an FPGA with programmable logic blocks consisting of 4-to-l multiplexers. Two of the three-input variables can be connected to the multiplexer select lines. Then we have to provide appropriate sig-nals to the multiplexer data input lines in order to realize the function. To derive these inputs, we will first construct a truth table of the function as shown below:

A	B	C	F	*Mux Input in Terms* $\{0, 1, C, C\}$
0	0	0	0	$\}C$
0	0	1	1	
0	1	0	1	$\}C$
0	1	1	0	
1	0	0	0	$\}0$
1	0	1	0	
1	1	0	1	0

Let us assume that A and B are connected to the select inputs of the multiplexer. Next, we will derive values of inputs to provide to the multiplexer input lines in terms of the third variable in the function. The third variable is C, and by providing one of the four values $[C, C', 0, 1\}$, any three variable function can be expressed. Considering the first two rows of the truth table, it can be seen that $F = C$ when $AB = 00$. Similarly, considering the third and fourth rows of the truth table, $F = C$ when $AB = 01$. When $AB = 10$, $F = 0$ irrespective of the value of C. Similarly, when $AB = 11$, the value of the function equals 1.The last column in the truth table presents the required multiplexer inputs. Hence, one 4-to-l multi-plexer with the connections shown in Figure 6-32 can implement function F_1.

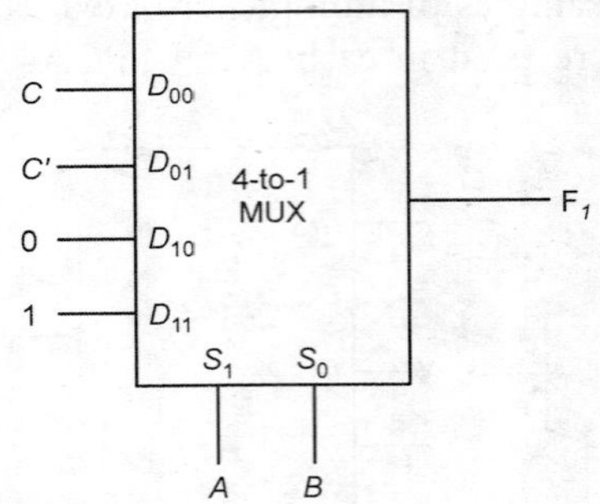

Figure 6.32: Multiplexer Implementing Function F_1

In the past three sections, we have provided an overview of the general architecture, logic block types, and programming technologies that can be used to build FPGAs. The general architecture, programming technology, and logic block types of several example commercial FPGAs are summarized in Table 6-8. LUT-based FPGAs are very common, especially for Xilinx and Altera. Actel is the manufacturer of multiplexer-based Fogs. SRAM programming technology, while expensive, is also common.

6.4.4 Programmable Interconnects

A key element of an FPGA is the general-purpose programmable interconnect interspersed between the programmable logic blocks. There are different types of interconnection resources in all commercial FPGAs. Every vendor has its own specific names for the different types of interconnects in its FPGA.

Table 6.8: Architecture, Technology, and Logic Block Types of Commercial FPGAs

Company	Device Names	General Architecture	Logic Block Type	Programming Technology
Actel	ProASIC/ProASO/ ProASIC^{plus}	Sea of Tiles	Multiplexers & Basic Gates	SRAM
	SX/SXA/eX/MX	Sea of Modules	Multiplexers & Basic Gates	Antifuse
	Accelerator	Sea of Modules	Multiplexers & Basic Gates	SRAM
	Fusion	Sea of Tiles	Multiplexers & Basic Gates	Flash, SRAM
Xilinx	Virtex	Symmetrical Array	LUT	SRAM
	Spartan	Symmetrical Array	LUT	SRAM
Atmel	AT40KAL	Cell Based	Multiplexers & Basic Gates	SRAM
QuickLogic	Eclipse II	Flexible Clock	LUT	SRAM
	PolarPro	Cell Based	LUT	SRAM
Altera	Cyclone II	Two-Dimensional Row and Column Based	LUT	SRAM
	Stratix **II**	Two-Dimensional Row and Column Based	LUT	SRAM
	APEX **II**	Row and Column, but Hierarchical Interconnect	LUT	SRAM

Interconnects in Symmetric Array FPGAs

In this section, we discuss some of the basic elements used for interconnection in symmetric array FPGAs.

General-Purpose Interconnect: Many FPGAs use switch matrices that provide interconnections between routing wires connected to the switch matrix. Figure 6.33(a) illustrates interconnecting logic blocks in an FPGA using switch matrices. Many FPGAs use this type of interconnect. A typical switch matrix is illustrated in Figure 6.33(b). where there is a switch at each intersection (i.e., wherever the lines cross). A switch matrix tha supports every possible connec-tion from every wire to every other wire is very expensive. The connectivity is often limited to some subset of a full crossbar connection; moreover, not all connections might be possible simul-taneously. In the switch matrix illustrated in Figure 6.33(b), each wire from a side of the switch can be routed to other wires using some combination of the switches. In order to support this type of a connection, each cross point in the switch matrix must support six possible interconnections as marked in Figure 6.33(*c*).

Depending on the programming technology, SRAM cells, flash memory cells, or antifuse connections control the configuration of the switches. The switch matrices interspersed between the logic blocks in an FPGA allow general-purpose interconnectivity between arbitrary points in the chip. However, the switch matrices are expensive in area and time (delay). If a signal passes through several of these switch matrices, it could contribute to a significant signal delay. Moreover, the delays are variable and unpredictable depending on the number of the switch matrices involved in each signal. In contrast, the interconnection resources in a CPLD are more restricted. However, interconnections in CPLDs result in smaller and more predictable delays.

Direct Interconnects: Many FPGAs provide special connections between adjacent logic blocks. These interconnects are fast because they do not go through the routing matrix. Many FPGAs provide direct interconnections to the four nearest neighbors: top, bottom, left, and right. Figure 3.34 illustrates examples of direct connections. In some cases, there are special interconnections to eight neighboring blocks, including the diagonally located logic blocks (Figure 6-34(b*)).* The direct interconnections do not go through the switch matrix but are implemented with dedicated switches, resulting in smaller delays. These types of direct interconnects are used in some Xilinx FPGAs.

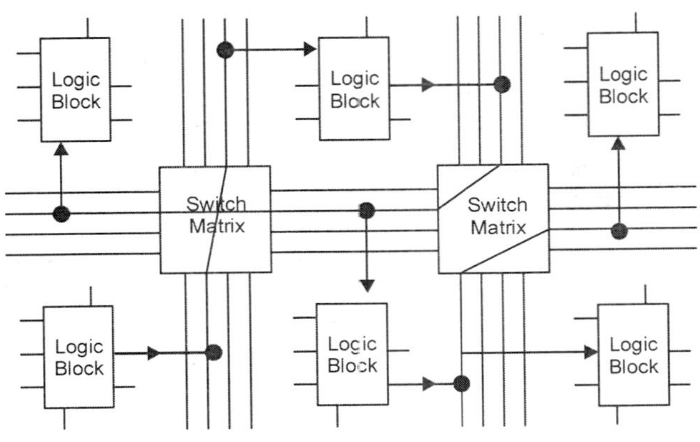

(a)

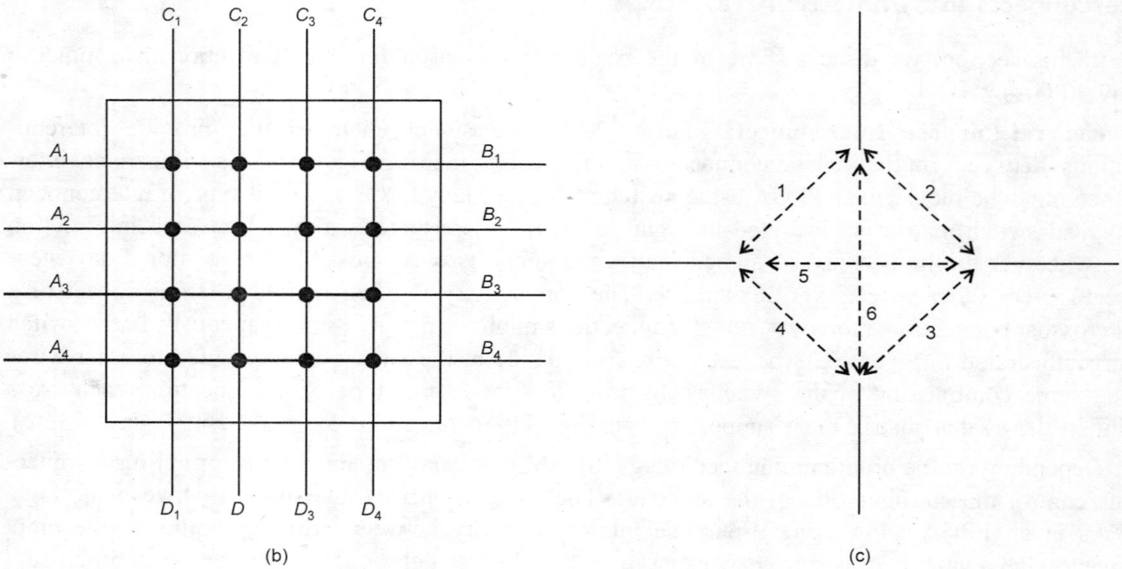

Figure 6.33: Routing Matrix for General-Purpose Interconnection in an FPGA

Global Lines: For purposes like high fan-out and low-skew clock distribution, most FPGAs provide routing lines that span the entire width of the device/height of the device. A limited number (two or four) of such global lines is provided by many FPGAs in the horizontal and vertical directions. Figure 6-35 llustrates **horizontal long lines** (global lines) in an example FPGA. The logic blocks often have tristate buffers to connect to the global lines.

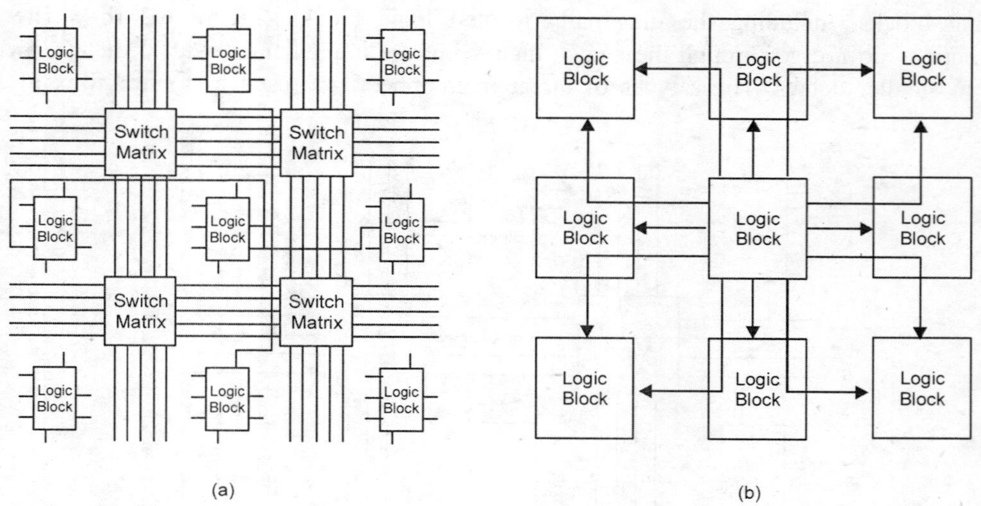

Figure 6.34: Direct Interconnects between Neighboring Logic Blocks

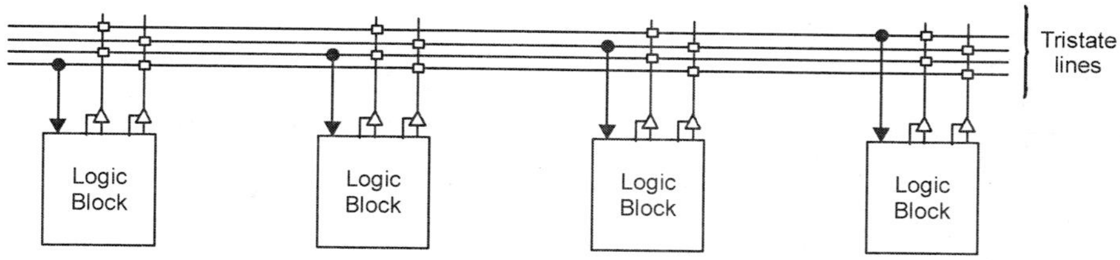

Figure 6.5: Global Lines

Clock Skew

There are several million gates in modern FPGA chips. When a clock is dis-tributed to various parts of such a large chip, the delays in the wire carrying the clock can result in the clock edge arriving at different times at different parts. This difference in the actual edge of the clock as it arrives at different flip-flops or other devices is called clock skew. Clock skew is a problem in large systems, including modern microprocessors. Carefully planned clock dis-tribution circuits are implemented in most systems in order to minimize the effect of clock skew. Modern FPGAs provide specialized clock distribution circuitry in order to create a clock of sufficient strength and low skew.

Interconnects in Row-Based FPGAs

Many of the interconnect resources mentioned previously are very characteristic of symmetric array devices with a two-dimensional array of logic blocks (e.g., Xilinx). In devices that are row based, there are rows of logic blocks, and there are channels of switches to enable connections between the logic blocks. Several switches are used to route a signal from a logic block in one row to another logic block elsewhere in the chip. There are arrays of switches in the routing channel between the rows of logic. The routing resources in these FPGAs are very similar to routing in traditional gate arrays.

The interconnects in row-based channeled architecture can be classified into two categories: non-segmented routing and segmented routing. In order to understand different types of channel routing, consider the connections *x, y,* and *z* in 6.36(a). Figure 6.36(b). indicates what is called as a **nonsegmented channel routing** architecture. There are three horizontal rows or tracks in this figure. There are several vertical wires and switches at the crosspoints. The switches technically can use any programming technology (SRAM, EPROM, or antifuse), although FPGAs that use this type of routing are typically antifuse FPGAs. Desired connectivity is obtained by programming the appropriate switches. Connectivity between the points marked *x* is obtained by the two switches at row 1, columns 1 and 4. Typically this is called net *x*. Net *x* simply means a wire that is named *x*. The connectivity for net y is obtained by programming the switches at row 2, columns 3 and 8. It may be noticed that row 1 cannot be used for any other connections other than net *x*. Similarly, row 2 is exclu-sively used for net *y*. Thus, a problem with this type of interconnect resource is that a full-length track (i.e., an entire row) is used even for a short net. The area overhead of this type of routing is very high for this reason.

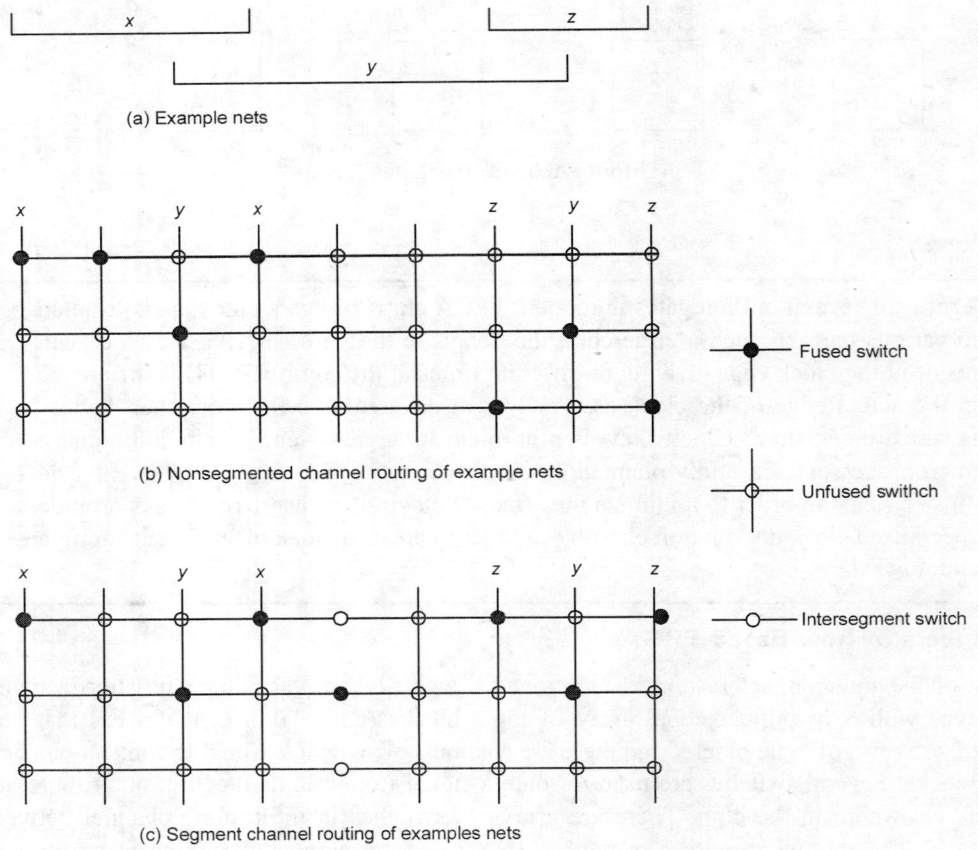

Figure 6.36: Typical Routing Resources in a Row-Based FPGA

In order to reduce the area overhead associated with using full-length tracks for each net, we can use **segmented tracks,** as in Figure 6.36(c). Instead of being full length, a track is divided into seg-ments. If a track in row 1 is segmentedinto two segments, we could use the same track for one more net. For example, nets *x* and *z* can both be routed on row 1 in Figure 6.36(c). That is the principle of segmented track routing. More nets can be routed using the same number of tracks; however, when long nets are desired, intersegment switches must be used to join the segments. These switches intro-duce more resistance and capacitance into the net. However, the overall routing resource area will reduce with segmented routing.

6.4.5 Programmable I/O Blocks in FPGAs

The I/O pads on an FPGA are connected to programmable input/output blocks, which facilitate con-necting the signals from FPGA logic blocks to the external world in desired forms and formats.

I/O blocks on modern FPGAs allow use of the pin as input and/or output, in direct (combinational) or latched iormson tristate true or inverted forms, and with a variety of I/O standards.

Figure 6.37 shows an example configurable input/output block (I/OB). Each I/OB has a number of I/O options, which can be selected by configuration memory cells, indicated by boxes with an M. The I/O pad can be programmed to be an output or an input. To use the cell as an output, the tris-tate buffer must be enabled. To use the cell as an input, the tristate control must be set to place the tristate buffer, which drives the output pin, in the high-impedance state.

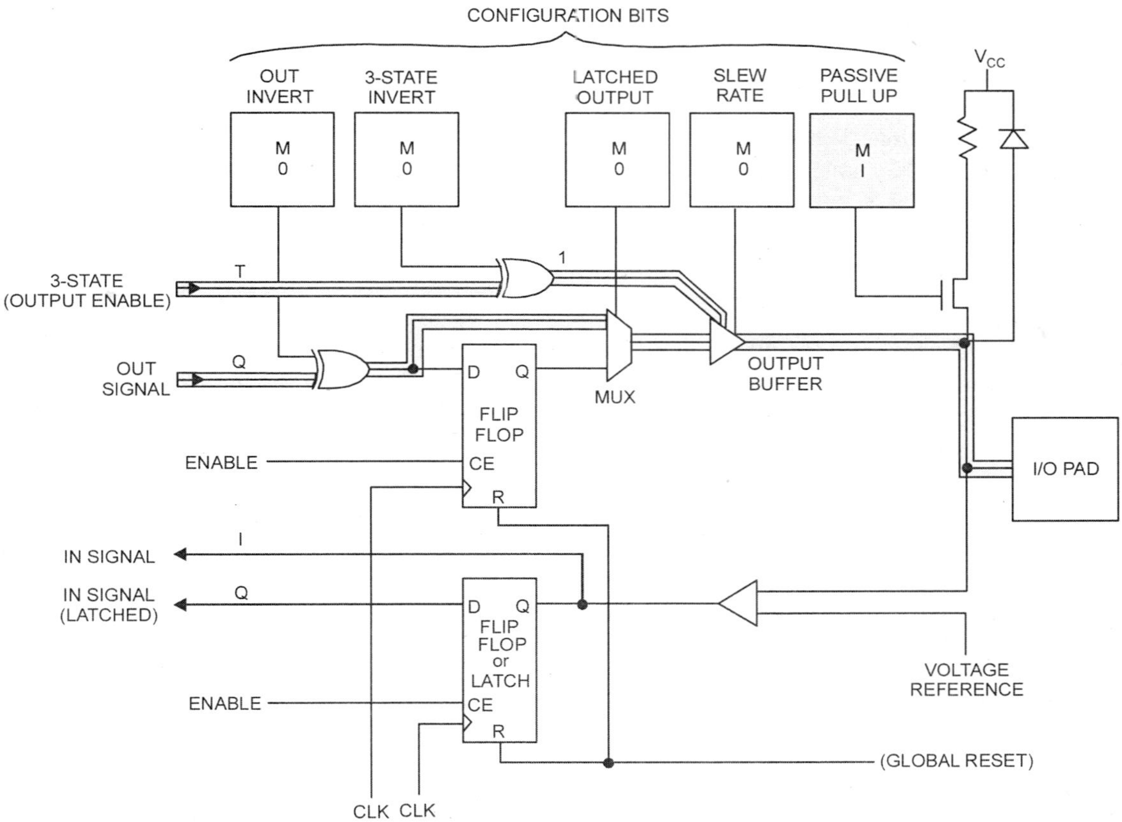

Figure 6.37: Programmable I/O Block for an FPGA

Flip-flops are provided so that input and output values can be stored within the I/O block. The flip-flops are bypassed when direct input or output is desired. The input flip-flop on many FPGAs can be programmed to act as an edge-triggered D flip-flop or as a transparent latch. Even if the I/O pin is not used, the I/O flip-flops can still be used to store data.

The configuration memory cells (marked M) allow control of various aspects associated with the I/O block. An output signal can be inverted by the I/O block if desired. The inversion is done using an XOR gate. The output signal goes through an exclusive-OR gate, where it is either complement-ed or not, depending on the contents of the configuration bit in the OUT-INVERT cell. The 3-STATE INVERT configuration bit allows one to create an active high or active low tristate control signal. If the *3-STATE* signal is 1 and the *3-STATE INVERT* bit is 0 (or if the *3-STATE* signal is 0 and the *3 STATE INVERT* bit is 1), the output buffer has a high-impedance output. Otherwise, the buffer drives the output signal to the I/O pad. When the I/O pad is used as an input, the output buffer must be in the high-impedance state. An external signal coming into the I/O pad goes through a buffer and then to the input of a D flip-flop. The buffer output provides a *DIRECT IN* signal to the logic array. Alternatively, the input signal can be stored in the D flip-flop, which provides the *LATCHED IN* signal to the logic array.

The *LATCHED OUTPUT* configuration bit allows one to provide the output in latched or combinational form. Depending on how the *LATCHED OUTPUT* bit is programmed, either the *OUT* signal or the flip-flop output goes to the output buffer. The *SLEW RATE* bit controls the rate at which the output signal can change. When the output drives an external device, reduction of the slew rate is desirable to reduce the induced noise that can occur when the output changes rapidly. When the *PASSIVE PULL-UP* bit is set, a pull-up resistor is connected to the I/O pad. This internal pull-up resistor can be used to avoid floating inputs. The highlighted path indicates the I/O block in an output configuration, with tristate enabled and with a passive pull-up resistor.

I/O Standards

Early FPGAs provided TTL and CMOS signal compatibility, but nowadays there are many more standards for input/output signals. I/O blocks on mod-ern FPGAs allow transforming signals to a variety of I/O signal standards, some of which are as follows:

LVTTL:	low-voltage transistor-transistor logic
PCI:	peripheral component interconnect
LVCMOS:	low-voltage complementary metal-oxide semiconductor
LVPECL:	low-voltage positive emitter-coupled logic
SSTL:	stub-series terminated logic
AGP:	advanced graphics port
CTT:	center tap terminated
GTL:	gunning transceiver logic
HSTL:	high-speed transceiver logic

Some of these standards use 5 volts whereas some use 3.3 volts or even 1.5 volts. The LVTTL is an example of a 3.3-V standard that can tolerate 5-V signals. The LVCMOS2 is a 2.5-V signal standard which can tolerate 5-V signals. The PCI standard has 5-V and 3.3-V versions. Some standards need an input voltage reference.

6.4.6 Dedicated Specialized Components in FPGAs

In the early days, FPGAs were simply logic blocks of medium or low complexity, integrated with pro-grammable I/O and interconnect. More recently, FPGA vendors have incorporated embedded processors, digital signal processing (DSP) processors, dedicated multipliers, dedicated memory, analog-to-digital (A/D) converters, and so on into FPGAs. These specialized components help to efficiently achieve the provided special-purpose functionality. For instance, if dedicated multipliers are not provided, we will have to implement multipliers using general-purpose logic blocks, albeit in an inefficient manner.

Dedicated Memory

A key feature of modern FPGAs is the embedding of dedicated memory blocks (RAM) onto the chip. The embedded RAM can be used to implement the memory needs of the circuit being designed. It could be a table storing constants/coefficients during processing, or it could be imple-menting memory for an embedded processor that you are designing using the FPGA. Modern FPGAs include 16K to 10M bits of memory. The width of the embedded RAM often can be adjust-ed. Let us assume that there are 32K of SRAM bits provided as blocks of RAM. This RAM can be used as 32K X 1,16K X 2, 8K X 4, or 4K X 8. Essentially there are several tiles or blocks of memo-ry. They can be placed in different ways to achieve different aspect ratios. The number of address lines and data lines get adjusted according to the aspect ratio, as illustrated in Table 6.9

Table 6.9: Variable-Width RAM Aspect Ratios

Width	Depth	Addr Bus	Data Bus
1	32K	15 bits	1 bit
2	16K	14 bits	2 bits
4	8K	13 bits	4 bits
8	4K	12 bits	8 bits
16	2K	11 bits	16 bits

Dedicated Arithmetic Units

Many users of FPGAs use them to implement arithmetic logic. When logic is implemented in FPGA logic blocks, the implementation generally takes more area and power and is slower than custom implementations. Hence, if most of the target users use arithmetic units such as adders and multipli-ers, it is beneficial to provide support for such dedicated operations inside the chip. Most FPGAs provide dedicated fast-carry logic to create fast adders. Nowadays, many FPGAs also contain dedi-cated multipliers (see Table 6.10). Thus, instead of mapping a multiplier into several logic blocks,dedicated multipliers provided on the FPGA fabric can be used. These dedicated multipliers are more efficient than a multiplier we could implement using the programmable logic in the FPGA. As indicated in Table 6.10 many Xilinx and Altera FPGAs provide 18 bit X 18 bit multipliers.

Table 6.10: Examples of FPGAs with Dedicated Multipliers

FPGA	Dedicated Multipliers
Xilinx Virtex-4, Virtex-ll Pro/X, Spartan-3E, Spartan 3/3L	18 × 18 multipliers
Altera Stratix II Cyclone II	18 x 18 multipliers

Digital Signal Processing Blocks

Multiplication is a common operation in DSP. Hence the dedicated multipliers help DSP applications. Similar to multipliers, an FPGA vendor can provide DSP building blocks such as hardware for fast Fourier transforms (FFTs), finite impulse response (FIR) filters, infinite impulse response (IIR) filters, and so on. Encryption/decryption, compression/decompression, and security functions can also be pro-vided. Once a large amount of specialized components are provided, a large part of an FPGA may be unused in applications that do not warrant such specialized components. In some FPGAs, DSP support is limited to the dedicated multipliers.

Embedded Processors

Many modern FPGAs contain an entire processor core (see Table 6.11 This is extremely useful when designers use hybrid solutions, where part of a system is in a programmable processor, but part of the sys-tem is implemented in hardware. Circuitry that needs a large amount of flexibility can be implemented in the microprocessor, but circuit parts that need better performance than that of a programmable processor can be implemented in the FPGA logic blocks. Some FPGAs include the core of a small MIPS processor such as the MIPS R 4000, and some include an embedded version of the IBM PowerPC proces-sor. Some FPGAs include custom processors designed by the FPGA vendors such as the MicroBlaze from Xilinx and the Nios processor from Altera.

Table 6.11: Examples of FPGAs with Embedded Microprocessors

FPGA	Embedded Processor
Xilinx Virtex-4, Virtex-ll Pro/X	IBM 400 MHz PowerPC
Xilinx Spartan-3E, Spartan 3/31	MicroBlaze PicoBlaze
Altera Stratix II Cyclone II	Nios II
Altera APEX APEX II	ARM, MIPS Nios
Altera Excalibur	ARM 9
Actel Fusion	ARM7

Content Addressable Memories

In some FPGAs, the memory blocks can be used as **content addressable memories** (CAMs). The general concept of a memory is that the user provides a memory address and the memory unit responds with the content. A CAM is a special kind of memory in which the content, not the address, is used to search the memory. We provide a data element, and the CAM responds with addresses where that data was found. CAMs contain more logic than RAMs because all locations of the memory have to be searched simultaneously to see whether the particular content is in any of the locations. Some FPGAs allow embedded CAM (e.g., Altera APEX II).

The Actel Fusion architecture, shown in Figure 3-40, provides several specialized components, including embedded RAM, decryption, and A/D converters. At the core of the chip are tiles of logic blocks (Versailles in Actel terminology). The embedded RAM is in the form of rows of SRAM blocks above and below the tiles of logic blocks. Several specialized components appear below the SRAM blocks in the bottom. There is a dedicated decryption unit that imple-ments the AES decryption algorithm. (AES stands for Advanced Encryption Standard, which has been the cryptograhic standard for the U.S. government since 2001.) There is an analogto digital converter (ADC) that accepts inputs from several analog quads, which are circuitry to condition analog signals received by the FPGA. The analog quads contain circuitry to monitor and condition signals according to voltage, current, and temperature.

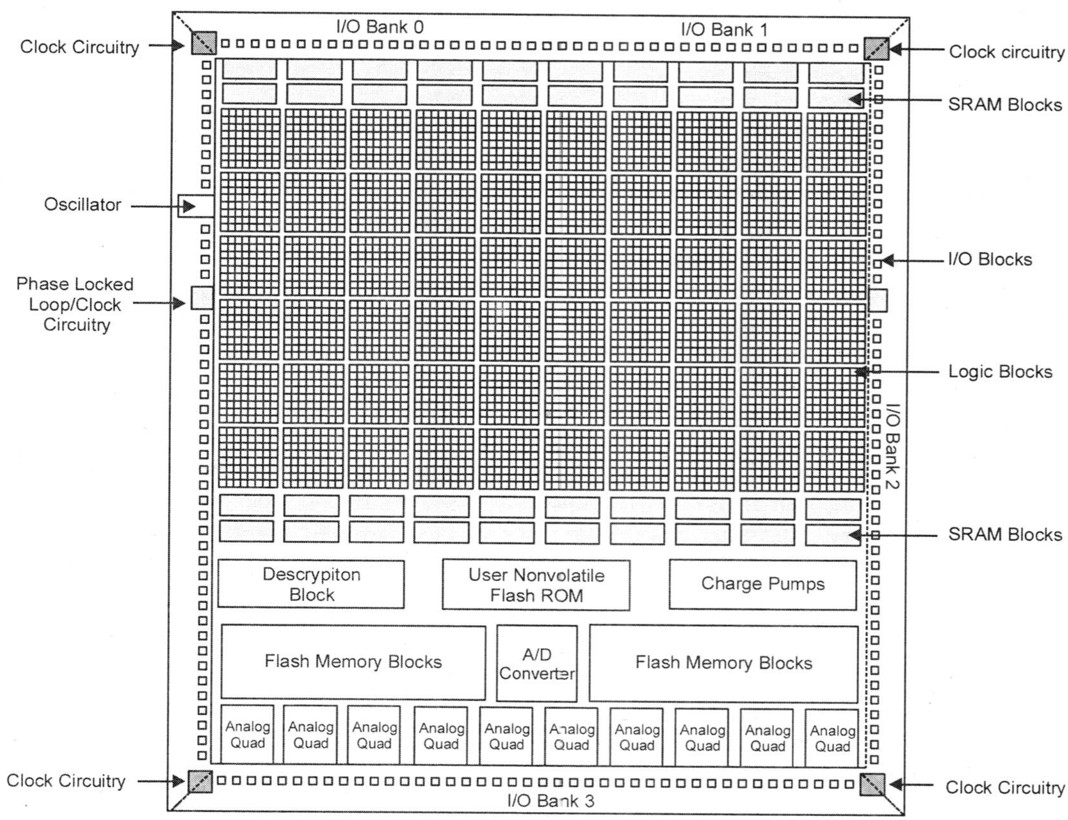

Figure 6.38: Overview of the Actel Fusion Chip (© 2006 Actel Corporation)

6.4.7 Applications of FPGAs

FPGAs have become a popular mode of circuit implementation for various applications:

Rapid Prototyping

FPGAs are very useful for building rapid prototypes of large systems. A designer can build proof-of-concept systems very quickly using field programmable gate arrays. Since FPGAs are large enough to contain 5 million or more gates, many large real-world systems can be prototyped using a single FPGA. If a single FPGA will not suffice, multiple FPGAs can be interconnected to realize large systems. Rapid prototyping of large systems is done by using boards with multiple FPGAs and plugging multiple boards into a backplane (motherboard).

As Final Product in Medium-Speed Systems

Circuits realized using FPGAs typically operate in the 150-200-MHz clock rate. For applications where this speed is sufficient, FPGAs can be used for the final product itself as opposed to the prototype. When an FPGA is used as the final product, enhancements to the system can be done as soft-ware updates rather than hardware changes. Modern FPGA speeds are adequate for many applications.

Reconfigurable Circuits and Systems

The reprogrammability of FPGAs lends itself to building dynamically reconfigurable circuits and systems. SRAM-based FPGAs make it possible to implement "soft" hardware. FPGAs have been used to design circuits and systems that need multiple functionalities at various times.

As an example, consider a reprogrammable Tomahawk missile that the Navy designed using FPGAs. The Navy designed a reconfigurable Tomahawk, which can operate in one of two modes, depending on the mission at hand. Rather than designing separate logic for each mode, the missile designers used FPGAs so that the configuration for each mode can be kept on-board in ROM. Depending on the mode of operation, the FPGA could be configured in midflight.

Glue Logic

FPGAs have become the medium of choice for implementing interface or glue logic between modules and components. Small changes in interface protocols or formats would conventionally necessitate build-ing new interface logic. With SRAM FPGAs, the new interface logic can be implemented on the same FPGA as in a software update.

Hardware Accelerators/Coprocessors

A software application running on a conventional system can be accelerated if a coprocessor/accelera-tor can implement some key routines/kernels from the application in hardware. An FPGA can be used to implement the key kernel. A SRAM-based, reconfigurable FPGA is well suited for this type of use because depending on the application running, different kernels can be dynamically programmed into the FPGA. This approach has been demonstrated for applications, such as pattern matching. FPGA-based hardware is used for several applications, including computer architecture simulator acceleration, emulation boards, hardware test/verification, and so on.

6.4.8 Design Flow for FPGAs

Sophisticated CAD tools are available to assist with the design of systems using programmable gate arrays. Designs can be entered in many ways.

In the early days of FPGAs, designs were entered using schematic entry or even lower levels of design entry tools. Low-level design entry means less abstraction, whereas high-level means entering designs at a higher level of abstraction (e.g., behavioral VHDL/Verilog description). Early FPGA tools allowed low-level utilities to enter logic equations, Karnaugh maps, and so on into specific logic blocks in the FPGA. Schematic capture technique means that the designer develops a schematic of the design. Schematic diagrams utilizing standard hardware components are created and entered into the CAD software.

Nowadays, automatic synthesis tools are available that will take a VHDL description of the system as an input and generate an interconnection of gates and flip-flops to realize the system. Behavioral models can be translated into design implementations reasonably efficiently. Synthesis tools have advanced significantly in the last decade.

One method of designing a digital system with an FPGA uses the following steps:

1. Create a behavioral, register-transfer level (RTL), or structural model of the design in a hardware description language such as VHDL or Verilog.
2. Simulate and debug the design.
3. Synthesize the design targeting the desired device.
4. Run a mapping/partitioning program. This program will break the logic diagram into pieces that will fit into the configurable logic blocks.
5. Run an automatic place and route program. This will place the logic blocks in appropriate places in the FPGA and then route the interconnections between the logic blocks.
6. Run a program that will generate the bit pattern necessary to program the FPGA.
7. Download the bit pattern into the internal configuration cells in the FPGA, and test the operation of the FPGA.

Steps 3, 4, and 5 are often integrated in modern CAD tools. However, the processes mentioned in the steps are happening whether presented as one step or several steps. This is analogous to how gen-eral-purpose compilers have integrated compiling and assembling steps. In the early days of high-level language compilers, the term *compiling* only meant translation into an assembly language format. Converting from assembly language to machine language code was considered the assem-bler's job. Nowadays, the steps are integrated in most high-level language compilation environments.

In SRAM-based FPGAs, when the final system is built, the bit pattern for programming the FPGA is normally stored in an EPROM and automatically loaded into the FPGA when the power is turned on. The EPROM is connected to the FPGA, as shown in Figure 6.39. The FPGA resets itself after the power has been applied. Then it reads the configuration data from the EPROM by supplying a sequence of addresses to the EPROM inputs and storing the EPROM output data in the FPGA internal configuration memory cells. This is not required in flash memory based FPGAs because the flash technology is nonvolatile. In antifuse FPGAs, the configuration bits permanently alter the switches.

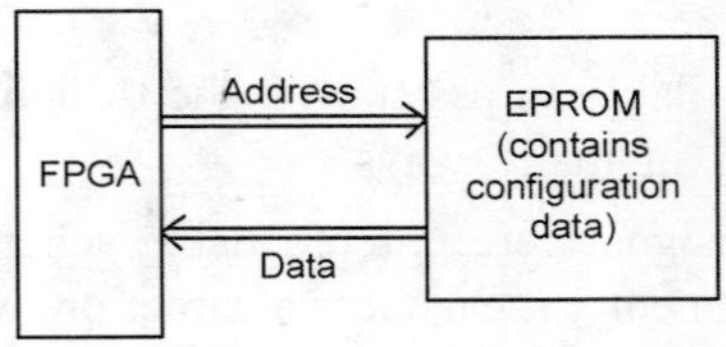

Figure 6.39: EPROM Connections for SRAM FPGA Initialization

In this chapter we have introduced several different types of programmable logic devices and used them for designing circuits. The technology underlying early programmable logic devices, such as ROMs, PALs, and PLAs, was presented first. Simple PLDs and GALs were presented next. Examples were presented to illustrate implementations of simple logic functions in these devices. CPLDs and FPGAs were presented next. The discussion on FPGAs was limited to an overview of the general technology underlying this class of devices. General organization of FPGAs, general structure of logic blocks, typical programming techniques, and so on were discussed. More details on FPGAs will be presented in Chapter 6.

PROBLEMS

6.1. What is the size of the smallest ROM that is needed to implement the following?

(*a*) An 8-bit full adder (assume carry-in and carry-out)
(*b*) A BCD-to-binary converter (2 BCD digits)
(*c*) A4-to-lMUX
(*d*) A 32-bit adder (adds two 32-bit numbers to give a 33-bit sum)
(*e*) A 3-to-8 decoder
(*f*) A 32-bit adder (no carry in or carry out)
(*g*) A 16 X 16 bit multiplier
(*h*) A 16-bit full adder (with carry-in and carry-out)
(*i*) An 8-to-3 priority encoder
(j) A 10-to-4 priority encoder
(*k*) An 8-to-l multiplexer

6.2. Given $F = A'B' + BC$ and $G = AC + B'$, write a complete VHDL module that realizes the functions F and G using an 8-word X 2-bit ROM. Include the array type declaration and the constant declaration that defines the contents of the ROM.

6.3. Implement the following state table using a ROM and two D flip-flops. Use a straight binary state assignment.

(*a*) Show the block diagram and the ROM truth table. Truth table column headings should be in the order Q_1 Q_0 X D_1 D_0 Z.
(*b*) Write VHDL code for the implementation. Use an array to represent the ROM table, and use two processes.

Present State	Next State X = 0	Next State X = 1	Output (Z) X = 0	Output (Z) X = 1
S_0	S_0	S_1	0	1
S_1	S_2	S_3	1	0
S_2	S_1	S_3	1	0
S_3	S_3	S_2	0	1

6.4 Find a minimum-row PLA table to implement the following sets of functions.

(*a*) $f_1(A, B, C, D) = \Sigma m(0,2,3,6,7,8,9,11,13)$,
$f_2(A, B, C, D) = \Sigma m(3,7,8,9,13)$,
$f_3(A, B, C, D) = \Sigma m(0,2,4,6,8,12,13)$

(*b*) $f_1(A, B, C, D) = cd + ad + a'bc'd'$
$f_2(A, B, C, D) = bc'd' + ac' + cd'$

6.5. An N-bit bidirectional shift register has *N* parallel data inputs, *N* outputs, a left serial input (*LSI*), a right serial input (*RSI*), a clock input, and the following control signals:

Load: Load the parallel data into the register (load overrides shift).
Rsh: Shift the register right (*LSI* goes into the left end).
Lsh: Shift the register left (*RSI* goes into the right end).

(*a*) If the register is implemented using a 22V10, what is the maximum value of *N*?
(*b*) Give equations for the rightmost two cells.

(*a*) What is the difference between a traditional gate array and an FPGA? 6.6 (b) What are the different types of FPGAs based on architecture (organization)?
(*c*) What are the different programming technologies for FPGAs?
(*d*) What is the main advantage of SRAM FPGAs?
(*e*) What is the main advantage of antifuse FPGAs?
(*f*) What are the major programmable elements in an FPGA?
(g) What are the disadvantages of SRAM FPGAs? (h) What are the disadvantages of antifuse FPGAs?
(*i*) How many transistors are typically required to make an SRAM cell?
(*j*) What is an MPGA?
(*k*) What is the difference between a CPLD and an FPGA?
(*l*) What is an advantage of a CPLD over an FPGA?
(*m*) What is the advantage of an FPGA over a CPLD?
(*n*) Name three vendors of CPLDs.
(*o*) Name three vendors of FPGAs.

6.7. (*a*) In what type of applications should a designer use a CPLD rather than an FPGA?
(b) In what type of applications should a designer use an MPGA rather than an FPGA?
(*c*) In what type of applications should a designer use an FPGA rather than an MPGA?
(*d*) A company is designing an experimental product, which is in version 1 now. It is expected that the product will undergo several revisions. The company's plan is to use an FPGA for the actual design. What type of FPGA (SRAM or antifuse) should be used?

(*e*) A company is designing a product using an FPGA. The company's plan is to use an FPGA for the actual design. The product has undergone several revisions and is fairly stable. Minimizing area, power, and cost is important for the company. What type of FPGA (SRAM or antifuse) should be used?

(*f*) A company is designing a product. It expects to sell 1000 copies of it. Should the company use an MPGA or FPGA for this product?

(*g*) A company is designing a product. It expects to sell 100 million copies of it. Should the company use an MPGA or an FPGA for this product?

6.8. (a) Route the '*w*', '*x*', '*y*', and '*z*' nets on the nonsegmented tracks shown below. Use the min-6.8 imum number of tracks possible.

(*b*) Route the '**w**', '*x*', '*y*', and '*z*' nets on the segmented tracks shown below. Use the minimum number of tracks possible.

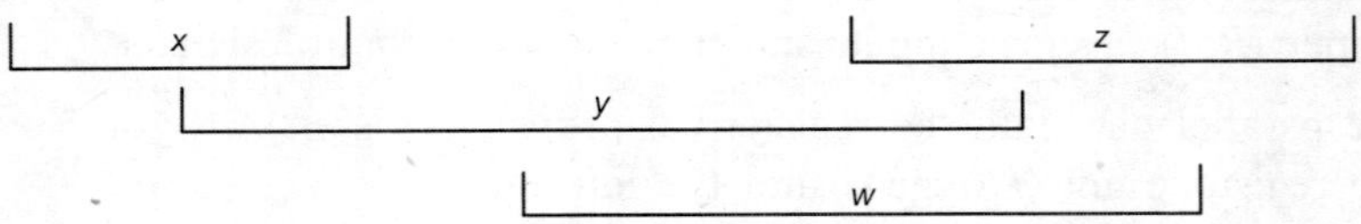

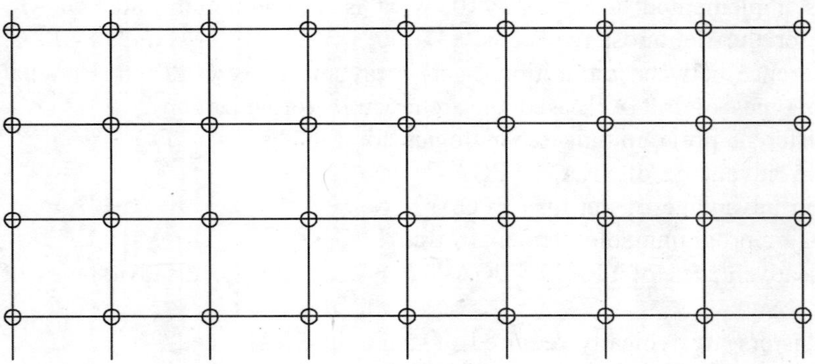

Nonsegmented tracks

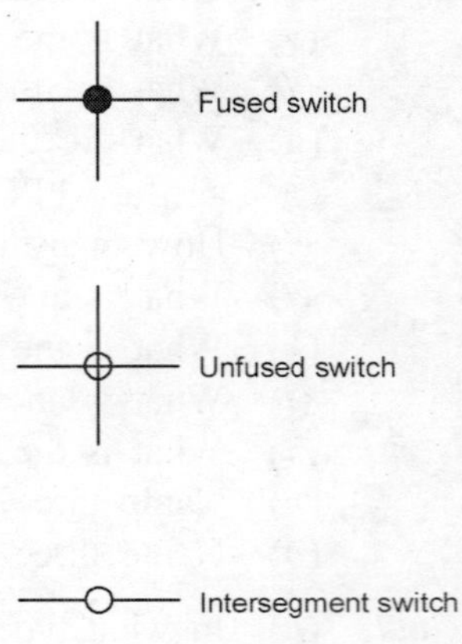

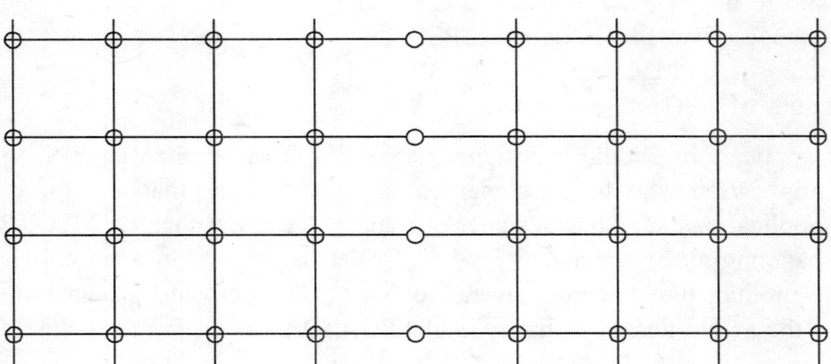

Segmented tracks

6.10. Write short notes on:

(*a*) GAL
(*b*) PEEL
(*c*) FPGA
(*d*) CPLD
(*e*) PAL
(*f*) Advavatages of PLD over fixed
(*g*) PROM
(*h*) PLA
(*i*) Designing using PLD's

UNIT 7 VHDL PROGRAMS

Example 7.1: *For 2-to-4 Decoder write down:*

(*a*) Behavior Model
(*b*) Dataflow Model
(*c*) Structural Model

Solution: Circuit diagram of 2- to-4 Decoder is shown in figure 7.1 and truth table is shown in table 7.1

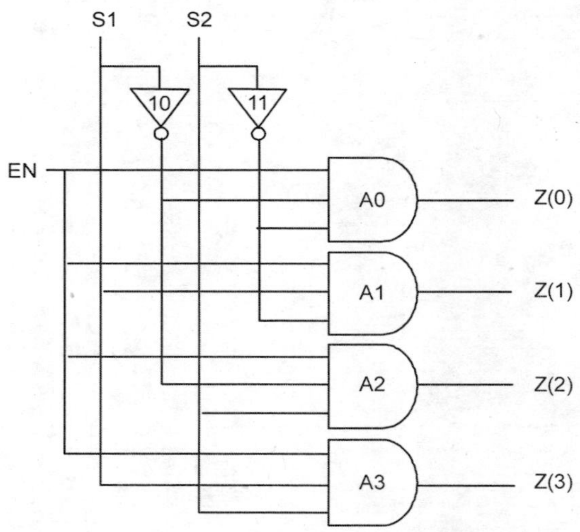

Figure 7.1: Circuit Diagram of 2-of-4 Decoder.

Table 7.1: Truth table of 2-to-4 Decoder

Inputs			Outputs			
EN	S1	S0	Z(0)	Z(1)	Z(2)	Z(3)
0	X	X	0	0	0	0
1	0	0	1	0	0	0
1	0	1	0	1	0	0
1	1	0	0	0	1	0
1	1	1	0	0	0	1

(a) Behavior model of 2-to-4 Decoder

```
LIBRARY ieee;
USE ieee.std_logic_1164.all;
USE ieee.std_logic_arith.all;

ENTITY decoder_2_to_4 IS
    port(s0,sl,en:in bit; z:out bit_vector(0 to 3));
END decoder_2_to_4;

ARCHITECTURE decoder_2_to_4_behaviour OF decoder_2_to_4 IS
BEGIN
  process(s0,sl,en)
    variable s0bar,slbar:bit;
    begin
        s0bar := not s0;
        slbar := not s1;
        if en='l'then
          z(0) <=s0bar and slbar;
          z(l) <=s0 and slbar;
          z(2) <=s0bar and s1;
          z(3) <=s0 and s1;
else
          z <="0000"; end if; end process; END decoder_2_to_4_behaviour;
```

(b) Dataflow model of 2-to-4 Decoder

```
LIBRARY ieee;
USE ieee.std_logic_1164.all;
USE ieee.std_logic_arith.all;
ENTITY decoder_2_to_4 IS
  port(s0,s1,en:in bit; z:out bit_vector(0 to 3));
END decoder_2_to_4;

ARCHITECTURE decoder_2_to_4_dataflow OF decoder_2_to_4 IS
    signal s0bar,slbar:bit;
    BEGIN
      s0bar <= not s0 ;
      slbar <= not s1;
      z(0) <=s0 bar and slbar and en;
      z(l) <=s0 and slbar and en;
      z(2) <=s0bar and s1 and en;
      z(3) <=s0 and s1 and en;
END decoder_2_to_4_dataflow; •
```

(c) Structural model of 2-to-4 Decoder

```
LIBRARY ieee;
USE ieee.std_logic_1164.all;
USE ieee.std_logic_arith.all;
```

```
ENTITY decoder_2_to_4 IS
  port(s0,sl,en:in bit; z:out bit_vector(0 to 3));
END decoder_2_to_4;
--

ARCHITECTURE decoder_2_to_4_structural OF decoder_2_to_4 IS
  component and3
    port(inl,in2,in3:in bit; outl:out bit);
  end component;
  component inv
      port(in4:in bit; out2:out bit);
  end component;
  signal s0bar,slbar:bit;
BEGIN
    a0:and3 port map(s0ar,slbar,en,z(0)) ;
    al:and3 port map(s0,slbar,en,z(l)) ;
    a2:and3 port map(s0bar,sl,en,z(2)) ;
    a3:and3 port map(s0,sl,en,z(3));
    inv0:inv port map(s0s0bar);
    invl:inv port map(sl,slbar);
END decoder_2_to_4_structural;
```

Example 7.2: *For a Full Adder write down*

(*a*) Behavior Model
(*b*) Dataflow Model
(*c*) *Structural Model*

Solution: Circuit diagram of Full Adder is shown in figure 7.3 waveforms are shown in figure 7.2 and truth table is shown in Table 7.2

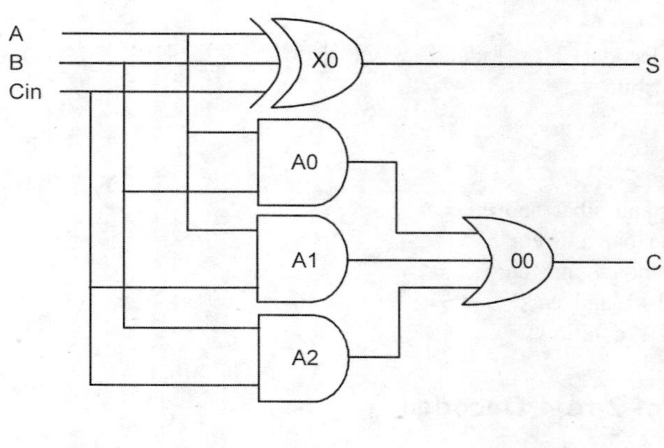

Figure7.3: Circuit Diagram of Full Adder

Table 7.2: Truth table of Full Adder

Inputs			Outputs	
A	B	Cin	S	C
0	0	0	0	0
0	0	1	1	0
0	1	0	1	0
0	1	1	0	1
1	0	0	1	0
1	0	1	0	1
1	1	0	0	1
1	1	1	1	1

(a) Behavior model of Full Adder

```
LIBRARY ieee;
USE ieee.std_logic_1164.all;
USE ieee.std_logic_arith.all;

ENTITY full_adder IS
    port(a,b,cin:in bit; s,c:out bit);
END full_adder;

ARCHITECTURE full_adder_behaviour OF full.adder IS
BEGIN
process(a,b,cin)
begin
  s <=a xor b xor cin;
  c <=(a and b)or(a and cin)or(b and cin) ;
end process;
END full_adder_behaviour;
```

(b) Dataflow model of Full Adder

```
LIBRARY ieee;
USE ieee.std_logic_1164.all;
USE ieee.std_logic_arith.all;

ENTITY ful_adder IS
    port(a,b,cin:in bit; s,c:out bit);
ENDfull_adder;
--
ARCHITECTURE full_adder_dataflow,OF full_adder IS
BEGIN
    s <=a xor b xor cin;
    c <=(a and b)or(a and cin)or(b and cin);
END full adder dataflow;
```

(c) Structural model of Full Adder

```
LIBRARY ieee;
USE ieee.std_logic_1164.all;
```

```
USE ieee.std_logic_arith.all;

ENTITY full_adder IS
    port(a,b,cin:in bit; s,c:out bit);
END ful_adder;

ARCHITECTURE full_adder_structure OF full_adder IS
  component xor3
        port(inl,in2,in3:in bit; outl:out bit);
  end component;

  component and2
        port(in4,in5:in bit; out2:out bit);
end component;

component or3
      port(in6,in7,in8:in bit; out3:out bit);
end component;
signal x,y,z:bit;
begin
  x0:xor3 port map(a,b,cin,s);
  a0:and2 port map(a,b,x);
  al:and2 port map(b,cin,y);
  a2:and2 port map(a,cin,z);
  o0:or3 port map(x,y,z,c);

END full_adder_structure;
```

Example 7.3: *For a Multiplexer of size 4-to-1 write down*

(*a*) *Behavior Model*

(*b*) *Dataflow Model*

(*c*) *Structural Model*

Solution: Circuit diagram of 4-to-1 Multiplexer is shown in figure 7.5 waveformsare shown in figure 7.6 and truth table is shown in Table7.3

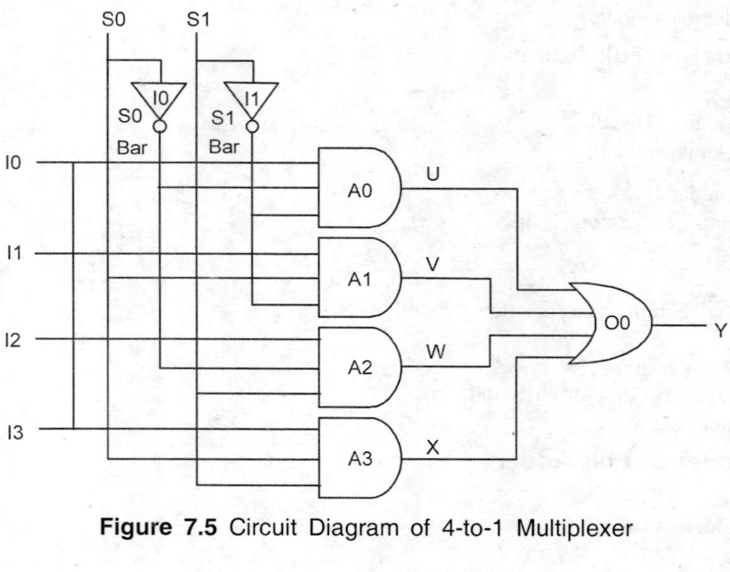

Figure 7.5 Circuit Diagram of 4-to-1 Multiplexer

Table 7.3: Truth table of 4-to-l Multiplexer

Select Input's		Output
S1	**SO**	**Y**
0	0	I0
0	1	11
1	0	12
1	1	13

(a) Behavior model of 4-to-1 Multiplexer

```
LIBRARY ieee;
USE ieee.std_logic_1164.all;
USE ieee.std_logic_arith.all;
ENTITY mux_4tol IS
    port(i0,il,i2,i3,s0,sl:in bit; y:out bit); END mux_4tol;

ARCHITECTURE mux_behaviour OF mux_4tol IS
BEGIN
    PROCESS(i0,il,i2,i3,s0,sl)
      BEGIN
           IF (sl='0' and s0='0') THEN
                y<=i0;
           ELSIF (sl='0' and s0'l') THEN
                y <= il; ELSIF (sl='l' and s0='0') THEN
                               y <= i2;
           ELSIF (sl='1nd s0'l') THEN
                               y<=i3;
           END IF;
    END PROCESS;
END mux_behaviour;
```

(b) Dataflow model of 4-to-1 Multiplexer

```
LIBRARY ieee;
USE ieee.std_logic_1164.aIl;
USE ieee.std_logic_arith.all;

ENTITY mux_4tol IS
    port(i0,il,i2,i3,s0,sl:in bit; y:out bit);
END mux_4tol;

ARCHITECTURE mux_dataflow OF mux_4tol IS
BEGIN
    Y<=i0when sl='0' and s0'0' else
    il when sl='0' and s0'l' else
    i2 when sl='l' and s0='0' else
    i3;
END mux_dataflow;
```

(c) Structural model of 4-to-1 Multiplexer

```
LIBRARY ieee;
USE ieee.std_logic_1164.all;
USE ieee.std_logic_arith.all;
ENTITY mux IS
    port(i0,il,i2,i3,s0,sl:in bit; y:out bit);
END mux;

--

ARCHITECTURE mux_structure OF mux IS
    component and3
            port(inl,in2,in3:in bit; outl:out bit);
    end component;

    component or4
            port(in4,in5,in6,in7:in bit; out2:out bit);
    end component;
    signal s0bar,slbar,u,v,w,x:bit;
    BEGIN

      a0:and3 port map(i0,slbar,s0bar,u);
      al:and3 port map(il,slbar,s0,v);
      a2:and3 port map(i2,sl,s0bar,w);
      a3:and3 port map(i3,sl,s0,x);
      or0:or4 port map(u,v,w,x,y);

END mux_structure;
```

Example 7.4: *For a BCD-to-Seven Segment Decoder write down*

(*a*) *Behavior Model*
(*b*) *Dataflow Model*
(*c*) *Structural Model*

Solution : Block diagram of BCD-to-Seven Segment Decoder is shown in figure7.4 and truth table is shown in Table 7.4

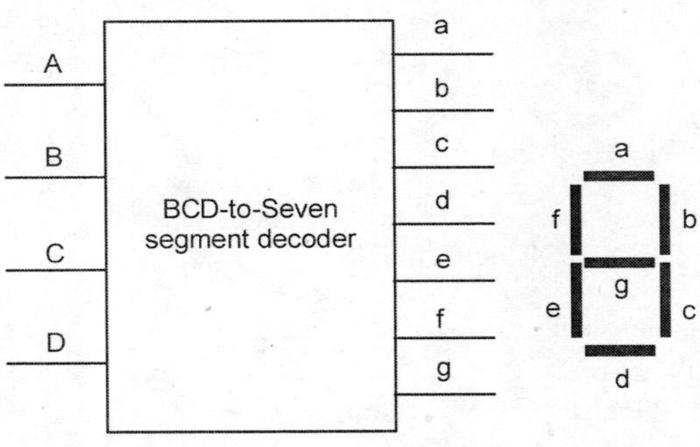

Fig.7.4: Block Diagram of BCD-to Seven Segment Decoder

Table 7.4: Trum table of BCD-to-Seven Segment Decoder

BCD Input's				Output's						
A	B	C	D	a	b	c	d	e	f	9
0	0	0	0	1	1	1	1	1	1	0
0	0	0	1	0	1	1	0	0	0	0
0	0	1	0	1	1	0	1	1	0	1
0	0	1	1	1	1	1	1	0	0	1
0	1	0	0	0	1	1	0	0	1	1
0	1	0	1	1	0	1	1	0	1	1
0	1	1	0	1	0	1	1	1	1	0
0	1	1	1	1	1	1	0	0	0	0
1	0	0	0	1	1	1	1	1	1	1
1	0	0	1	1	1	1	1	0	1	1

(a) Behavior model of BCD-to-Seven Segment Decoder

```
LIBRARY ieee;
USE ieee.std_logic_1164.all;
USE ieee.std_logic_arith.all;

ENTITY seven_seg_decoder IS
    port(bcd:in bit_vector(0 to 3); led:out bit_vector(0 to 6));
END seven_seg_decoder;

ARCHITECTURE seven_behaviour OF seven_seg_decoder IS
    BEGIN
      process(bcd)
begin
     case bcd is
          when"0000"=>
               led<="1111110";
          when"0001"=>
               led<="0110000";
          when"0010"=>
               led<="1101101";
          when"0011"=>*
               led<="1111001";
          when"0100"=>
               led<="0110011";
          when"0101"=>
               led<="1011011";
          when"0110"=>
               led<="1011110";
          when"0111"=>
               led<="1110000";
```

```
            when"1000"=>
                led<="lllllll";
            when "1001"=>
                led<="1111011";
            when others=>
                led<="lllllll";
        end case;
    end process;
END seven_behaviour;
```

(b) Dataflow model of BCD-to-Seven Segment Decoder

```
LIBRARY ieee;
USE ieee.std_logic_1164.all;
USE ieee.std_logic_arith.all;

ENTITY seven_seg_decoder IS
    port(bcd:in bit_vector(0 to 3); led:out bit_vector(0 to 6));
END seven_seg_decoder;

ARCHITECTURE seven_dataflow OF seven_seg_decoder IS
BEGIN
    Led<="llllilO" when bcd="0000" else
      "0110000" when bcd="0001" else
      "1101101" when bcd="0010" else
      "1111001" when bcd="0011" else
      "0110011" when bcd="0100" else
      "1011011" when bcd="0101" else
      "1011110" when bcd="0110" else
      "1110000" when bcd="0111" else
      "1111111" when bcd="1000" else
      "1111011" when bcd="1001" else
      "1111111" when others;
END seven_dataflow;
```

(c) Structural model of BCD-to-Seven Segment Decoder

```
LIBRARY ieee;
USE ieee.std_logic_1164.all;
USE ieee.std_logic_arith.all;

ENTITY seven_seg_decoder IS
    port(bcd:in bit_vector(0 to 3); led:out bit_vector(0 to 6));
END seven_seg_decoder;

--

ARCHITECTURE seven_structure OF seven_seg_decoder IS
    Component decoder
            Port(input:in bit_vector(0 to 3); output:out bit_vector(0 to 6));
    End component;
    Begin
Decl:decoder port map(bcd,led); END seven_structure;
```

Example 7.5: *For a 4-bit Parallel in Parallel Out Register write down*

(*a*) *Behavior Model*
(*b*) *Dataflow Model*
(*c*) *Structural Model*

Solution: Circuit diagram of 4-bit Parallel in Parallel out Register is shown in figure 7.5

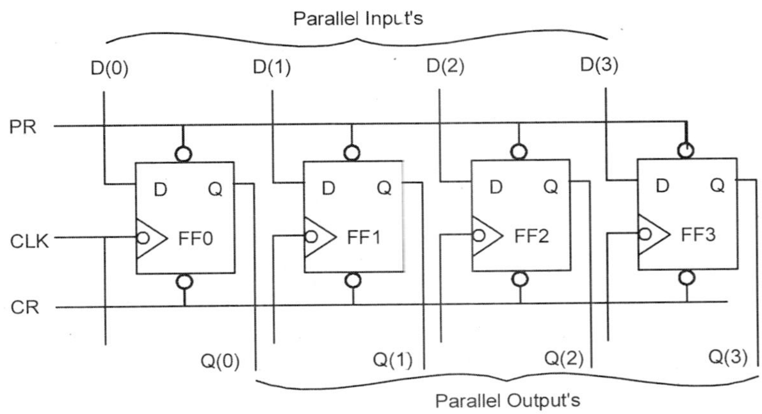

Figure 7.5: Circuit diagram of 4-bit Parallel In Parallel Out Register (or Parallel load Register)

(a) Behavior model of 4-bit Parallel In Parallel Out Register

```
LIBRARY ieee;
USE ieee.std_logic_1164.all; USE ieee.std_logic_arith.all;

ENTITY pipo IS
    port(d:in bit_vector(0 to 3) ;clk,pr,cr:in bit; q:out bit_vector(0 to 3));
END pipo;

ARCHITECTURE pipo_behaviour OF pipo IS
BEGIN
    process(pr,cr,clk)
     begin
         if (pr='0' and cr='1') then
              q<="1111";
        elsif (pr='1' and cr='0') then
              q <="0000";
        elsif (pr='0'and cr='0')then
              q<="1111";
        elsif (pr='1'and cr='1' and clk='0' and clk'event) then
              q<=d;
        end if;
  end process;
END pipo_behaviour;
```

(b) Dataflow model of 4-bit Parallel in Parallel Out Register

```
LIBRARY ieee;
USE ieee.std_logic_1164.all;
USE ieee.std_logic_arith.all;

ENTITY pipo IS
   port(d:in bit_vector(0 to 3); pr,cr,clk:in bit; q:out bit_vector(0 to 3));
END pipo;

ARCHITECTURE pipo_dataflow OF pipo IS
BEGIN
   q <="1111" when pr='0' and cr='l' else
   "0000" when pr='l' and cr='0' else
   "1111" when pr='0' and cr='0' else
   d when pr='l' and cr='l' and clk='0' and clk'event;
END pipo_dataflow;
```

(c) Structural mode! of 4-bit Parallel In Parallel Out Register

```
LIBRARY ieee;
USE ieee.std_logic_1164.all;
USE ieee.std_logic_arith.all;

ENTITY pipo IS
   port(d:in bit_vector(0 to 3); pr,cr,clk:in bit; q:out bit_vector(0 to 3));
END pipo;

--

ARCHITECTURE pipo_structure OF pipo IS
   component dff
         port(inl,in2,in3,in4:in bit; outl:out bit);
   end component;
   -- inl=d,in2=pr,in3=cr,in4=clk,outl=q
BEGIN
     d0:dff port map(d(0),pr,cr,clk,q(0));
     d1:dff port map(d(l),pr,cr,clk,q(l));
     d2:dff port map(d(2),pr,cr,clk,q(2));
     d3:dff port map(d(3),pr,cr,clk,q(3));
END pipo_structure;
```

Example 7.6: *For a 4-bit Serial In Serial Out Register write down*

(*a*) *Behavior Model*
(*b*) *Dataflow Model*
(*c*) *Structural Model*

Solution: Circuit diagram of 4-bit Serial in Serial out Register is shown in figures 7.6

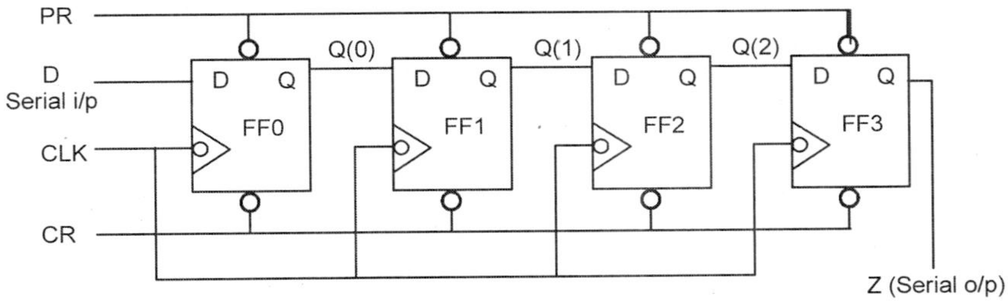

Fig. 7.11: Circuit Diagram of 4-bit Serial In Serial Out Register (or Right Shifter)

(a) Behavior model of 4-bit Serial In Serial Out Register

```
LIBRARY ieee;
USE ieee.std_logic_1164.all;
USE ieee.std_logic_arith.all;

ENTITY siso IS
    port(d,clk,pr,cr:in bit; z:out bit);
END siso;

ARCHITECTURE siso_behaviour OF siso IS
    signal q(0),q(l),q(2):bit;
BEGIN
  process(clk,pr,cr)
  begin
      if (pr='l'and cr='1' and clk='0' and clk'event) then
            q(0) <=d ;
            q(1) <=q(0) ;
            q(2) <=q(1) ;
            z <=q(2) ;
      elsif (pr='0' and cr='1') then
            q(0) <='1' ;
            q(1) <='i';
            q(2) <='1';
            z <='0' ;
 elsif (pr='1' and cr='0') then
            q(0) <='0';
            q(1) <='0';
            q(2) <='0';
            z <='0';
 elsif (pr='0' and cr='0') then
            q(0) <='1';
            q(1) <='1';
            q(2) <='l';
            z <='1';
     end if;
 end process;
END siso_behaviour;
```

(b) Dataflow model of 4-bit Serial In Serial Out Register

```
LIBRARY ieee;
USE ieee.std_logic_1164.all;
USE ieee.std_logic_arith.all;

ENTITY siso IS
    port(pr,cr,clk,d:in bit; z:out bit); END siso;

ARCHITECTURE siso_dataflow OF siso IS
    signal q:bit_vector(0 to 3);
BEGIN
    q<=d & q(0) & q(l) & q(2) whenpr='1' and cr='1' and clk= '0' and clk'event else
    "1111" when pr='0' and cr='1' else
    "0000" when pr='l' and cr= '0' else
    "1111" when pr='0' and cr='l';
      z <=q(3);
END siso_dataflow ;
```

(c) Structural model of 4-bit Serial In Serial Out Register

```
LIBRARY ieee;
USE ieee.std_logic_1164.all;
USE ieee.std_logic_arith.all;

ENTITY siso IS
    port(d,clk,pr,cr:in bit; z:out bit);
END siso;

ARCHITECTURE siso_structure OF siso IS
    component dff
            port(inl,in2,in3,in4:m bit; out1:out bit);
    end component;
      - - inl=d,in2=pr,in3=cr,in4=clk,outl=q
signal q:bit_vector(0 to 2);
BEGIN
        d0:dff port map(d,pr,cr,clk,q(0));
        dl:dff port map(q(0),pr,cr,clk,q(l)) ;
        d2:dff port map(q(l),pr,cr,clk,q(2));
        d3:dff port map(q(2),pr,cr,clk,z);
END siso_structure;
```

Example 7.7: *For a 4-bit Serial In Parallel, Out Register write down*

(*a*) *Behavior Model*

(*b*) Dataflow Model

(*c*) Structural Model

Solution: Circuit diagram of 4-bit Serial in Parallel out Register is shown in figures 7.7

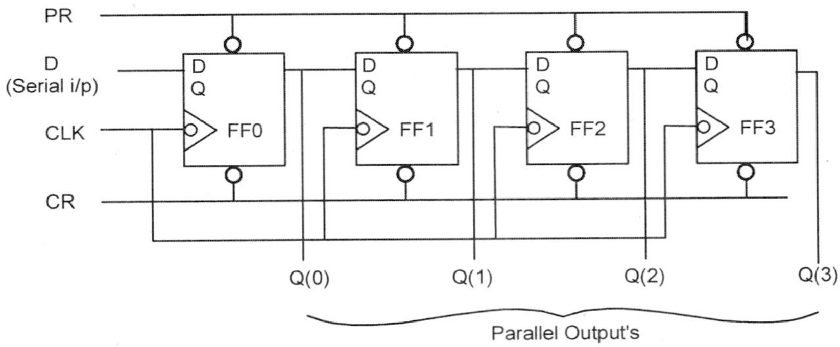

Figure 7.13: Circuit Diagram of 4-bit Serial In Parallel Out Register

(a) Behavior model of 4-bit Serial In Parallel Out Register

```
LIBRARY ieee;
USE ieee.std_logic_1164.all;
USE ieee.std_logic_arith.all;

ENTITY sipo IS
    port(d,pr,cr,clk:in bit; q:buffer bit_vector(0 to 3));
END sipo;

--

ARCHITECTURE sipo_behavioural OF sipo IS
BEGIN
        process (pr,cr,clk)
                begin
                    if (pr=’l’ and cr=’l’ and clk=’0' and clk’event) then
                            q(0)<=d;
                            q(1)<=q(0);
                            q(2)<=q(1);
                            q(3)<=q(2);
                    elsif (pr=‘l’ and cr=‘0')then
                            q(0)<=‘0';
                            q(1)<=‘0';
                            q(2)<=‘0';
                            q(3)<=‘0';
                    elsif (pr= ‘0’ and cr=‘l’) then
                            q(0)<=‘l’;
                            q(l)<=‘1’ ;
                            q(2)<=‘l’ ;
                            q(3)<=‘l’;
                    elsif (pr=’0' and cr=’0') then
                            q(0)<=‘l’;
                            q(l)<=‘l’;
                            q(2)<=‘l’;
                            q(3)<=‘l’;
```

```
            end if;
      end process;
END sipo_behavioural;
```

(b) Dataflow model of 4-bit Serial In Parallel Out Register

```
LIBRARY ieee;
USE ieee.std_logic_1164.all;
USE ieee.std_logic_arith.all;

ENTITY sipo IS
      port(d,pr,cr,clk:in bit; q:buffer bit_vector(0 to 3));
END sipo;

- -

ARCHITECTURE sipo_dataflow OF sipo IS
   BEGIN
     q<= "0000" when(pr='l' and cr='0')else
     "1111" when(pr='0' and cr=T)else
     "1111" when(pr='0' and cr='0')else
     d & q(0) & q(l) & q(2)when(pr='l' and cr='l' and clk='0' and clk'event);

END sipo_dataflow;
```

(c) Structural model of 4-bit Serial In Parallel Out Register

```
LIBRARY ieee;
USE ieee.std_logic_1164.all;
USE ieee.std_logic_arith.all;

ENTITY sipo IS
    port(d,clk,pr,cr:in bit; q:buffer bit_vector(0 to 3));
END sipo;

ARCHITECTURE sipo_structure OF sipo IS

     component dff
           port(inl,in2,in3,m4:in bit; outl:out bit);
     end component;
       - - inl=d,in2=pr,in3=cr,in4=clk,outl=q
     BEGIN
       d0:dff port map(d,pr,cr,clk,q(0));
       d1:dff port map(q(0),pr,cr,clk,q(l));
       d2:dff port map(q(l),pr,cr,clk,q(2));
       d3:dff port map(q(2),pr,cr,clk,q(3));
END sipo_structure;
```

Example 7.8: *For a 3-bit UP Counter write down*

(*a*) *Behavior Model*

(*b*) *Dataflow Model*

(*c*) *Structural Model*

Solution: Block diagram of 3-bit UP Counter is shown in figure 7.8

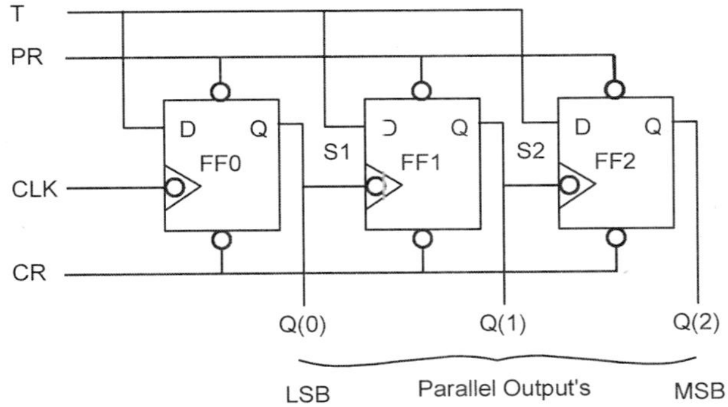

Figure 7.8: A 3-bit UP Counter *T input of T flip-flops is always set to '1'

(a) Behavior model of 3-bit UP Counter

```
LIBRARY ieee;
USE ieee.std_logic_1164.all;
USE ieee.std_logic_arith.all;

ENTITY counter IS
    port(pr,cr,clk,t:in bit; q:inout bit_vector(0 to 2));
END counter;

- -
ARCHITECTURE behaviour OF counter IS
    function "+"(a,b:bit_vector(0 to 2))return bit_vector is
      variable cout:bit;
      variable cin:bit:='0';
      variable sum:bit_vector(0 to 2):="000";
      begin
          for i in 0 to 2 loop
                cout:=(a(i) and b(i)) or (b(i) and cin) or (a(i) and cin)
                sum(i):=a(i) xor b(i) xor cin ;
                cin:=cout;
          end loop;
          return sum;
    end "+" ;
    BEGIN
    process(clk,pr,cr)
      begin
          if(pr='0' and cr='l')then
                q <=111";
          elsif(pr='l' and cr='0')then
                q<="000";
          elsif(pr='l' and cr='l'and clk='0' and clk'event)then
                q<=q + "100";
          end if;
  end process;
END behaviour;
```

(b) Dataflow model of 3-bit UP Counter

```
LIBRARY ieee;
USE ieee.std_logic_1164.all;
USE ieee.std_logic_arith.all;

ENTITY counter_up IS
   port(pr,cr,clk,t:in bit; qdnout bit_vector(0 to 2));
END counter_up ;

--

ARCHITECTURE dataflow OF counter_up IS
    -- function for addition of two bit arrays
   function "+"(a,b:bit_vector(0 to 2))return bit_vector is
     variable coutbit;
     variable cin:bit:='0';
     variable sum:bit_vector(0 to 2):="000" ;
     begin
        for i in 0 to 2 loop
              cout:=(a(i) and b(i)) or (b(i) and cin) or (a(i) and cin)
              sum(i):=a(i) xor b(i) xor cin;
              cin:=cout;
        end loop;
      return sum;
 end "+" ;

BEGIN

   q<="lll" when(pr='0' and cr='l')else
   "000" when(pr='l' and cr='0')else
   q + "100"when(pr='1' and cr='l'and clk='0' and clk'event);
end dataflow;
```

(c) Structural model of 3-bit UP Counter

```
LIBRARY ieee;
USE ieee.std_logic_1164.all;
USE ieee.std_logic_arith.all;

ENTITY counter_up IS
   port(pr,cr,clk,t:in bit; q:inout bit_vector(0 to 2));
END counter_up;

--

ARCHITECTURE counter_up_structure OF counter_up IS
   component tff
          port(inl,in2,in3,in4:in bit; outl:inout bit);
   end component;
   BEGIN
      ff0:tff port map(t,clk,pr,cr,q(0));
      ffl:tff port map(t,q(0),pr,cr,q(l));
      ff2:tff port map(t,q(l),pr,cr,q(2));
END counter_up_structure;
```

Example 7.9: *For a 3-bit DOWN Counter write down*

(*a*) *Behavior Model*

(*b*) *Dataflow Model*

(*c*) *Structural Model*

Solution: Block diagram of 3-bit DOWN Counter is shown in figure 7.17 and waveforms are shown in figure 7.18

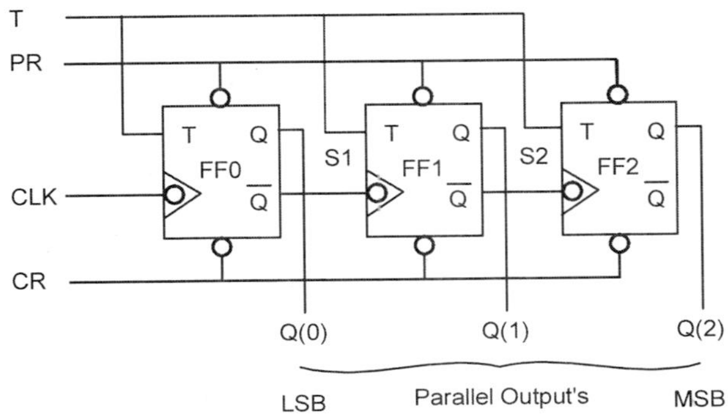

Fig. 7.9: A 3-bit DOWN Counter

(a) Behavior model of 3-bit DOWN Counter

```
LIBRARY ieee;
USE ieee.std_logic_1164.all;
USE ieee.std_logic_arith.all;

ENTITY counter_down IS
    port(pr,cr,clk,t:in bit; q:inout bit_vector(0 to 2));
END counter down;

--

ARCHITECTURE behaviour OF counter_down IS
    function "-"(a,b:bit_vector(0 to 2))return bit_vector is
      variable cout:bit;
      variable cin:bit:='0';
      variable dif:bit_vector(0 to 2):="000";
      begin
      for i in 0 to 2 loop
            cout:=((not a(i)) and b(i)) or (b(i) and cin) or ((not a(i)) and cin);
            dif(i):=a(i) xor b(i) xor cin;
            cin:=cout;
      end loop;
      return dif;
end "-";

BEGIN

process(clk,pr,cr)
      variable S1,S2 : bit;
```

```
            begin
               S1:=not q(0);
               S2:=not q(1);
               if(pr='0' and cr='l')then
                     q <="111";
             elsif(pr='l' and cr='0')then
                     q <="000";
             elsif(pr=T and cr= 'l')then
                  if (clk='0' and clk'event)then
                        q<=q - "100";
                  end if;
               end if;
         end process;
   END behaviour;
```

(b) Dataflow model of 3-bit DOWN Counter

```
LIBRARY ieee;
USE ieee.std_logic_1164.all;
USE ieee.std_logic_arith.all;

ENTITY counter_down IS
     port(pr,cr,clk,t:in bit; q:inout bit_vector(0 to 2));
END counter_down;

- -

ARCHITECTURE dataflow OF counter_down IS
   - - function for subtraction of two bit arrays
   function "-" (a,b:bit_vector(0 to 2))return bit_vector is
         variable cout:bit;
         variable cin:bit:='0';
         variable dif:bit_vector(0 to 2):="000" ;
         begin
         for i in 0 to 2 loop
               cout:=((not a(i)) and b(i)) or (b(i) and cin) or ((not a(i)) and cin);
               dif(i):=a(i) xor b(i) xor cin ;
               cin:=cout;
         end loop;
         return dif; end "-" ;
      signal S1,S2 : bit;

      BEGIN
         S1 <=not q(0);
         S2 <=not q(l);
         q <="111" when(pr='0' and cr='l')else
         "000" when(pr='l' and cr='0')else
         q - "100"when(pr=T and cr='l'and clk='0' and clk'event);
      end dataflow;
```

(c) Structural model of 3-bit DOWN Counter

```
LIBRARY ieee;
USE ieee.std_logic_1164.all;
```

```
USE ieee.std_logic_arith.all;

ENTITY counter_down IS
    port(t,clk,pr,cr:in bit; q:out bit_vector(0 to 2));
END counter_down;
ARCHITECTURE counter_down_structure OF counter_down IS
    component tff
          port(inl,in2,in3,in4:in bit; z,zbar:buffer bit);
    end component;

    signal sl,s2:bit;
    BEGIN
          t0:tff port map(t,clk,pr,cr,q(0),sl);
          tl:tff port map(t,sl,pr,cr,q(l),s2);
          t2:tff port map(t,s2,pr,cr,q(2),open);
END counter_down_structure;
```

Example 7.10: *For a 3-bit UP/DOWN Counter write down*

(*a*) *Behavior Model*

(*b*) *Dataflow Model*

(c) *Structural Model*

Solution: Block diagram of 3-bit DOWN Counter is shown in figure 7.10

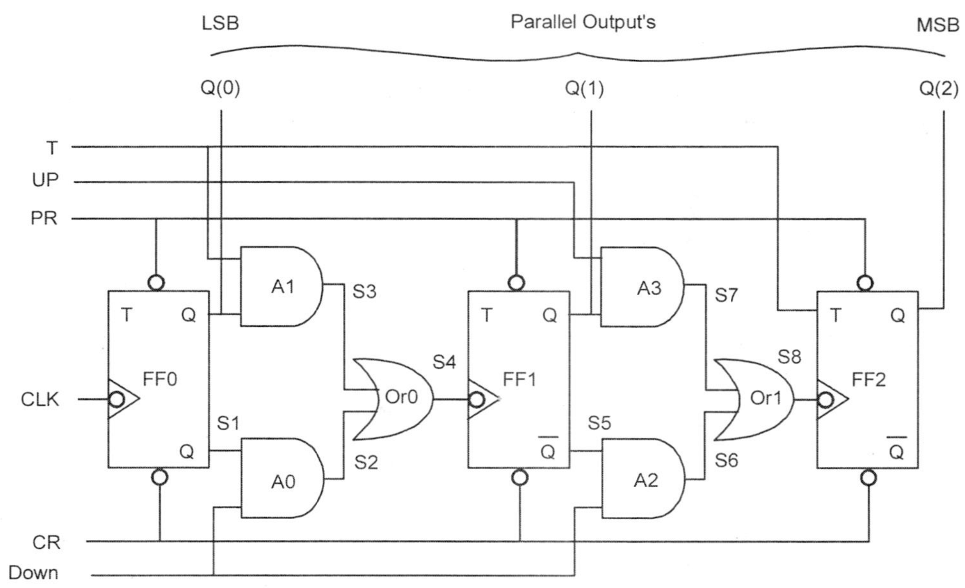

Fig.7.10: A 3-bit UP/DOWN Counter (Asynchronous)

- When UP = '1', circuit works as UP counter.
- When DOWN = '1', circuit works as DOWN counter.

(a) Behavior model of 3-bit UP/DOWN Counter

```
LIBRARY ieee;
USE ieee.std_logic_1164.all;
USE ieee.std_logic_arith.all;

ENTITY counter_updown IS
      port(pr,cr,up,dn,clk,t:in bit; q:inout bit_vector(0 to 2));
END counter_updown;

--

ARCHITECTURE behaviour OF counter_updown IS
    -- function for addition of two bit arrays
    function "+"(a,b:bit_vector(0 to 2))return bit_vector is
            variable cout:bit;
            variable cin:bit:='0';
            variable sum:bit_vector(0 to 2):="000";
            begin
            for i in 0 to 2 loop
                  cout:=(a(i) and b(i)) or (b(i) and cin) or (a(i) and cin);
                  sum(i):=a(i) xor b(i) xor cin ;
                  cin:=cout; end loop; return sum;
            end "+" ;
            -- function for subtraction of two bit arrays
            function "-"(a,b:bit_vector(0 to 2))return bit_vector is
                  variable cout:bit;
                  variable cin:bit:='0';
                  variable dif:bit_vector(0 to 2):="000";
                  begin
                  for i in 0 to 2 loop
                        cout:=((not a(i)) and b(i)) or (b(i) and cin) or ((not a(i)) and cin);
                        dif(i):=a(i) xor b(i) xor cin;
                        cin:=cout;
                  end loop;
                  return dif;
            end "-";

            BEGIN
            process(clk,pr,cr)
                  variable s1,s2,s3,s4,s5,s6,s7,s8:bit;
                  begin
                  s1:=not q(0);
                  s2:=up and q(0);
                  s3:=sl and dn;
                  s4:=s2 or s3;
                  s5:=not q(l);
                  s6:=up and q(l);
                  s7:=s5 and dn;
                  s8:=s6 or s7;

                  if(pr='0' and cr='l')then
                      q<="111";
```

```
                elsif(pr='1' and cr= '0')then
                    q<="000";
                elsif(pr='l' and cr='l'and up='l'and dn='0'
                                            and clk='0' and clk'event)then
                    q<=q + "100";
                elsif(pr='l' and cr='l'and up='0' and dn='l' and clk='0'
                                            and clk'event)then
                    q<=q - "100";
                end if;
            end process ;
        end behaviour;
```

(b) Dataflow model of 3-bit UP/DOWN Counter

```
LIBRARY ieee;
USE ieee.std_logic_1164.all;
USE ieee.std_logic_arith.all;

ENTITY counter_updown IS
    port(pr,r,up/dn,clk,t:in bit; q:inout bit_vector(0 to 2));
END counter_updown;

- -

ARCHITECTURE dataflow OF counter_updown IS
    - - function for addition of two bit arrays
    function "+"(a,b:bit_vector(0 to 2))return bit_vector is
            variable cout:bit;
            variable cin:bit:='0';
            variable sum:bit_vector(0 to 2):= "000";
            begin
            for i in 0 to 2 loop
                cout:=(a(i) and b(i)) or (b(i) and cin) or (a(i) and cin);
                sum(i):=a(i) xor b(i) xor cin ;
                cin:=cout;
            end loop ;
            return sum;
end "+" ;
- - function for subtraction of two bit arrays
function "-"(a,b:bit_vector(0 to 2))rerum bit_vectcr is
            variable cout:bit;
            variable cin:bit:='0';
            variable dif:bit_vector(0 to 2):="000" ;
            begin
            for i in 0 to 2 loop
                cout:=((not a(i)) and b(i)) or (b(i) and cin) or ((not a(i)) and cin);
                dif(i):=a(i) xor b(i) xor cin ;
                cin:=cout;
            end loop;
            return dif;
end "-" ;

signal sl,s2,s3,s4,s5,s6,s7,s8:bit;
```

```
BEGIN
    begin
s1 <=not q(0);
s2 <=up and q(0);
s3 <=sl and dn;
s4 <=s2 or s3 ;
s5 <=not q(l);
s6 <=up and q(l);
s7 <=s5 and dn;
s8 <=s6 or s7;
q<= "111" when(pr= '0' and cr= 'l')else
"000" when(pr='1' and cr= '0')else
q+ "100"when(pr='l' and cr= 'l'and up='l'and dn= '0' and clk='0'
                                        and clk'event)else
q - "100"when(pr='l' and cr='l'and up= '0'and dn='l' and clk='0'
                                        and clk'event);
end dataflow;
```

(c) Structural model of 3-bit UP/DOWN Counter

```
LIBRARY ieee;
USE ieee.std_logic_1164.all;
USE ieee.std_logic_arith.all;

ENTITY counter_up_down IS
     port(t,clk,pr,cr,up,dn:in bit; q:buffer bit_vector(0 to 2));
END counter_up_down;

ARCHARCHITECTURE counter_up_down_structure OF counter_up_down
ISITECTURE counter_up_down_structure OF counter_up_down IS
   component tff
          port(inl,in2,in3,in4:in _bit; z,zbar:buffer bit);
   end component;

   component and2
          port(in5, in6:in bit; outl:out bit);
   end component;
   component or2
          port(in7,in8:in bit; out2:out bit);
   end component;
   signal sl,s2,s3,s4,s5,s6,s7,s8:bit;
   BEGIN
        t0:f port map(t,clk,pr,cr,q(0),sl);
        tl:tff port map(t,s4,pr,cr,q(l),s5);
        t2:tff port map(t,s8,pr,cr,q(2),open);
        a0:and2 port map(sl,dn,s2);
        al:and2 port map(q(0),up,s3); a2:and2 port map(s5,dn,s6); a3:and2 port map(q(l),up,s7);
        or0:or2 port map(s2,s3,s4); orl:or2 port map(s6,s7,s8);
END counter_up_down_structure;
```

Example 7.11: *For a ALU(74381) write down Behavior Model.*

Solution: Block diagram of a ALU (74381) is shown in figure 7.11 and Function table is shown in Table7.11

Table 7.5 Function Table of ALU(74381)

Select Input			Operation	Output
S2	S1	S0		F
0	0	0	CLEAR	0 0 0 0
0	0	1	Q – P	Q – P
0	1	0	P – Q	P – Q
0	1	1	ADD	P + Q
1	0	0	XOR	P XOR Q
1	0	1	OR	P OR Q
1	1	0	AND	P AND Q
1	1	1	PRESET	1 1 1 1

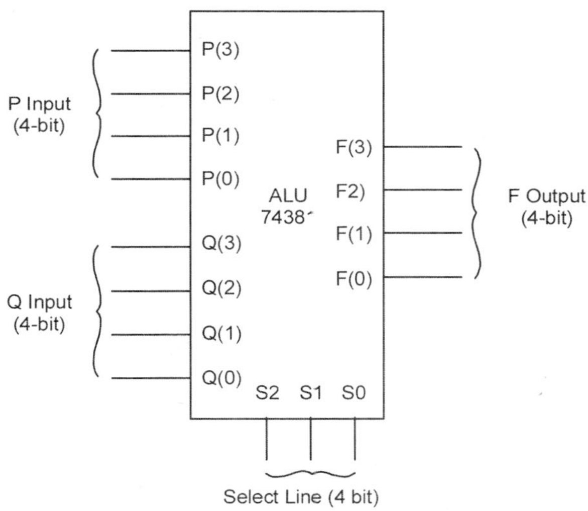

Figure 7.21: Block Diagram of ALU(74381)

Behavior Model of ALU(74381):

```
LIBRARY ieee;
USE ieee.std_logic_1164.all;
USE ieee.std_logic_arith.all;

ENTITY alu IS
   port(p,q:inbit_vector(0 to 3); s:in bit_vector(0 to 2); f:out bit_vector(0 to 3));
END alu;
--
ARCHITECTURE behavior OF alu IS
```

```
    - - function for addition of two bit arrays
    function "+"(a,b:bit_vector(0 to 3))return bit_vector is
       variable cout:bit;
       variable cin:bit:='0';
       variable sum:bit_vector(0 to 3):="0000";
       begin
       for i in 0 to 3 loop
              cout:=(a(i) and b(i)) or (b(i) and cin) or (a(i) and cin);
              sum(i):=a(i) xor b(i) xor cin;
              cin:=cout;
       end loop ;
       return sum;
  end "+" ;
  - - function for subtraction of two bit arrays
  function "-"(a,b:bit_vector(0 to 3))return bit_vector is
       variable cout:bit;
       variable cin:bit:='0'; variable dif :bit_vector(0 to 3) :="0000";
       begin for i in 0 to 3 loop
                    cout:=((not a(i)) and b(i)) or (b(i) and cin) or ((not a(i)) and cin);
                    dif(i):=a(i) xor b(i) xor cin;
                    cin:=cout; end loop; return dif;
       end "-";
  begin
  process(p,q,s)
    begin
       case s is
            when "000"=>
                  f<="0000";
            when "001 "=>
                  f<=q-p;
            when"010"=>
                  f<=p-q;
            when "011"=>
                  f<=p + q;
            when "100"=>
                  f <=p xor q;
            when "101 "=>
                  f <=p or q;
            when "110"=>
                  f <=p and q;
            when"111"=>
                  f<="1111";
       end case;
   end process;
END behavior;
```

Example: 7.12: *For a D flip-flop write down Behavior Model.*

Solution: A D flip-flop is shown in figure 7.12 and Truth table is shown Table 7.12

7.12: Truth table of D flip-flop. Boolean Expression of D flip-flop- Q(t+1) = D

PR	CR	D(i/p)	Output Q (t + 1)
0	0	X	1
0	1	X	1
1	0	X	0
1	1	0	0
1	1	1	1

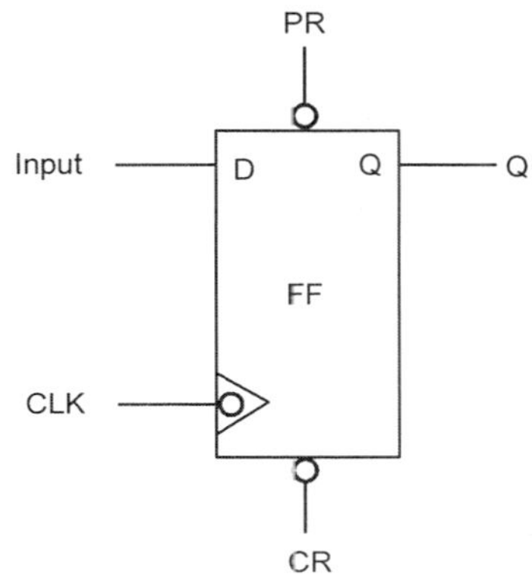

Fig. 7.12: D flip-flop.

Behavior Model of D flip-flop:

```
LIBRARY ieee;
USE ieee.std_logic_1164.all;
USE ieee.std_logic_arith.all;
ENTITY dff IS
    port(clk,pr,cr,d:in bit; q:out bit);
END dff,;
ARCHITECTURE dff_behaviour OF dff IS
begin
    process(clk,pr,cr)
    begin
      if (pr ='0' and cr ='1') then
            q<='1';
      elsif (pr ='1' and cr ='0') then
            q<= '0';
      elsif (pr = '1' and cr ='1' and clk = '0' and clk'event) then
            q<=d;
      end if;
   end process;
end dff_behaviour;
```

Example 7.13: *For a T flip-flop write down Behavior Model*

Solution: A D flip-flop is shown in figure 7.13 and Truth table is shown in Table 7.13

Table 7.13: Truth table of T flip-flop

PR	CR	T(i/p)	Output
			Q (t + 1)
0	0	X	1
0	1	X	1
1	0	X	0
1	1	0	Q
1	1	1	$\overline{Q}$

Boolean Expression of T flip-flop : $Q(t+1) = T\overline{Q} + \overline{T}Q$

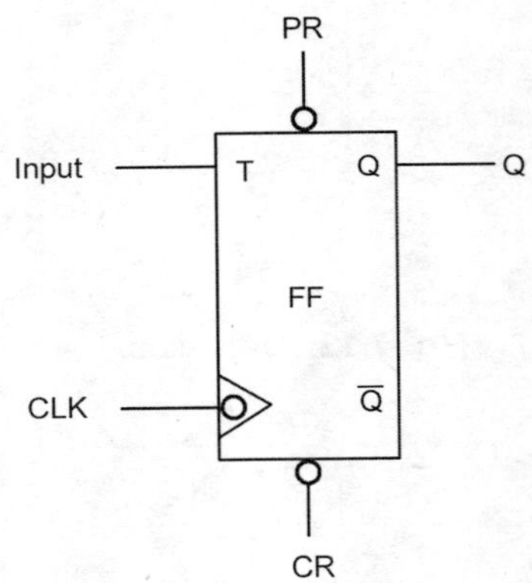

Behavior Model of T flip-flop :

```
LIBRARY ieee;
USE ieee.std_logic_1164.all;
USE ieee.std_logic_arith.all;
ENTITY tff IS
    port(clk,pr,cr,t:in bit; q:inout bit; qbanout bit) ;
END tff;
ARCHITECTURE tff_behaviour OF tff IS
begin
    process(clk,pr,cr)
    begin
        if (pr ='0' and cr ='1') then
            q <='1';
        elsif (pr ='1' and cr ='0') then
            q<='0';
```

```
                elsif (pr ='1' and cr ='1' and clk = '0' and elk'event) then
                      q <=(not t and q) or (not q and t);
                end if;
          end process;
                qbar <= not q;
      end tff_behaviour;
```

Example 7.14: *For a 1-Bit comparator do the following:*

(*a*) *Design the ciruit*

(*b*) *Behavior Model*

(*c*) *Dataflow Model*

(*d*) *Structural Model*

Solution: (*a*) A 1-Bit comparator is a logic circuit which compares two bits and generates three outputs.

The block diagram of a 1-Bit comparator is given in figure 7.14 Truth table of 1-Bit comparator is given in Table 7.8

Table 7.8 : Truth Table of 1 -Bit Comparator

I/P		O/P		
A	**B**	**AGB**	**AEB**	**ALB**
0	0	0	1	0
0	1	0	0	1
1	0	1	0	0
1	1	0	1	0

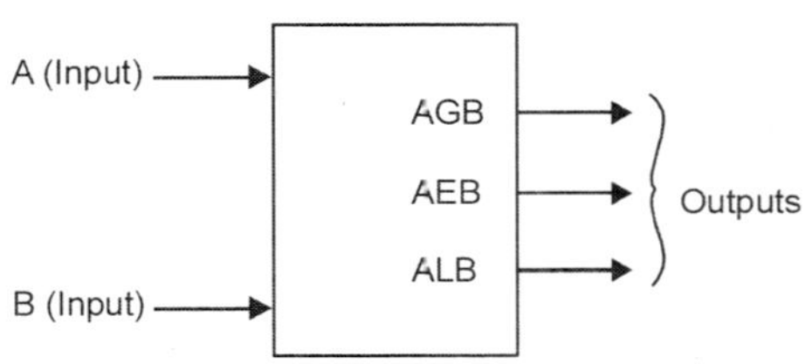

Figure 7.14: Block Diagram of 1-Bit Comparator.

Now by using K-map find out AGB, AEB and ALB.

(I) K-map for AGB

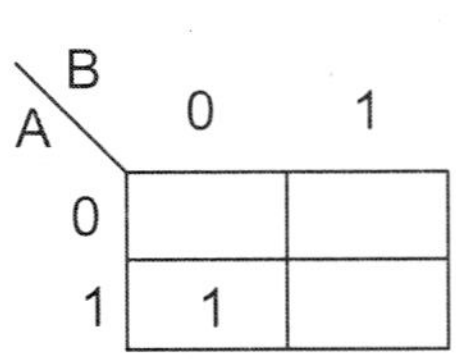

AGB = $A\bar{B}$

(I) K-map for AEB

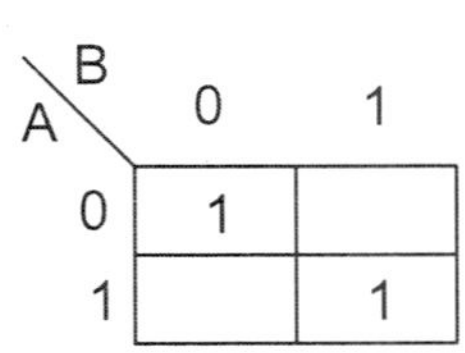

AEB = $AB + \bar{A}\bar{B}$

(III) K-map for ALB

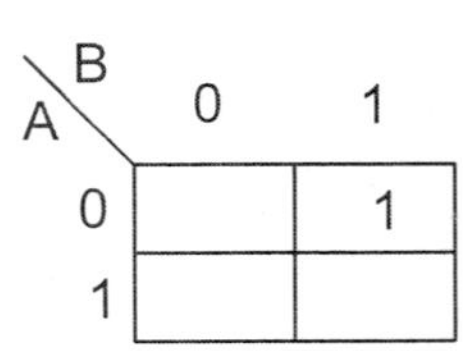

ALB = $\bar{A}B$

(b) Behavior Model of 1 -Bit comparator:

```
LIBRARY ieee;
USE ieee.std_logic_1164.all;
USE ieee.std_logic_arith.all;
ENTITY comparator_l_bit IS
      port(A,B:in bit; AGB. AEB, ALB:out bit);
END comparator_l_bit;
ARCHITECTURE comparator_l_bit_behaviour OF comparator_l_bit IS
begin
      process(A, B) begin
      if (A>B) then
            AGB <='1';
            AEB <='0';
            ALB <='()';
      elsif (A = B)then
            AGB <='0';
            AEB <='1';
            ALB <='()';
      elsif (A < B) then
            AGB <='{)';
            AEB <='0';
            ALB <='1';
      end if;
  end process;
end comparator_l_bit_behaviour;
```

(c) Dataflow Model of 1-Bit comparator :

```
LIBRARY ieee;
USE ieee.std_logic_1164.all;
USE ieee.std_logic_arith.all;

ENTITY comparator_l_bit IS
      port(A,B:in bit; RESULT:out bit_vector (2 downto 0))
END comparator_l_bit;

ARCHITECTURE comparator_l_bit_behaviour OF comparator_l_bit IS
begin
      RESULT <= "100" when (A > B) else
                "010" when (A =B) else
                "001";
end comparator_l_bit_dataflow;
- - here RESULT(2) is AGB, RESULT(l) is AEB and RESULT(O) is ALB.
```

(d) Structural Model of 1-Bit comparator:

```
LIBRARY ieee;
USE ieee.std_logic_1164.all;
USE ieee.std_logic_arith.all;

ENTITY comparator_l_bit IS
    port(A,B:in bit; AGB, AEB, ALBrout bit);
```

```
END comparator_l_bit;

ARCHITECTURE comparator_l_bit_structure OF comparator_l_bit IS
    component AND2
        port (IN0, INI : in bit; OUTO : out bit) ;
    end component;
    component XNOR2
        port (IN2, IN3 : in bit; OUT1: out bit);
    end component;

    component NOT1
        port (IN4 : in bit; OUT2 : out bit)l;
    end component;
    signal Ul, U2: bit;

    begin

     A0 : AND2 port map (A, U2, AGB) ;
     A1 : AND2 port map (Ul, B, ALB);
     N0 : NOT1 port map (A, Ul);
     N1 : NOT1 port map (B, U2);
     X0 : XNOR2 port map (A, B, AEB);

end comparator_l_bot_structure;
```

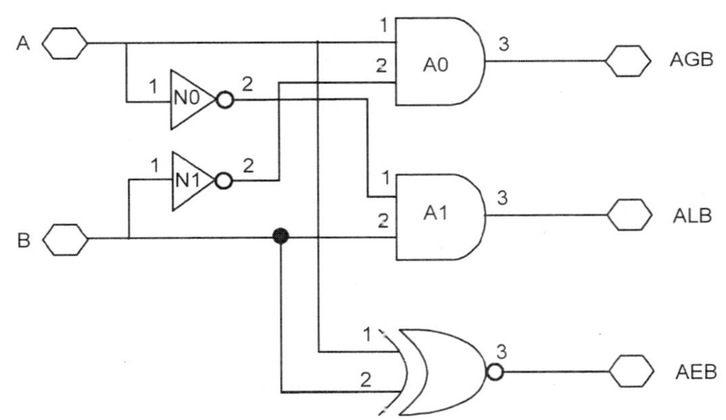

Fig. 7.15: Circuit diagram of 1-Bit comparator.

Example:1.15: *For a 4-Bit comparator do the following:*

(*a*) Behavior Model
(*b*) Dataflow Model
(*c*) Structural Model

Solution: Block diagram of a 4-Bit comparator is given in figure 7.16. This 4-Bit comparator compares two 4-bit numbers and accordingly generate outputs AGB (A greater than B), AEB (A equal than B) and ALB (A less than B).

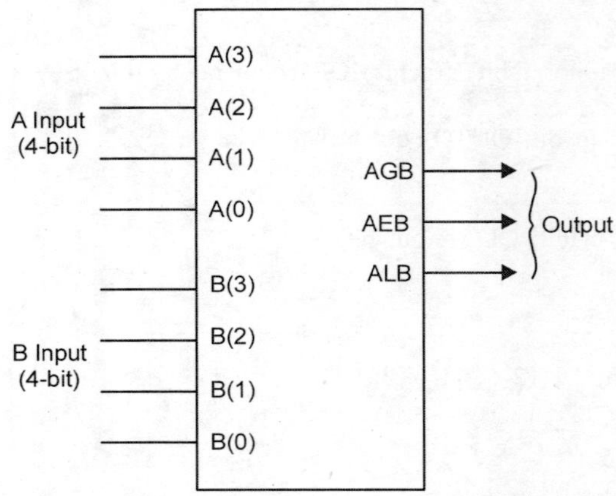

Fig. 7.16: Circuit Diagram of 4-Bit Comparator.

(a) Behavior Model of 4-Bi.t comparator

```
LIBRARY ieee;
USE ieee.std_logic_1164.all;
USE ieee.std_logic_arith.all;

ENTITY comparator_4_bit IS
   port(A,B:in bit_vector (3 downto 0); AGB, AEB, ALBrout bit);
END comparator_4_bit;

ARCHITECTURE comparator_4_bit_behaviour OF comparator_4_bit IS
begin
   process(A, B)
   begin
         if (A > B) then
               AGB <='1';
               AEB <='0';
               ALB <='0';
         elsif (A = B) then
               AGB <='0';
               AEB<= '1';
               ALB<= '0';
         elsif (A < B) then
               AGB <='0';
               AEB<='0';
               ALB<='1';
         end if;
   end process;
end comparator_4_bit_behaviour;
```

(b) Dataflow Model of 4-Bit comparator

```
LIBRARY ieee;
USE ieee.std_logic_1164.all;
```

```
USE ieee.std_logic_arith.all;

ENTITY comparator_4_bit IS
    port(A,B:in bit_vector (3 downto 0); RESULT: out bit_vector (2 downto 0));
END comparator_4_bit;

ARCHITECTURE comparator_4_bi .dataflow OF comparator_4_bit IS
begin
    RESULT <=  "100" when (A > B) else
               "010" when (A = B) else
               "001";
end comparator_4_bit_dataflow;
-- here RESULT(2) is AGB, RESULT(l) is AEB and RESULT(O) is ALB
```

(c) Structural Model of 4-Bit comparator

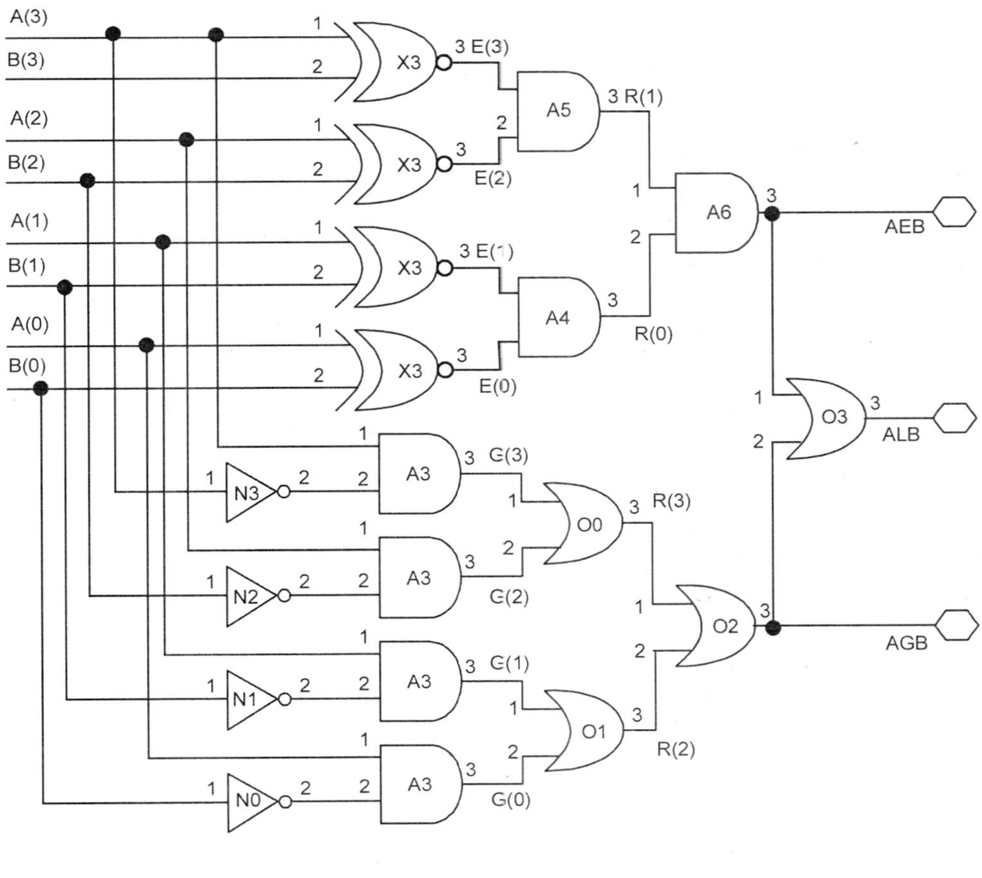

Fig. 7.30: Circuit Diagram of 4-Bit Comparator

```
LIBRARY ieee;
USE ieee.std_logic_1164.all;
USE ieee.std_logic_arith.all;

ENTITY comparator_4_bit IS
    port(A,B:in bit_vector (3 downto 0); ALB: out bit; AGB, AEB : inout bit);
END comparator_4_bit;

ARCHITECTURE comparator_4_bit_dataflow OF comparator_4_bit IS

component AND2
      port (IN0, INI : in bit; OUT0 : out bit);
end component;
component XNOR2
    port (IN2, IN3 : in bit; OUT1: out bit);
end component;
component OR2
    port (IN4, IN5 : in bit; OUT2 : out bit);
end component;

component NOT1
    port (IN6 : in bit; OUT3 : out bit);
end component;

signal E, G, X, R : bit_vector (3 downto 0) ;

begin
    X3 : XNOR2 port map (A(3), B(3), E(3));
    X2 : XNOR2 port map (A(2), B(2), E(2));
    X1 : XNOR2 port map (A(l), B(l), E(l));
    X0 : XNOR2 port map (A(0), B(0), E(0));
    A6 : AND2 port map (R(l), R(0), AEB);
    A5 : AND2 port map (E(3), E(2), R(l));
    A4 : AND2 port map (E(l), E(0), R(0));
    A3 : AND2 port map (A(3), X(3), G(3));
    A2 : AND2 port map (A(2), X(2), G(2));
    A1 : AND2 port map (A(l), X(l), G(l));
    A0 : AND2 port map (A(0), X(0), G(0));
    03  : OR2 port map (AEB, AGB, ALB);
    02  : OR2 port map (R(3), R(2), AGB));
    01  : OR2 port map (G(l), G(0), R(0));
    00  : OR2 port map (G(3), G(2), R(3));
    N3 : NOT1 port map (B(3), X(3));
    N2 : NOT1 port map (B(2), X(2));
    N1 : NOT1 port map (B(l), X(l)) ;
    N0 : NOT1 port map (B(0), X(0));
end comparator_4_bit_structure;
```

Example 7.16: *For a 3-Bit Binary to Gray code converter*:

(*a*) *Design the circuit* (*b*) *Write Behavior Model*

(*c*) *Write Dataflow Model* (*d*) *Write Structure Model*

Solution: (a) Truth table of a 3-Bit Binary to Gray code converter is given in Table 7.9

Table 7.9: Truth Table of 3-Bit Binary to Gray converter

Inputs			Outputs		
B(2)	**B(1)**	**B(0)**	**G(2)**	**G(1)**	**G(0)**
0	0	0	0	0	0
0	0	1	0	0	1
0	1	0	0	1	1
0	1	1	0	1	0
1	0	0	1	1	0
1	0	1	1	1	1
1	1	0	1	0	1
1	1	1	1	0	0

Now by using K-map find out G(2), G(1) and G(0).

(*i*) K-map for G(2)

B(2) \ B(1) B(0)	00	01	11	10
0				
1	1	1	1	1

G(2) = B(2)

(*ii*) K-map for G(1)

B(2) \ B(1) B(0)	00	01	11	10
0			1	1
1	1	1		

G(1) = B(2) × orB (1)

(*iii*) K-map for G(0)

B(2) \ B(1) B(0)	00	01	11	10
0		1		1
1		1		1

G(0) = B(1) × orB (0)

(b) Write Behavior Model of 3-Bit Binary to Gray code converter

```
LIBRARY ieee;
USE ieee.std_logic_1164.all;
USE ieee.std_logic_arith.all;

ENTITY binary_to_gray IS
    port(B:in bit_vector (2 downto 0); G: out bit_vector (2 downto 0));
END binary_to_gray;

ARCHITECTURE binary_to_gray_behaviour OF binary_to_gray IS

begin
    process(B)
    begin
            G(2) <=B(2);
            G(2) <=B(2)xor B(l);
            G(0) <=B(l)xorB(0);
    end process;
end binary_to_gray_behaviour;
```

(c) Write Dataflow Model of 3-Bit Binary to Gray code converter

```
LIBRARY ieee;
USE ieee.std_logic_1164.all;
USE ieee.std_logic_arith.all;

ENTITY binary_to_gray IS
    port(B:in bit_vector (2 downto 0); G: out bit_vector (2 downto 0));
END binary_to_gray;

ARCHITECTURE binray_to_gray_dataflow OF binray_to_gray IS
begin
    G(2) <=B(2);
    G(1) <=B(2) xor B(1);
    G(0) <=B(1) xor B(0);
end binary_to_gray_dataflow;
```

(d) Structure Model of 3-Bit Binary to Gray code converter

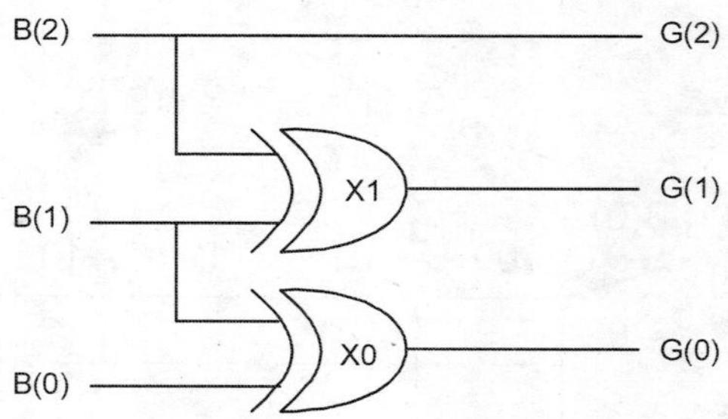

Fig. 7.18 Circuit diagram of 3-Bit Binary to Gray Converter.

```
LIBRARY ieee;
USE ieee.std_logic_1164.all;
USE ieee.std_logic_arith.all;

ENTITY binary_to_gray IS
    port(B:in bit_vector (2 downto 0); G: out bit_vector (2 downto 0));
END binary_to_gray;

ARCHITECTURE binary_to_gray_structure OF binary_to_gray IS

    component XOR2
            port (IN0, IN1: in bit; OUTO: out bit);
    end component;
begin
  Xl:XOR2 port map (B(2), B(l), G(l));
  X0:XOR2 port map (B(l), B(0), G(0));
end binary_to_gray_structure;
```

Example 7.17: *For a 3-Bit Even Parity Checker:*

(*a*) *Design the circuit.*
(*b*) *Write Behavior Model*
(*c*) *Write Dataflow Model*
(*d*) *Write Structural Model*

Solution: *(a)* Truth table of a 3-Even Parity Checker is given in Table 7.16. Three bits are received by even parity checker. Out of these three bits **two** bits A and B message bits and C is parity bit send along with message bits. If the number of '1' in input bits are even then PE = '0' and if the number of '1' in input bits are odd then PE = '1'.

Table 7.16: Table of a 3-Even Parity Checker

I/P			O/P
A	**B**	**C (Parity bit)**	**PE**
0	0	0	0
0	0	1	1
0	1	0	1
0	1	1	0
1	0	0	1
1	0	1	0
1	1	0	0
1	1	1	1

Now by using K-map find out PE

K-map for G(2)

A \ BC	00	01	11	10
0		1		1
1	1		1	

PE = A xor B xor C

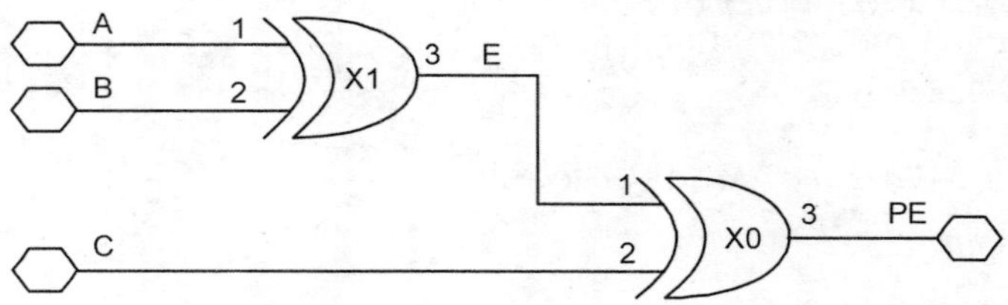

Fig. 7.19: Circuit Diagram of 3-Bit Even Parity Checker

(b) Behaviour Model of a 3-Bit Even Parity Checker

```
LIBRARY ieee;
USE ieee.std_logic_1164.all;
USE ieee.std_logic_arith.all;

ENTITY parity_checker_IS
    port(A, B, Gin bit; PE: out bit);
END parity_checker;

ARCHITECTURE parity_checker_behaviour OF parity_checker IS
begin
    process (A, B, C)
    begin
        PE <= AxorBxorC;
    end process;
end parity_checker_behaviour;
```

(c) Write Dataflow Model of a 3-Bit Parity Checker

```
LIBRARY ieee;
USE ieee.std_logic_1164.all;
USE ieee.std_logic_arith.all;

ENTITY parity_checker_IS
      port(A, B, Gin bit; PE: out bit);
END parity_checker;
```

```
ARCHITECTURE parity_checker_dataflow OF parity_checker IS
begin
   PE <= A xor B xor C ;
end parity_checker_dataflow;
```

(d) Write Structural Model of a 3-Bit Even Parity Checker

```
LIBRARY ieee;
USE ieee.std_logic_1164.all ;
USE ieee.std_logic_arith.all;

ENTITY parity_checker_IS
   port(A, B, Gin bit; PE: out bit);
END parity_checker;

ARCHITECTURE parity_checker_structure OF parity_checker IS

component XOR2
   port (INO, IN1: in bit ; OUT0 : out bit);
end component;
signal E:bit;

begin

   X1 :XOR2 port map (A, B, E);
   X0:XOR2 port map (E, C, PE);

end parity_checker_structure;
```

Example 7.18: *For a 4-to-2 Encoder*:

(*a*) *Explain the operation*
(*b*) *Write Behaviour Model*
(*c*) *Write Dataflow Model*
(*d*) *Write Structural Model*

Solution: (*a*) Encoder is a logic circuit which converts decimal input into its equivalent binary output. Block diagram of a 4 to 2 Encoder is shown in figure 7.20 Truth table of 4 to 2 Encoder is given in Table.

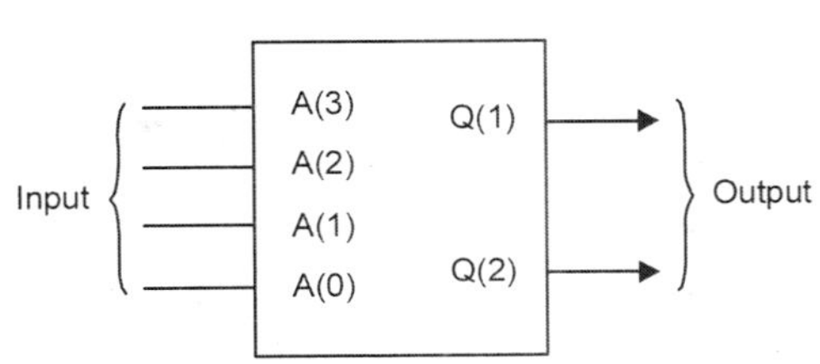

Fig. 7.20: Block Diagram of 4 to 2 Encoder.

Table 7.11: Truth table of 4-to-2 Encoder

Input	Output Q(1)	Output Q(0)
A(0)	0	0
A2)	0	1
A2)	1	0
A(3)	1	1

Operation of this decoder:

- when A(0) = '1' then Q(1) = '0' and Q(0) = '0'.
- when A(1) = '1' then Q(1) = '0' and Q(0) = '1'.
- when A(2) = '1' then Q(1) = '1' and Q(0) = '0'.
- when A(3) = '1' then Q(1) = '1' and Q(0) = '1'.

Disadvantages of this encoder :

(i) When A(0) = '1' or when all inputs are at '0' then Q(1) = '0' and Q(0) = '0'. So it is not clear that Q(1) = '0' and Q(0) = '0' are due to 'HIGH' on A(0). This problem can be solved by providing an additional output that specifies the condition that none of the inputs are active.

(ii) Only one input can be active at any given time. If two inputs are active simultaneously, the output produces an undefined combi-nation. For example, when A(1) and A(2) both are at '1' simul-taneously then according to figure 7.21 Q(1) = '1' and Q(0) = '1'. But both outputs should be 'HIGH' when A(3) is 'HIGH'.

To resolve this problem, the encoder must establish a priority to ensure that only one input is encoded. To overcome this disadvantage we have to design a Priority Encoder.

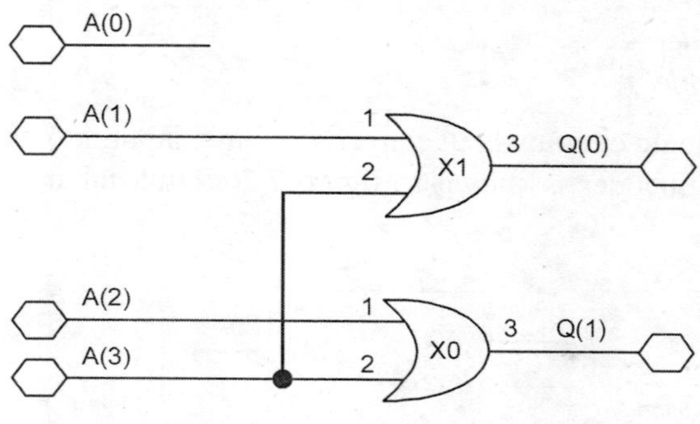

Fig 7.21: Circuit Diagram of 4 to 2 Encoder

(b) Behaviour Model of a 4 to 2 Encoder

```
LIBRARY ieee;
USE ieee.std_logic_1164.all;
USE ieee.std_logic_arith.all;
```

```
ENTITY encoder_IS
    port(A:in bit_vector (3 downto 0); Q: out bit_vector (1 downto 0));
END encoder;

ARCHITECTURE encoder_behaviour OF encoder IS
begin
    process(B)
    begin
            case A is
            when "0001" =>
            Q <= "00";
            when "0010" =>
            Q <= "01";
            when "0100" =>
            Q <= "10";
            when "1000" =>
            Q<= "11";
    end process;
end encoder_behaviour;
```

(c) Write Dataflow Model of a 4 to 2 Encoder

```
LIBRARY ieee;
USE ieee.std_logic_1164.all;
USE ieee.std_logic_arith.all ;

ENTITY encoder_IS
    port(A:in bit_vector (3 downto 0); Q: out bit_vector (1 downto 0));
END encoder;

ARCHITECTURE encoder_dataflow OF encoder IS
begin
    Q <=  "00" when A = "0001" else,
          "01" when A = "0010" else,
          "10" when A = "0100" else,
          "11";
end encoder_dataflow;
```

(d) Write Structural Model of a 4 to 2 Encoder

```
LIBRARY ieee;
USE ieee.std_logic_1164.all;
USE ieee.std_logic_arith.all;

ENTITY encoder_IS
    port(A:in bit_vector (3 downto 0); Q: out bit_vector (1 downto 0));
END encoder;

ARCHITECTURE encoder_structure OF encoder IS

    component OR2
        port (INO, IN1: in bit ; OUTO: out bit);
    end component;
begin
```

```
    X1:OR2 port map (A(1), A(3), Q(0));
    X0:OR2 port map (A(2), A(3), Q(l));
end encoder_structure;
```

Example 7.19: *For a 4-to-2 Priority Encoder:*

(a) Explain the operation
(b) Write Behaviour Model
(c) Write Dataflow Model
(d) Write Structural Model

Solution, *(a)* A Priority Encoder is a encoder circuit that includes priority function. The operation of priority encoder is such that if two or more inputs are equal to '1' at the same time, the input having the highest priority will take precedence. Block diagram of a 4 to 2 priority encoder is shown in figure 7.22 truth table of 4 to 2 priority encoder is given in Table 7.12 The xs are Don't Care conditions. The input A(3) has the highest pnority,'so regardless of the values of the other inputs, when this input is '1', the output for Q(1) and Q(0) is "11". A(2) has next priority level. The output is "10" if A(2) = '1', A(3) = '0' and A(1) and A(2) are V. The input for A(1) is generated if A(3) = '0' and A(2) = '0'.

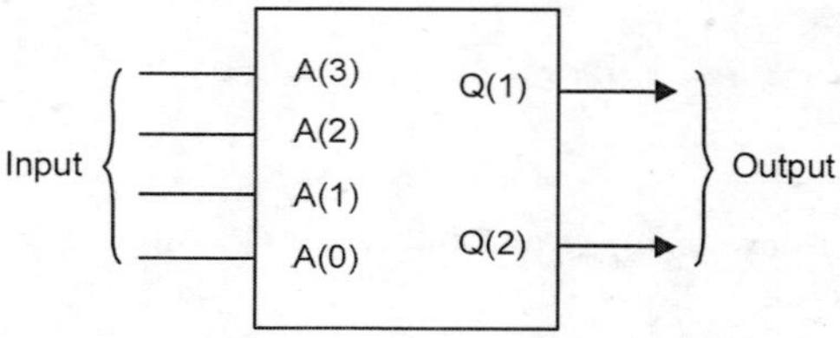

Fig. 7.22: Block Diagram of 4 to 2 Priority Encoder

Another output 'Valid' is also incorporated in the circuit. The function of 'Valid' is to check at least one input is '1'

Table7.12: Truth Table of 4 to 2 Priority Encoder

Inputs				**Outputs**		
A(3)	**A(2)**	**A(l)**	**A(0)**	**Q(1)**	**Q(O)**	**Valid**
0	0	0	0	0	0	0
0	0	0	1	0	0	1
0	0	1	X	0	1	1
0	1	X	X	1	0	1
1	X	X	X	1	1	1

Now by using K-map find out Q(1), Q(0) and Valid.

(i) K-map for Q(1)

A(3) A(2) \ A(1) A(0)	00	01	11	10
00				
01	1	1	1	1
11	1	1	1	1
10	1	1	1	1

$$Q(1) = A(3) + A(2)$$

(ii) K-map for Q(0)

A(3) A(2) \ A(1) A(0)	00	01	11	10
00			1	1
01				
11	1	1	1	1
10	1	1	1	1

$$Q(1) = A(3) + \overline{A(2)}\,A(1)$$

(iii) K-map for 'Valid'

A(3) A(2) \ A(1) A(0)	00	01	11	10
00		1	1	1
01	1	1	1	1
11	1	1	1	1
10	1	1	1	1

Valid = A(3) + A(2) + A(1) + A(0)

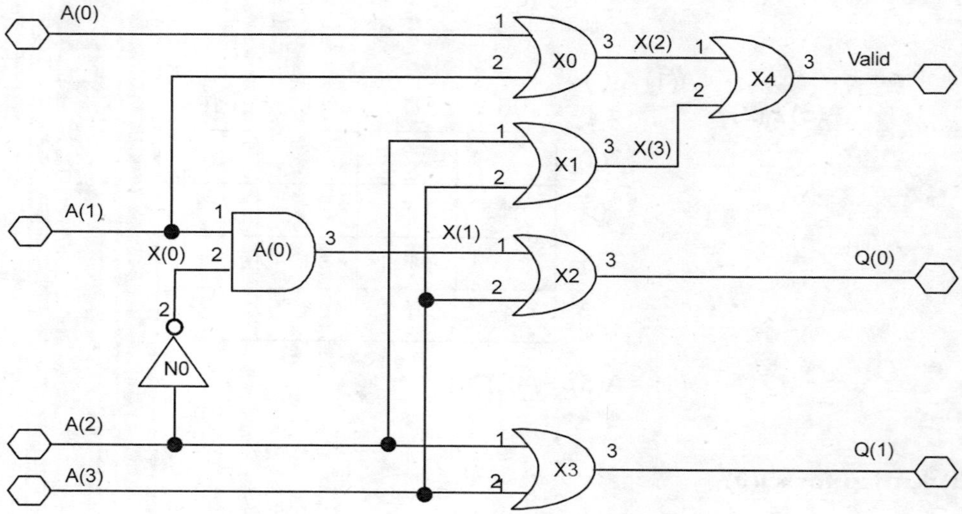

Fig. 7.23: Circuit Diagram of 4 to 2 Priority Encoder

(b) Behavior Model of a 4 to 2 priority Encoder

```
LIBRARY ieee;
USE ieee.std_logic_1164.all;
USE ieee.std_logic_arith.all;

ENTITY priority_encoder_IS
   port(A:in bit_vector (3 downto 0); VALID: out bit;
                              Q: out bit_vector (1 downto 0));
END priority_encoder;
ARCHITECTURE priority_encoder_behaviour OF priority_encoder IS
begin

process(A)
begin
    if(A(3)='l')then
        Q<="11";
        VALID <= '1';
    elsif(A(2) = 'l')then
        Q <= "10";
        VALID <='1';
    elsif(A(l)='l')then
        Q<="01";
        VALID <='l';
    elsif(A(0)='l')then
        Q<="00";
        VALID <='1';
    else
      Q<="00";
```

```
            VALID <='0';
         End if;
      end process;
   end priority_encoder_behaviour;
```

(c) Dataflow Model of a 4 to 2 priority Encoder

```
LIBRARY ieee;
USE ieee.std_logic_1164.all;
USE ieee.std_logic_arith.all;

ENTITY priority_encoder_IS
   port(A:in bit_vector (3 downto 0); VALID: out bit;
        Q: out bit_vector (1 downto 0));
END priority_encoder;

ARCHITECTURE priority_encoder_dataflow OF priority_encoder IS
begin
   process(A)
   begin
      Q <=  "11" when (A = "1 --- "else
            "10" when (A = "01-- "else
            "01" when (A = "001- "else
            "00" when (A = "0001" else
            "00"
         VALID <='1' when (A(3)='l'or '1'(2)='1' or A(1)='1' or A(0)=' l') else
                  '0';
end priority_encoder_dataflow;
```

(d) Write Structural Model of a 4 to 2 priority Encoder

```
LIBRARY ieee;
USE ieee.std_logic_1164.all;
USE ieee.std_logic_arith.all;

ENTITY priority_encoder_IS
    port(A:in bit_vector (3 downto 0); VALID: out bit;
              Q: out bit_vector (1 downto 0));
END priority_encoder;

ARCHITECTURE priority_encoder_structure OF priority_encoder IS

   component OR2
         port (INO, IN1: in bit; OUT0 : out bit) ;
   end component;

   component AND2
         port (IN2, IN3: in bit; OUT1: out bit);
   end component;

   component NOT1
         port (IN4: in bit; OUT2: out bit);
   end component;
```

```
begin
      A0:AND2 port map (A(l), X(0), X(l));
      X0:OR2 port map (A(0), A(l), X(2));
      Xl:OR2  port map (A(2), A(3), X(3));
      X2:OR2 port map (X(l), A(2), Q(0)) ;
      X3:OR2 port map (A(3), A(2), Q(l));
      X4:OR2 port map (X(2), X(3), VALID);
      N0:NOT1 port map (A(2), X(0));

end priority_encoder_structure;
```

PROBLEMS

7.1. Using two half adders design a full adder and do the following:
(*a*) Entity Declaration. (*b*) Structural modeling.

7.2. Design a 4-bit comparator and do the following:
(*a*) Entity Declaration. (*b*) Structural modeling,
(*c*) Behavior Modeling. (*d*) Dataflow Modeling.

7.3. Design a I-bit comparator and do the following:
(*a*) Entity Declaration. (*b*) Structure Modeling.
(*c*) Behavior Modeling. (*d*) Dataflow Modeling.

7.4. Write the VHDL code for a 4×4 Bit RAM.

7.5. Write down the Behavior Model VHDL code for a JK flip-flop.

7.6. For a 8-to-3 Encoder gate do the following:
(*a*) Entity Declaration. (*b*) Dataflow Model.
(*c*) Behavior Model.

7.7. For a 3-bit Binary-to-Gray code converter do the following:
(a) Entity Declaration. (*b*) Dataflow Model.
(c) Behavior Model.

7.8. For a 3-bit Gray Binary-to-code converter do the following:
(*a*) Entity Declaration. (*b*) Dataflow Model,
(*c*) Behavior Model.

7.9. For a 2-bit comparator do the following:
(*a*) Entity Declaration. (*b*) Dataflow Model,
(*c*) Behavior Model.

7.10. For figure 7.24 do the following:
(*a*) Entity Declaration. (*b*) Structural modeling.
(*c*) Data flow modeling. (*f*) Behavioral modeling.

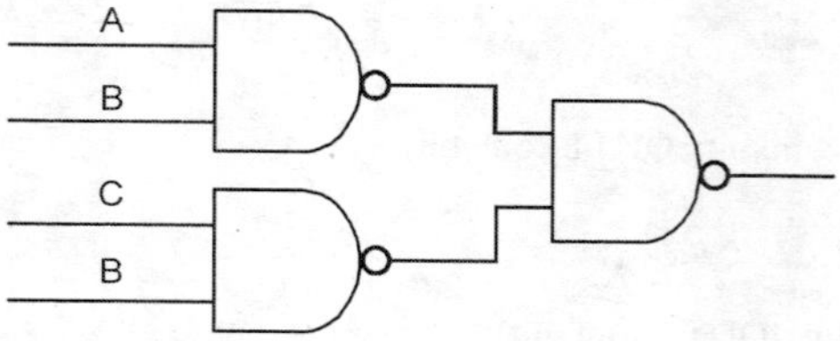

Fig. 7.24

7.11. For figure7.38 do the following:

(*a*) Entity Declaration

(*b*) Structural modeling

(*c*) Data flow modeling

(*d*) Behavioral modeling

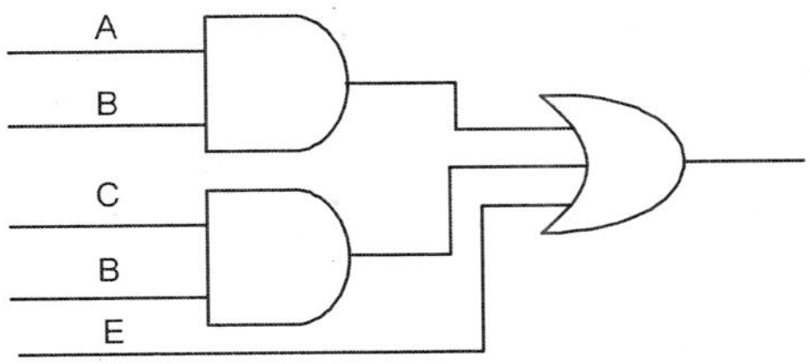

Figure 7.25

7.12. For the figure 7.39 the 'following:

(*a*) Entity Declaration.

(*b*) Structural modeling.

(*c*) Behavioral Modeling

(*d*) Dataflow Model.

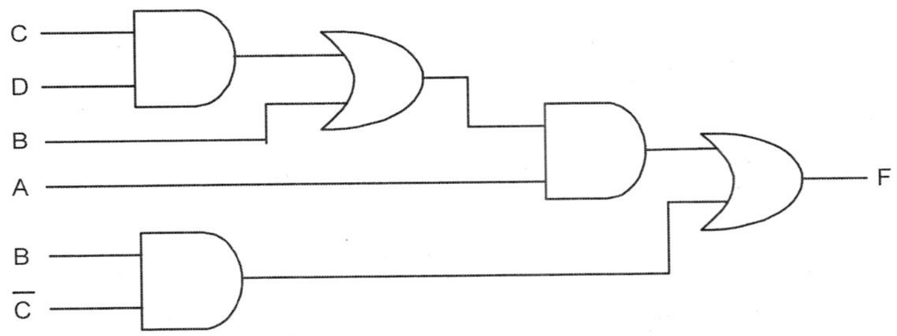

Fig. 7.26

MODEL TEST PAPERS-I

Time: 3 Hours **Maximum Marks : 100**

Note : *Attempt Five Questions in all. Question No. 1 is compulsory. Attempt one question from each section. Each question carries 20 marks.*

1. Answer the following question:

(*a*) In hardware simulation, what is the role of stimuli.
(*b*) What is the use of " LIBRARY" & " USE " clause?
(*c*) Discuses in brief about:
  (*i*) Enumeration Type
  (*ii*) Real type
(*d*) What is the significance of "Wait for O ns" statement?
(*e*) Write the syntax for " Conditional signal assignment statement".
(*f*) What is Operator Overloading?
(*g*) What are generics?
(*h*) What is VHDL?
(*i*) What do you understand by semi- conductor memory?
(*j*) Explain PEEL.

SECTION- A

2. While designing a digital system, What possibly could be Various design stages. Also prone that how VHDL is capable of realizing the purpose of each design stage.
3. What are identifiers? How many types of identifiers? Explain with example.

SECTION- B

4. (*a*) Explain the difference between concurrent and sequential statements.
(*b*) Differentiate between a process and wait statement. Can they be used simultaneously in a program.
5. (*a*) How will you compare component declaration and component installation.
(b) What are technology libraries?

SECTION-C

6. (*a*) Write VHDL code for 3 bit binary to excess- 3 code converter in date flow modeling.
(*b*) Write VHDL code for modulo- 3 counter in behavioral modeling.
7. Draw a complete structural model for a 9- bit parits generator circuit in structural modeling.

SECTION- D

8. (*a*) Implement a simple micro computer using VHDL.
(b) Write a short notes on:
  (*i*) FPGA (ii) CPLD.

MODEL QUESTION PAPER-2

Time: 3 Hours **Maximum Marks : 100**

Note : Attempt Five Questions in all. Question No.1 compulsory. Attempt one question from each section. Each question carries 20 Marks.

1. Answer the following questions:

(*a*) What are the capabilities of VHDL.
(*b*) Write short note on packages.
(*c*) Write down types of flexible design methodologies.
(*d*) What are the advantages pf PLDs ?
(*e*) Write down the entity for the circuit given below?
(*f*) What are the identifiers?
(*g*) What is the difference between arrays and records?
(*h*) What is Subprogram Overloading.
(*i*) Write short note a test bench.
(*j*) Define Delay and its types.

SECTION- A

2. (*a*) What is the VHDL ?Who for umlauted rules for VHDL? What are its specific capabilities?
(*b*) In Behavioural Modeling explain transport and interial delays.

3. (*a*) Differentiate types of data types.
(*b*) What is the role of HDL in digital system design? How do HDL differ from other software language.

SECTION- B

4. How case statement is different from loop statement .

5. What are generics? Specify with example and explain. Why do me need configuration?

6. Using structural modeling for exampling the circuit of 9- bit parity generator.

7. (*a*) Write VHDL code for any comparator.
(*b*) Draw and explain look ahead carry circuit.

SECTION- D

8. How do you implement a micro computer system using VHDL.

9. Write short note on:
(*i*) FPGA (*ii*) PLD designing
(*iii*) PEEL

QUESTION BANK

1. How many types of over loadings are there? Explain.
2. What are data objects? Explain all classes and data types.
3. What is the purpose of Writing a Test Bench? Explain with example.
4. Differentiate between component declaration and component installation.
5. Why do you use ARRAYS AND Loops ? Give explain of each.
6. Write a programs for full adder circuit to differentiate between structural and data flow modeling.
7. What are generics? Specify with example and explain. Why do me need configuration?
8. Write a program to implement micro processor system.
9. Differentiate between concurrent and Assignment statements.
10. What is a micro cell and how can be design it (memory circuit) using PLD's and macro cell.
11. Write short note on:-
    (*a*) PEEL
    (*b*) FPGA
    (*c*) 22 V10
12. Write a program to implement a shift register in Mixed style of modeling.
13. Design a Ex - OR gate using only four NAND gates. Write a program for the same using structural style of modeling.
14. Draw and explain look ahead carry circuit.
15. What is VHDL? Who formulated rules for VHDL? What are its specific capabilities?
16. What is entity declaration. Explain it for 2-4 decoder kit.
17. How is Behavioural style of modeling different from mixed style of modeling ?
18. How many types of identifiers are there?
19. Differentiate between sub and scalar types of data types.
20. In Behavioural modeling What are signal drivers? What is the effect of transport delay on signal drivers?
21. Draw a complete structural model for a 9- bit parity generator ckt in structural modeling.
22. What are attributes? Explain it with the help of any digital circuit?
23. What are technology libraries? How delay effects are used in delay models?
24. Draw a block diagram of CPU and implement it using synthesis description.
25. Use IF control OR gate.
26. Write short notes on:
    (*a*) PLD designing
    (*b*) FPGA
    (*c*) Simulation of Johnson counter.
27. Sate differences between Latch and Flip Flop.
28. Compare symbols versus Entities.
29. Compare schematics versus Architecture
30. In Behavioural modeling explain transport and inertial delays.
31. Using Process statement implement input MUX symbol and its functions.
32. What are EXIT and ASSERT statements in sequential processing?

33. What are data types? Explain all its types with program examples.

34. Differentiate between value types and value array attributes. Why attributes are being used?

35. Write short notes non:

(*i*) CPLDs

(*ii*) Packages and libraries

(*iii*)Test Bench

36. What is meant by operator overloading? Discuss with suitable examples.

37. Compare the following:

(i) Concurrent statement and sequential statement

(ii) Behavioural modeling and structural modeling.

38. Write VHDL code for 16:1 multiplexer using 4:1 multiplexer using 4:1 multiplexer as a component.

39. Discuss in detail implementation of simple micro computer system using VHDL.

40. How can a ROM be used as a:

(*i*) PAL

(*ii*) PLA

41. Explain function and procedure used in VHDL with suitable examples.

42. Discuss Various types of delays.

43. Discuss various types of data objects and operator used in VHDL.

44. Write VHDL for 4:1 multiplexer using:

(*i*) Case statement

(*ii*) If then else statement.

45. Explain the concept of delta delay transport delay and overloading with examples.

46. What are subprograms? Explain with examples.

47. Discuss packages and libraries used in VHDL .

48. Write VHDL code for any comparator.

49. Write short notes on:

(*i*) FPGA.

(*ii*) PAL

(*iii*) PEEL

50. Explain the use of the following VHDL, with examples:

(*i*) Generics

(*ii*) Resolution functions

(*iii*) Procedures

(*iv*)Concurrent and sequential statements

51. Give architecture of simple micro computer system.

52. How do you implement a micro computer system using VHDL.

53. Draw circuit diagrams write VHDL code for shift right register.

54. What do you mean by hardware description languages? Discuss in brief history of VHDL.

55. Discuss various types of classes and operators in VHDL.

APPENDIX A

KEYWORDS

A

abs
access
after
alias
all
and
architecture
array
assert
attribute

B

begin
block
body
buffer
bus

C

case
component
configuration
constant
disconnect
downto

D

else
elsif
end
entity exit

F

file
for
function

G

generate
generic group
guarded

I

if
impure in
inertial
inout
is

L

label library
linkage
literal
loop

M

map
mod

N

nand
new
next
nor
not
null

O

of
on
open
or
others
out

P

package
port
postponed
procedure
process
pure

R

range
record
register
reject
rem
report
return
rol
ror

S

select
severity
shared
signal
sla
sll
sra
srl
subtype

T

then
to
transport
type

U

unaffected
units
until
use

V

variable

W

wait
when
while
with

X

xnor
xor

APPENDIX B

PACKAGE STD_LOGIC_L 164

This package shall be compiled into a design library symbolically named IEEE.

```
      package STD_LOGIC_1164 is
logic State System (unresolved)

        type STD_ LOGIC is
        (
                'U'   — Uninitialized
                'X,   — Forcing Unknown
                '0',  — Forcing 0
                '1',  — Forcing 1
                'Z'   — High Impedance
                'W,   — Weak   Unknown
                'L',  — Weak   0
                'H'   — Weak   1
                '-'   — don't care
        );
Unconstrained array of std_ulogic for use with the resolution function
        type STD_ULOGIC_VECTOR is array ( NATURAL range )
        of STD-ULOGIC ;
resolution function
      function RESOLVED (S : STD_ULOGIC_VECTOR)
        return STD_ULOGIC;
*** industry standard logic type ***
        subtype STD-LOGIC is RESOLVED STD_ULOGIC;
Unconstrained array of std_logic for use in declaring signal arrays
        type STD_LOGlC_VECTOR is array (NATURAL range )
          of STD_LOGIC;
common  subtypes
        subtype X01 is RESOLVED STD_ULOGIC range 'X' to '1';
                    - - ('X' '0', '1')
        subtype X01Z is RESOLVED STD_ULOGIC range 'X' to 'Z';
                    - - ('X', '0', 'V, 'Z ;)
        subtype UX01 is RESOLVED STD_ULOGIC range 'U' to '1';
                    - - ('U', 'X', '0', '1')
        subtype UX01Z is RESOLVED STD_ULOGIC range 'U' to 'Z';
```

```
                    - - ('U', 'X', '0', '1', 'Z')
```

overloaded logical operators

```
function "and" ( L : STD_ULOGIC ; R : STD_ULOGIC ) return UX01;
function "nand" (L : STD_ULOGIC ; R : STD_ULOGIC )
        return UX01;
function "or" (L : STD_ULOGIC ; R : STD_ULOGIC ) return UX01;
function "xor" (L : STD_ULOGIC ; R : STD_ULOGIC ) return UX01;
function "xnor" (L : STD_ULOGIC ; R : STD_ULOGIC )
        return UX01;
function "not" (L : STD_ULOGIC ) return UX01;
```

vectorized overloaded logical operators

```
function "and" (L, R: STD_LOGIC_VECTOR)
        return STD_LOGIC_VECTOR;
function "and" (L, R : STD_ULOGIC_VECTOR)
        return STD_ULOGIC_VECTOR;
function "nand" ( L, R : STD_LOGIC_VECTOR)
        return STD_LOGIC_VECTOR;
function "nand" ( L, R: STD_ULOGIC_VECTOR )
        return STD_ULOGIC_VECTOR;
function "or" ( L, R : STD_LOGIC_VECTOR )
        return STD_ULOGIC_VECTOR;
function "or" (L, R : STD _ULOGIC_VECTOR )
        return STD_ULOGIC_VECTOR;
function "nor" (L, R : STD_LOGIC_VECTOR)
        return STD_UOGIC_VECTOR;
function "nor" (L, R: STD_ULOGIC_VECTOR)
        return STD_ULOGIC_VECTOR;
function "xor" (L, R: STD_LOGIC_VECTOR )
        return STD_ULOGIC_VECTOR;
function "xor" (L, R : STD _ULOGIC_VECTOR)
        return STD_ULOGIC_VECTOR;
function "xnor" ( L, R : STD_LOGIC_VECTOR)
        return STD LOGIC VECTOR;
function "xnor" ( L, R : STD_ULOGIC_VECTOR)
        return STD_ULOGIC_VECTOR;
function "not" (L : STD_LOGIC_VECTOR)
        return STD_LOGIC_VECTOR;
function "not" (L : STD_ULOGIC_VECTOR)
        return STD_ULOGIC_VECTOR;
```

conversion functions

```
function TO_BIT ( S : STD_ULOGIC ; XMAP : BIT := '0'
        return BIT ;
function TO_BITVECTOR (S : STD_LOGIC_VECTOR ; XMAP : BIT :='0')
        return BIT_VECTOR
```

```
function TO_BITVECTOR  (S: STD_ULOGIC_VECTOR; XMAP: BIT :='0')
        return BIT_VECTOR
function TO_STD_ULOGIC (B:BIT)
        return STD_ULOGIC;
function TO_STD_LOGIC_VECTOR ( B : BIT_VECTOR)
        return STD_ULOGIC_VECTOR;
function TO_STD_LOGIC_VECTOR ( S : STD_ULOGIC_VECTOR )
        return STD_LOGIC_VECTOR;
function TO_STD_ULOGIC_VECTOR ( B : BIT_VECTOR )
        return STD_ULOGIC_VECTOR ;
function TO_STD_ULOGIC_VECTOR ( S : STD_ULOGIC_VECTOR )
        return STD_ULOGIC _VECTOR;
```

strength strippers and type converters

```
function TO_X01 ( S : STD_LOGIC_VECTOR)
        return STD_LOGIC _VECTOR;
function TO_X01 ( S : STD_ULOGIC_VECTOR)
        return STD_ULOGIC_VECTOR;
function TO_X01 ( S : STD _ULOGIC    )
        return X01;
function TO_K01 (B : BIT_VECTOR    )
        return STD_LOGIC_VECTOR;
function TO_X01 ( B : BIT_VECTOR    )
        return STD _ULOGIC_VECTOR;
function TO_X01 ( B : BIT )
        return X01;
function TO_X01Z ( S : STD_LOGIC_VECTOR)
        return STD_LOGIC_VECTOR;
function TO_X01Z ( S : STD_ULOGIC_VECTOR)
        return STD_ULOGIC_VECTOR;
function TO_X01Z ( S : STD_ULOGIC    )
        return X01Z;
function TO_X01Z ( B : BIT_VECTOR    )
        return STD_LOGIC_VECTOR;
function TO_X01Z (B : BIT_VECTOR    )
        return STD_ULOGIC_VECTOR;
function TO_X01Z (B : BIT)
        return X01Z;
function TO_UX01 (S : STDJLOGIC.VECTOR)
       return STD_LOGIC_VECTOR;
function TO_UX01 (S : STD_ULOGIC_.VECTOR)
       return STD_ULOGIC_VECTOR;
function TO_UX01 (S: STD_ULOGIC    )
       return UX01;
function TO_UX01 (B : BIT_VECTOR    )
      return STD_LOGIC_VECTOR;
function TO_UX01 (B : BIT_VECTOR    )
      return STD_ULOGIC_VECTOR;
function TO_UX01 (B:BIT)
      return UX01;
```

edge detection

```
function RISING_EDGE (signal S : STD_ULOGIC)
        return BOOLEAN;
function FALLING_EDGE (signal S : STD_ULOGIC)
        return BOOLEAN;
```

object contains an unknown

```
function IS_X ( S : STD_ULOGIC_VECTOR) return BOOLEAN ;
function IS_X (S : STD_LOGIC_VECTOR) return BOOLEAN;
function IS_X ( S : STD_ULOGIC     ) return BOOLEAN;
end STD_LOGIC_1164;
```

APPENDIX C

PACKAGE STANDARD

- This package shall be predefined in the compiler.
- The name of the package is STANDARD.
- This design unit contains some special tokens, which are only recognized by the analyzer when it is in special "bootstrap" mode.

```
package STANDARD is

-- predefined enumeration types:
type BOOLEAN is (FALSE, TRUE);
type BIT is ('0', '1');
type CHARACTER is
(
NUL,    SOH,    STX,    ETX,    EOT,    ENQ,    ACK,    BEL,
BS,     HT,     LF,     VT,     FF,     CR,     SO,     SI,
DLE,    DC1,    DC2,    DC3,    DC4,    NAK,    SYN,    ETB,
CAN,    EM,     SUB,    ESC,    FSP,    GSP,    RSP,    USP,
' ',    '!',    '"',    '#',    '$'     '%''    '&',    ''',
'(',    ')',    '*'     '+',    ',',    '-',    '/',
'0',    '1',    '2',    '3',    '4',    '5',    '6',    '7',
'8',    '9',    ':',    ';',    '<',    '=',    '>',    '?',


'©',    'A',    'B',    'C,     'D',    'E',    'F,     'G',
'H',    T,      ']',    'K',    'L',    'M',    'N',    'O',
'P,     'Q',    'R',    'S',    T,      'U',    'V',    'W',
'X',    'Y'     'Z',    '['.    '\',    ']',    '^',    '_',


'`'     'a',    'b',    'c',    'd',    'e',    'f,     'g',
'h',    'i',    'j',    'k',    'm',    'n',    'o',    'o'.
'p',    'q,'    'r',    's',    't',    'u',    'v',    'W,
'x',    'y',    'z',    '{',    'l',    '}',    '~',    DEL,
```

C128,	C129,	C130,	C131,	C132,	C133,	C134,	C135,
C136,	C137,	C138,	C139,	C140,	C141,	C142,	C143,
C144,	C145,	C146,	C147,	C148,	C149,	C150,	C151,
C152,	C153,	C154,	C155,	C156,	C157,	C158,	C159,

‘ ’,	‘!’,	‘¢’,	‘£’,	‘¤’,	‘¥’,	‘¦’,	‘§’,
‘¨’,	‘©’,	‘ª’,	‘«’,	‘¬’,	‘’,	‘®’,	‘¯’
‘°’	‘±’,	‘2’,	‘3’	‘´’,	‘µ’,	‘¶’,	‘·’,
‘¸’,	‘1’,	‘º’	‘»’,	‘¼’,	‘½’,	‘¾’,	‘¿’,
‘À’,	‘Á’	‘Â’,	‘Ã’,	‘Ä’,	‘Å’,	‘Æ’,	‘Ç’,
‘È’,	‘É’,	‘Ê’,	‘Ë’,	‘Ì’,	‘Í’,	‘Î,	‘Ï’,
‘Ð’,	‘Ñ’,	‘Ò’,	‘Ó’,	‘Ô’,	‘Õ’,	‘Ö’,	‘×’,
‘Ø’	‘Ù’,	‘Ú’,	‘Û’,	‘Ü’,’	‘Ý’,	‘Þ’,	‘ß’
‘à’,	‘á’,	‘â’,	‘ã’,	‘ä’,	‘å’,	‘æ’,	‘ç’,
‘è’,	‘é’,	‘ê’,	‘ë’,	‘ì’,	‘í’,	‘î’,	‘ï’,
‘ð’,	‘ñ’,	‘ò’,	‘ó’,	‘ô’,	‘õ’,	‘ö’,	‘÷’,
‘ø’,	‘ù’,	‘û’,	‘ü’,	‘ý’,	‘þ’	‘ÿ’	‘ÿ’

);

type SEVERITY_LEVEL is (NOTE, WARNING, ERROR, FAILURE);

-- predefined numeric types :

-- Do INTEGER first to aid implicit declarations of “”.**

type INTEGER is range -2147483647 to 2147483647;

type $UNIVERSAL_INTEGER is range $- to $–;

type $UNIVERSAL_REAL is range $-. to $+.;

function “*” ($LEFT: $UNIVERSAL_REAL ; $RIGHT: $UNIVERSAL_INTEGER)
return $UNIVERSAL_REAL;

function “*” ($LEFT: $UNIVERSAL_INTEGER; $RIGHT: $UNIVERSAL_REAL)
return $UNIVERSAL_REAL;

function “/” ($LEFT: $UNIVERSAL_REAL; $RIGHT: $UNIVERSAL_INTEGER)
return $UNIVERSAL_REAL;

type REAL is range $-. to $+.;

-- predefined type TIME :

type TIME is range $- to $+

```
units

        fs ;                        -- femtosecond
        ps      = 1000 fs;          -- picosecond
        ns      = 1000 ps;          -- nanosecond
        us      = 1000 ns;          -- microsecond
        ms      = 1000 us;          -- millisecond
        sec     = 1000 ms;          -- second
        min     = 60 sec;           -- minute
        hr      = 60 min;           -- hour;

end units;

-- subtype used internally for checking time expressions for non-negativness:

subtype $NATURAL_TIME is TIME range 0 sec to TIME'HIGH;

subtype DELAY_LENGTH is TIME range 0 fs to TIME'HIGH;

-- function that returns the current simulation time:

impure function NOW return TIME ;

-- predefined numeric subtypes :

subtype NATURAL is INTEGER range 0 to INTEGER'HIGH;

subtype POSITIVE is INTEGER range 1 to INTEGER'HIGH;

-- predefined array types:

type STRING is array (POSITIVE range ) of CHARACTER;

type BITJVECTOR is array (NATURAL range ) of BIT ;

-- type FILE_OPEN_KIND is (READ_OPEN, WRITE_OPEN, APPEND_OPEN);

type FILE_OPEN_KIND is (READ_MODE, WRITE_MODE, APPEND_MODE);

type FILE_OPEN_STATUS is (OPEN.OK, STATUS_ERROR,
                          NAME_ERROR, MODE_ERROR);

attribute FOREIGN: STRING;

-- The rest of this package is SVEN specific stuff required to make
-- this implementation go.
-- Note that all things are declared use leading $ characters, so we don't
-- trample the user's name space.

attribute $BUILTIN: BOOLEAN;
```

```
procedure $RTINDEX (I: NATURAL; FIRSTARG: INTEGER;
                            PASSAP,SAVEFREGS: BOOLEAN);

procedure $RTSYMBOL (S: STRING; FIRSTARG: INTEGER;
                            PASSAP,SAVEFREGS: BOOLEAN);

attribute $BUILTIN of all: function is TRUE;

attribute $BUILTIN of all: procedure is TRUE;

end STANDARD;
```

APPENDIX D

BIBLIOGRAPHY

1. Aggarwal, A., Lewis, D., Routing Architecture for Hierarchical Field Programmable Gate Arrays, IEEE Conference on Custom Integrated Circuits, 1994, pp. 475-478.
2. Armstrong, James and Gray, Gail, Structured Logic Design with VHDL, Prentice-Hall, 1993.
3. Ashenden, Peter, J., The Students Guide to VHDL, Morgan Kaufman Pub. Co.
4. Berge, F., Maginot, R., J., VHDL Designer's Reference, Kluwer Academic Publishers, 1992.
5. Bhasker, J., A VHDL Primer, Revised Edition Prentice-Hall, 1992.
6. Breeding, K.J., Digital Design Fundamentals, Prentice-Hall, 1989.
7. Brown, S., Vranesic, Z., Fundamentals of Digital Logic with VHDL Design, McGraw-Hill, 200C.
8. Chang, K., C, Digital Design and Modeleing with VHDL and Synthesis, IEEE Comput. Soc. Press, 1997.
9. Coelho, D., HELIX, A Tool for Multi-Level Simulation of VLSI Systems, Conference on Semi-Custom IC, 1993.
10. Coelho, David, R., The VHDL Handbook, Kluwer Academic Publishers, 1989.
11. Dewey, Allen., Analysis and Design of Digital Systems with VHDL, PWS Pub. Co., 1997.
12. Douglas, Ott, E., Wilderotter., Thomas, J., A., A Designer's Guide to VHDL Synthesis, Kluwer Academic Publishers, 1994.
13. Fletcher, W., I., An Engineering Approach to Digital Design, Prentice-Hall, 1979.
14. Floyd, L., Thomas, Digital Fundamentals, Pearson Education, 2004.
15. Friedman, A, D., Fundamentals of Logic Design and Switching Theory, Computer Science Press, 1986.
16. Givone, D., D., Introduction to Switching Circuit Theory, McGraw-Hill, 1970.
17. Hill, F., J., Peterson, G., R., Introduction to Switching Theory and Logical Design, 3rd ed., John Wiley, 1981.
18. Hill, J., F., Peterson, R., G., Digital Systems: Hardware Organization and Design, 3nd ed., Wiley, 1987.
19. IEEE Standard VHDL Reference Manual (ANSI/IEEE Std 1076-1993), IEEE, 1994.
20. Jain, R., P., Digital Electronics and Microprocessors - Problems and Solutions, Tata McGraw-Hill Publishing Company Limitedi 1987.
21. Jain, R., P., Modern Digital Electronics, Tata McGraw-Hill Publishing Company Limited, 2005.
22. Kitson, B., Programmable Array Logic Handbook. Sunnyvale, CA: Advanced Micro Devices, 1983.
23. Lala, P., K., Digital System Design using Programmable Logic Devices, Prentice-Hall, 1990.

24. Gupta Slipra, Chaudhary Neelu, Digital Circuit System using VHDL, SK- Kateria & Sons
25. Mishra Yogesh, Digital System Design using VHDL, Dhanpat Rai& Co. (Pvt) Ltd.
26. Lipsett, Roger., VHDL: Hardware Description and Design, Kluwer Academic Publishers, Boston, 1989.
27. Mano, M., Digital Design, 2nd ed., Prentice-Hall of India,2001. 28. Mano, M., Kime, C, Logic and Computer Design Fundamentals, Prentice-Hall, 2000. 29. Mano, M., M., Computer Engineering: Hardware Design, Prentice-Hall, 1988. 30.Cluskey, E., J., Logic Design Principles, Prentice-Hall, 1986. 31.Micdheli, De, G., Synthesis and Optimization of Digital Circuits, McGraw-Hill, 1994. 32. Navabi, Z., VHDL: Analysis and Modeling of Digital Systems, McGraw-Hill, 1993.
33. Nemani, M., Najm, F. N., High-Level Area and Power Estimation for VLSI Circuits, IEEE International Conference on Computer Aided Design, 1997.
34. Pandey, Sujata., Manoj Pandey., VLSI Design, 1st ed., Dhanpat Rai & Co., 2003.
35. Peatman, J., B., Digital Hardware Design, McGraw-Hill, 1980.
36. Pellerin David and Taylor., Douglas., VHDL Made Easy!, Prentice-Hall PTR Pub. Co., 1997.
37. Perry, Douglas L, 2nd ed., McGraw-Hill, 1994.
38. Phadke, K., Anuradha., VLSI Design, 2nd ed., Nirali Prakashan, 2003. 39.Programmable Logic Databook. Dallas: Texas Instruments, 1988.
40. Prosser, Franklin., Winkel, David., The Art of Digital Design- An Introduction To Top-Down Design, 2nd ed., Prentice-Hall, 1987. 41. Quick-HDL User's and Reference Manual, Mentor Graphics, 1996. 42.RabaeyJ., M.,DigitallntegratedCircuits-adesignperspective, Prentice-Hall of India, 2001.
43. Rose, J. S., Brown, S., Flexibility of Interconnection Structures for Field Programmable Gate Arrays, IEEE J. Solid-State Circuits, Vol.26, No.3 mar. 1991, pp.277-282.
44. Roth, Jr., Charles H., Digital System design Using VHDL, PWS Pub. C-o., 1998.
45. Salcic, C, Smailagic, A., Digital System Design and Prototyping Using Field Programmable Logic, Kluwer Academic Publishers, 1997.
46. Sandige, R., S., Digital Concepts using Standard Integrated Circuits, McGraw-Hill, 1978.
47. Scarpino, Frank,. VHDL and AHDL Digital System Implementation, Prentice-Hall PTR Pub. Co., 1998.
48. Shahdad, M,, An Overview of VHDL Language and Technology, 23rd ACM/IEEE Conference on Design Automat, 1986.
49. Skahill, K., VHDL for Programmable Logic, Addison Wesley, 1996.
50. Tocci, R., J., Digital Systems Principles and Applications, Prentice-Hall, 1988.
51. Uycmura, P., J., A First Course In Digital System Design: An Integrated Approach, Brooks/ Cole Publishing Co., 2000.
52. Wakerly, J., Digital Design: Principles & Practices, Prentice-Hall, 2001.
53. Walsh, Peter., Integrating VHDL Into A First Course In Logic Design, IEEE Conference on Electrical and Computer Engineering, 1997, pp. 1531-1534.
54. Yalamanchili, Sudhakar., VHDL Starter's Guide, Prentice-Hall PTR Pub. Co., 1998.
55. Yeap, G., Practical Low Power Digital VLSI Design, Kluwer Academic Publishers, 1998.
56. Zwolinski, Mark., Digital System Design with VHDL, 2nd ed., Pearson Education, 2004.